FRANCIS BACON AND THE REFIGURING OF EARLY MODERN THOUGHT

Literary and Scientific Cultures of Early Modernity

Series editors:

This series provides a forum for groundbreaking work on the relations between literary and scientific discourses in Europe, during a period when both fields were in a crucial moment of historical formation. We welcome proposals for books that address the many overlaps between modes of imaginative writing typical of the sixteenth and seventeenth centuries – poetics, rhetoric, prose narrative, dramatic production, utopia – and the vocabularies, conceptual models, and intellectual methods of newly emergent 'scientific' fields such as medicine, astronomy, astrology, alchemy, psychology, mapping, mathematics, or natural history. In order to reflect the nature of intellectual inquiry during the period, the series is interdisciplinary in orientation and publishes monographs, edited collections, and selected critical editions of primary texts relevant to an understanding of the mutual implication of literary and scientific epistemologies.

Other titles in the series:

Giordano Bruno and the Geometry of Language
Arielle Saiber

Staging Anatomies
Dissection and Spectacle in Early Stuart Tragedy
Hillary M. Nunn

Disease, Diagnosis and Cure on the Early Modern Stage
Edited by Stephanie Moss and Kaara L. Peterson

Francis Bacon and the Refiguring of Early Modern Thought

Essays to Commemorate *The Advancement of Learning* (1605–2005)

Edited by

JULIE ROBIN SOLOMON and CATHERINE GIMELLI MARTIN

ASHGATE

Published by

Ashgate Publishing Limited
Gower House
Croft Road
Aldershot
Hants GU11 3HR
England

Ashgate Publishing Company
Suite 420
101 Cherry Street
Burlington
Vermont, 05401–4405
USA

Ashgate website: http://www.ashgate.com

British Library Cataloguing in Publication Data
Francis Bacon and the Refiguring of Early Modern Thought
Essays to Commemorate The Advancement of Learning (1605–2005). – (Literary and Scientific Cultures of Early Modernity).
1. Bacon, Francis, 1561–1626 – Knowledge and learning. 2. Bacon, Francis, 1561–1626 – Influence. 3. Learning – Philosophy. 4. Knowledge, Theory of. 5. England – Intellectual life – 17th century. 6. England – Intellectual life – 16th century. I. Solomon, Julie Robin. II. Martin, Catherine Gimelli.
192

US Library of Congress Cataloging in Publication Data
Francis Bacon and the Refiguring of Early Modern Thought
Essays to Commemorate The Advancement of Learning (1605–2005) / edited by Julie Robin Solomon and Catherine Gimelli Martin.
p. cm. – (Literary and Scientific Cultures of Early Modernity)
Includes bibliographical references and index.
1. Bacon, Francis, 1561–1626. 2. Bacon, Francis, 1561–1626. Advancement of learning. 3. Logic. 4. Knowledge, Theory of. 5. Learning – Philosophy, 6. Science – Methodology. I. Solomon, Julie Robin. II. Martin, Catherine Gimelli. III. Series.
B1194.F72 2006
192–dc22 2005002769

ISBN 0 7546 5359 5

This book is printed on acid free paper.
Typeset by Saxon Graphics Ltd, Derby
Printed and bound in Great Britain by MPG Books, Bodmin, Cornwall

Table of Contents

Contributors

Reid Barbour is Gillian T. Cell Distinguished Professor of English at the University of North Carolina at Chapel Hill. His books include studies of Elizabethan fiction, of Epicureanism and Stoicism in the seventeenth century, of the religious imagination in the age of Charles I, and of John Selden. He is currently at work on an intellectual biography of Sir Thomas Browne and finishing up his edition of Lucy Hutchinson's Lucretius for Oxford.

John C. Briggs is Professor of English at the University of California, Riverside. He is the author of *Francis Bacon and the Rhetoric of Nature* (Harvard, 1989), winner of that press's Thomas J. Wilson prize. His essay on Bacon's science and religion appears in *The Cambridge Companion to Bacon* (Cambridge, 1996). He has contributed articles on Bacon's life and works for the *Dictionary of Literary Biography* (Gale, 1995) and the *Philosophy of Education: An Encyclopedia* (Garland, 1996) and has published articles on Shakespeare and the history of rhetoric. His new book–*Lincoln's Speeches Revisited* (Johns Hopkins, forthcoming) –examines Lincoln's idea and practice of persuasion.

Daniel R. Coquillette received a B.A. from Williams College, and earned a Fulbright Scholarship to study law at University College, Oxford University, England, where he earned a B.A. and an M.A. He acquired a J.D. from Harvard Law School, and went on to clerk for justice Robert Braucher of the Supreme Judicial Court of Massachusetts and Chief Justice Warren E. Burger of the Supreme Court of the United States. He has taught at Boston University Law School, Cornell Law School and Harvard Law School. He was Dean of Boston College Law School from 1985–1993, and was named J. Donald Monan University Professor in 1996. In 2001 he was also named the Lester Kissel Visiting Professor of Law at Harvard Law School. Among his many outside activities, Professor Coquillette is Reporter to the Committee on Rules of Practice and Procedure of the Judicial Conference of the United States.

Guido Giglioni works at the Warburg Institute in London where he specializes in early modern history of medicine with a focus on sixteenth- and seventeenth-century Italy and England. He is the author of *Immaginazione e malattia. Saggio su Jan Baptiste van Helmont* (Angeli, 2000) and *Latin Manuscripts of Francis Glisson (1). Philosophical Papers* (Cambridge Wellcome Unit, 1996). He has published articles on Cardano, Comenius, and the place of women in early modern natural philosophical thought.

Fritz Levy is Professor of History Emeritus at the University of Washington. He is the author of *Tudor Historical Thought* (Huntington Library, 1967; Toronto, 2004), and of articles on Sidney, Spenser, Bacon, and Jonson among others.

William T. Lynch is Associate Professor in the Department of Interdisciplinary Studies at Wayne State University, Detroit, MI. He is the author *of Solomon's Child: Method in the Early Royal Society of London* (Stanford University Press, 2001). He has an article entitled "Surveying and the Cromwellian Reconquest of Ireland" forthcoming in Steven A. Walton, ed., *Instrumental in War: Scientific Instruments and Research between Knowledge and the World,* Brill Academic Publishers. He has published articles in *Minerva, Philosophy of the Social Sciences*, and *Scientia Poetica.*

Catherine Gimelli Martin is the author of a prize-winning book entitled *The Ruins of Allegory: "Paradise Lost" and the Metamorphosis of Epic Convention* (Duke, 1998). She has recently completed another book-length study of Milton, *Milton among the Puritans.* Her essay collection, *Milton and Gender* (2004) is available from Cambridge University Press. She has so far produced two articles on Bacon, one on the *New Atlantis* in *Faultlines in the Field,* and another in the *John Donne Journal.* Her other articles deal with a wide range of seventeenth-century poets and playwrights, while her current book project, "Proteus Unbound," examines the influence of Baconian rhetoric throughout the period.

Michael McCanles is Emeritus Professor of English, Marquette University. He has published widely in Renaissance Literature and Literary Theory. Current projects are a book on sexual harassment jurisprudence and a novel with William Tecumseh Sherman at its center.

Timothy J. Reiss is Professor of Comparative Literature at New York University. His most recent books are *Against Autonomy: Global Dialectics of Cultural Exchange* (2002), *Mirages of the Self: Patterns of Personhood in Ancient and Early Modern Europe* (2003), and the edited collections *Sisyphus and Eldorado Magical and Other Realisms in Caribbean Literature*, revised edition (2002), and *Music, Writing, and Cultural Unity in the Caribbean* (2005). He is completing a book on Descartes and the beginnings of the modern era, a collection on Ngugi in the Americas, and working on rethinking the Renaissance as a process of cultural exchange between Africa, the Americas, and Europe.

Julie Robin Solomon is a scholar-in-residence at The American University in Washington, D.C. She is the author of *Objectivity in the Making: Francis Bacon and the Politics of Inquiry* (Johns Hopkins University Press, 1998) as well as essays on Calvin, Thomas Harriot, Shakespeare, and the passions of the soul. Currently she is working on issues surrounding fixated thought in the early modern period as well as a memoir entitled *Strange Mother*.

Jerry Weinberger is Professor of Political Science at Michigan State University. He is Director of the LeFrak Forum and Co-Director of the Symposium on Science, Reason, and Modern Democracy. From 1997 until 2001 he was Chair of the Department of Political Science. Professor Weinberger has written widely on the political thought of Francis Bacon. His latest paper is "Pious Princes and Red-Hot Lovers: The Politics of Shakespeare's *Romeo and Juliet*" (in *The Journal of Politics*,

Vol. 65, 2003), and his latest book, *Benjamin Franklin Unmasked: On the Unity of His Moral, Religious, and Political Thought* (University Press of Kansas), will be available in August 2005. He is currently at work on a new book on the political philosophy of Martin Heidegger and the problem of technology.

Acknowledgments

We thank the following people for encouraging our intellectual work and supporting this project: Michael J. B. Allen, Gloria Anderson, Jim Berg, Mary Thomas Crane, Erika Gaffney, Tobias Gregory, Ed Morman, Tim Reiss, Joshua Scodel, Bill Sessions, Beverly Solomon, and Arlene Stidel. We would also express our gratitude to the staff and intellectual communities of The Folger Shakespeare Library and The Huntington Library.

List of Abbreviations

LL — *The Letters and Life of Francis Bacon*, ed. James Spedding, 7 vols. (Longman: London, 1861–74).

OFB — *The Oxford Francis Bacon*, gen. eds. Graham Rees and Lisa Jardine, 15 vols. (Oxford: Clarendon Press, 1996–).

SEH — *The Works of Francis Bacon*, eds. James Spedding, Robert Leslie Ellis, and Douglas Denon Heath, 7 vols. (Longman: London, 1859–64).

In Praise of

Dominique Adrea Britton
&
Gordon Frank Martin Michels
1912–2004

HEERE Learning ſits, a comely Dame in yeares;
Vpon whoſe head, a heavenly dew doth fall:
Within her lap, an opened booke appeares:
Her right hand ſhewes, a ſunne that ſhines to all;
* Blind Ignorance, expelling with that * light:
The Scepter ſhewes, her power and ſoveraigne might.

Her out * ſpread Armes, and booke her readines,
T' imbrace all men, and entertaine their loue:
The ſhower, thoſe ſacred graces doth expreſſe
By Science, that do flow from heaven aboue.
Her age declares the ſtudie, and the paine;
Of many yeares, ere we our knowledge gaine.

Via ad Deum eſt Scientia quæ ad inſtitutionem recte et honeſte vivendi pertinet.

* Exempla omnia iacerent in tenebris niſi literarum lumen Hiſtoriæ accederet. Cicero. pro Archia Poeta.

* Studiis ac literis res ſecundæ ornantur adverſæ iuvantur: Cic: ad Luceium Epiſt. 5 famil: vide plura in orat: pro Archia poeta.

Hugo.

F 1. *Sine*

"Doctrina (Learning)" from Henry Peacham's *Minerva Britanna* (London, 1612), 26.
By kind permission of the Folger Shakespeare Library;
STC 19511 copy 2 sig F1 (page 26).

Chapter 1

Introduction

Julie Robin Solomon

Setting the Agenda

Natural philosopher, jurist, essayist, statesman, and England's Lord Chancellor, Francis Bacon (1561–1626), Viscount St. Albans and Baron Verulam, needs no introduction. This volume, however, requires something along those lines in order to explain why the editors felt that now was the moment to publish a collection of essays such as these. We both had been struck by the recent claim of sociobiologist E. O. Wilson that Bacon was the "grand architect" of the Enlightenment. Moreover, as Wilson sees him, Bacon was not culpable (as L. C. Knights claimed) for what T. S. Eliot described as the late seventeenth-century dissociation of sensibility between thought and feeling. Nor did Bacon contribute to what C. P. Snow had contended was the mid-twentieth-century antipathy between the "two cultures" of the sciences and the humanities.[1] Wilson justifies his position by pointing out that Bacon was intent on the logical organization of the intellectual disciplines and advocating induction as a method that could serve, and thus help unify, all branches of learning.[2] Wilson's large and programmatic contentions spurred us to ask specific questions about how Bacon might have or have not gone about unifying the sciences and the humanities, linking the various scientific disciplines, and forging connections between past intellectual traditions and emerging knowledge. Wilson names this state of unity consilience.

Opportunely, we found ourselves thinking about these and similar questions in time to pull together a collection commemorating the four hundredth anniversary of the publication of *The Advancement of Learning* in 2005. While no would-be editor should ignore the providence and promise that commemoration provides, we would like to believe that our decision to compile this collection owes more to intellectual convergence and vitality than to rank opportunism (if such a term can be used to describe the desire to publish a scholarly tome). Catherine Martin, my co-editor, worked her way back to Bacon from her study of the intertwined intellectual roots of Milton's political and scientific thought and its connection to emerging seventeenth-

---

[1] L. C. Knights, "Bacon and the Dissociation of Sensibility," *Explorations: Essays in Criticism mainly in the Literature of the Seventeenth Century* (New York: New York Univ. Press, 1964) ; T. S. Eliot, "The Metaphysical Poets," *Selected Prose of T. S. Eliot*, ed. Frank Kermode (New York: Harcourt Brace Jovanovich, 1975), 64; C.P. Snow, *The Two Cultures and the Scientific Revolution* (New York: Basic Books, 1993); E. O. Wilson, *Consilience: The Unity of Knowledge* (New York: Knopf, 1998), 40.

[2] Wilson, 23, 27.

century epistemologies in England. For Catherine, Milton needed to be rescued from those who misread him as adhering to a hidebound, Christian cosmology, instead of being the admirer of a largely Baconian empirical and rationalist intellectual project. And Bacon then needed to be rescued from critics who misprised his natural philosophy as tyrannical towards nature (and women), by ignoring the degree to which he saw nature as a full and equal, if not superior, partner to human art in his natural philosophical scheme. I, in contrast, approached Bacon by sidling up to him from my study of English Renaissance theater. I then jumped ship from the study of literature to intellectual history–retaining, however, much of my literary training (or baggage, depending on your perspective)–in theory and discourse analysis. The result was a book analyzing the intellectual-historical context and political purport of Bacon's notions of natural philosophical disinterestedness, or objectivity.[3] One thing that troubled both Catherine and me were the divisions that existed among scholars who approached Bacon from their own particular disciplinary perspective with little awareness of how their views might intersect those from other disciplines. We were even more disturbed by the rancorous divide between Bacon scholars working within intellectual history who explored Bacon's thought in terms of internalist intellectual traditions, and those scholars–often coming from the disciplines of literature and general history–who were interested in relating Bacon's ideas to his contemporary political, social, and economic context. We thought it was time to bring scholars together from a variety of disciplines to explore in one volume the ways in which Bacon constructed intellectual discourses. We did this with an eye to unifying the intellectual and socio-political field, while acknowledging the inevitable gaps and disunity that such a wide-reaching project would still encounter.

The editors of this volume believe that to establish Bacon's overarching importance to modern culture one must focus on how he sought to unify knowledge: diachronically, across historical time and, synchronically, across social and intellectual space. We read Bacon not as a "dissociator of sensibility," one who crudely distanced organic nature from a domineering inorganic technology, a magical worldview from a scientific one, the humanities from the sciences, or the feminine from the masculine. As a proto-Enlightenment thinker, Bacon sincerely struggled to unify–substantially and methodologically, and across both space and time–a range of philosophies, disciplines, practices, and intellectual institutions. Of course in the effort to unify he often recognized the difficulties, even the impossibility, of creating intellectual coherence. He thus tried to forge unity where none was clearly available and often appealed to the notion of what is "unknown" to human minds to beg the question, or to avoid premature formulation of that "unity." We thus raised the subject of Bacon's interest in the unity of knowledge to our essayists. Coming from a variety of disciplines, they would shed varied and rich light on this subject. We wanted our authors to consider such questions as how did Bacon seek to establish methodological or theoretical unity among the branches of learning? To what extent did Bacon succeed or not in weaving together threads of ancient intellectual

[3] Julie Robin Solomon, *Objectivity in the Making: Francis Bacon and the Politics of Inquiry* (Baltimore: Johns Hopkins Univ. Press, 1998).

tradition with the new concerns and insights of contemporary thinkers to create a unified discourse on natural philosophy? How did Bacon recast the popular image and rhetoric of natural philosophical learning so as to make its institutionalization conceivable?

The essays

Bacon's investment in diachronic continuity comes to the fore in his express interest in the *ars tradendi* or the arts of tradition or transmission (*SEH*, 4: 405, 438). Bacon's *ars tradendi* comprise the various disciplinary, rhetorical, pedagogical, and technological means of conveying tradition. A recent volume of *Critical Inquiry* hails Bacon as an important early user of the term, and includes essays devoted to the various modalities of information transmission from the ancients to the present.[4] But by *ars tradendi* Bacon meant not only the means and media of transmission but the way in which receivers of knowledge traditions would handle the content of the tradition–the way, in other words, knowers would behave in the presence of "received knowledge." In particular, Bacon feared that deliverers of traditional knowledge would do their best to hide its weaknesses and that slothful receivers of such knowledge would simply take it at face value. For as

> Knowledges are now delivered, there is a kinde of Contract of Errour, betweene the Deliverer, and the Receiver: for he that delivereth knowledge; desireth to deliver it in such fourme, as may be best beleeved; and not as may [be] best examined: and hee that receiveth knowledge, desireth rather present satisfaction, than expectant Enquiry, & so rather not to doubt, than not to erre: glorie making the Author not to lay open his weaknesse, and sloth making the Disciple not to knowe his strength. (*OFB*, IV: 123)

Rather than passively imbibing received knowledge, the Baconian knower would be interested in the "*progression*" of knowledge (and a natural philosophic thinker in his own right) and participate actively in determining the meaning of received knowledge, remaking that knowledge in the process of knowing it. He would remake traditional knowledge as "as a threade to bee spunne on" and "delivered and intimated, if it were possible, *In the same Methode wherein it was invented.*" Inductive knowledge is most easily delivered and received in this way, but even non-inductive knowledges should be handled proactively, so that

> a man may revisite, and descend unto the foundations of his Knowledge and Consent: and so transplant it into another, as it grewe in his own Minde. For it is in Knowledges, as it is in Plantes; … if you meane to remoove it to growe, then it is more assured to rest uppon rootes, than Slippes: So the deliverie of Knowledge (as it is nowe used) is as of faire bodies of Trees without the Rootes: good for the Carpenter, but not for the Planter: But if you will

[4] *Critical Inquiry* 31 (Autumn 31), a special volume on the "Arts of Tradition," eds. James Chandler, Arnold I. Davison, and Adrian Johns.

> have Sciences grewe: it is lesse matter for the shafte, or bodie of the Tree, so you looke well to the takinge up of the Rootes. (*OFB*, IV: 123)

Five of our essayists take up Bacon's more or less successful efforts to forge a coherent and creative relationship with past intellectual, rhetorical, or religious traditions.

Reid Barbour's essay takes on Bacon's complex relationship with the ancient tradition of atomism, and in so doing, demonstrates how Bacon struggled with the very form and meaning of the "particulate" matter upon which he would found his ratio-empirical natural philosophy. As a theory of matter, atomism helped stem superstitious natural theological notions concerning the divinity of nature itself, but it was also distastefully associated with atheism. In contrast, the Stoic theory of *pneuma,* or spirits animating matter, avoided the charge of atheism, but bordered on reintroducing the idea of natural divinity. At various points in his philosophical writings, Bacon finds himself uncertainly suspended between the superstition associated with natural divinity and the atheistical imposture that shadows the theory of atomism. Towards the end of his life, Bacon seems to have embraced pneumaticism as a hedge against charges of atheism, but he still expressed admiration for Democritus, the founder of atomism, as well as most aspects of the theory. Barbour's exploration suggests that despite the fact that Bacon shared with E. O. Wilson an intent to unify knowledge through the use of a common method of empirical observation and induction, Bacon himself remained fundamentally uncertain about the very nature of the particulate matter which would provide the material basis for natural philosophic induction.

Michael McCanles's essay delves into Bacon's debt to two strains of Franciscan philosophy: transcendental mysticism and Ockhamist nominalism. The mystical strain reinforced Bacon's ethical fear of the psychological impediments to knowledge posed by the human passions and the will. The nominalist strain led to Bacon's technical concerns with the intellectual distortions induced by human concept-formation, given that it is limited by the coordinates of human space-time, and discursive, fragmented, and incremental reason. Like the Franciscan mysticism and nominalism upon which he draws, Bacon forges a distinction between "apophatic" and "cataphatic" language. The latter is a first-order form of speech which claims to express, semantically, the truth value or content of that to which it refers. Apophatic language, in contradistinction, recognizes that it cannot express, mirror, or satisfactorily represent the intellectual content to which it refers–it can only point to or refer to such a content without in fact satisfactorily reproducing or representing it. Bacon's natural philosophy thus shares much with what McCanles labels a "negative theology" that points the way to God's truth but can never discursively express that truth. According to McCanles, Bacon views human concept-formation or philosophical universal claims as essentially apophatic in nature. In order to have any intellectual validity, apophatic statements must establish their detachment from the self's desires and from conceptual and semiotic human inventions; they must peel away those aspects of self and society that could corrode their referential validity.

This kind of philosophical apophaticism might have been sufficient for Ockham, who sought to rewrite Aristotelian epistemology, but Bacon's goal was to advance a natural philosophy that would produce real material effects on the world for the betterment of society. If scientific knowledge was only referential, rather than cataphatic or essential, then how could the new natural philosophy make claims to understand the world as "it really is," or at least, better than Aristotle or other ancient philosophies? McCanles argues that Bacon solved this dilemma by appealing to the superior value of pragmatically "testing" hypotheses. The problem of validating the relation of *ficta* or hypotheses to the extra-mental world could be resolved if one could demonstrate the real, pragmatic effects of an hypothesis on that world (66). While not addressing Wilson's concerns directly, McCanles's argument simultaneously supports and subverts aspects of Wilson's idea that the unity of knowledge is indeed a possibility. Insofar as McCanles shows how Bacon turned to experimental practice to insure the validity of natural philosophic knowledge claims, Baconian science foreshadows the workings of modern science along Wilsonian lines. Yet, insofar as Baconian philosophy bespeaks the kind of negative theology or apophatism that McCanles explores, it undermines Wilson's relatively naive belief in a simple empirical understanding of the world that would ensure consilience. Where Wilson assumes that our knowledge *is* simply equivalent with the data that we assemble and then analyze, McCanles suggests that Baconian knowledge is a much more circuitous set of signposts and indicators–what Bacon would sometimes refer to as "prerogative instances," gesturing towards truths that we can access, but whose essence we can never fully know.

Catherine Martin argues that we do Bacon a disservice in accepting the "monological readings" of recent feminists who claim that Bacon viewed "Mother Nature" simply as matter for conquest, or the "feminine mind" as that dangerous ingredient of the human mental apparatus that brings forth a teeming "brood of errors." As Martin shows, Bacon wrestled with traditional rhetorical tropes of feminine intellectual inadequacy. He frequently reverses his gendered terms, instead viewing nature as male and human art or science as female, as in his reading of the myth of Atalanta (art) and Hippomenes (nature). In fact, she contends that Bacon sought to promote a view of the marriage of art and nature that recognized and respected the rightful powers of each, while maintaining a productive equilibrium between them. The subtle and strenuous workings of nature must be matched by a feminized natural philosophical prudence and imagination that could, if properly restrained, heroically advance the human condition. Martin traces the fate of Bacon's gendered language in the hands of later Baconians like Sprat who eliminated Bacon's appreciation of the feminine, and Milton who championed it in his notions of natural philosophy as well as actual companionate marriage. Ultimately, Bacon's championing of a "feminine" prudential and pragmatic approach to knowledge makes Wilson's project to unify the arts, sciences, and humanities conceivable.

John Briggs takes up the difficult task of analyzing how E. O. Wilson and Bacon integrate transcendental traditions into their scientific perspectives and hence contribute to the unity of knowledge. Briggs focuses on the way in which Wilson and Bacon respectively "reject", but then "partially rehabilitate", the "persistence" of "certain [transcendental] *ideas*" (89–90) within their unifying projects. To make matters more complex, Briggs focuses on Bacon's and Wilson's notions about the status and effect of not any old ideas, but of the category of "metaphysical ideas" themselves. According to Briggs, Bacon borrows from ancient Neoplatonism the transcendental concept of "*eidos*" or transcendental "Form" and translates and rehabilitates it into his notion of a materially embedded "Law" of nature. This notion does not totally extinguish all of the transcendental meaning of the notion of "form", but lends to it the important weight of material and operational force, specifically the ability of human beings to transform any bit of matter into any other bit of it. For Bacon the origin of the idea of "Forms" or of other ideas–such as the essential nature of human beings–can only be explained as a coming forth from the "divine limbus" (93), that is, a border or spontaneous opening out from which divine truths come into human understanding. For Briggs, this "limbus" or opening can be unproblematically viewed as simultaneously divine and empirical. Although the origins of such ideas are thus in a sense logically undecidable (being either the consequence of divine will, mechanistic fate, or chance), they are nonetheless susceptible to empirical investigation, calculation, and testing. It is thus, for Briggs, that Bacon's metaphysical ideas can stand in a mutually interacting relationship with his empirical actions and perspectives. In a similar vein, Wilson's views about the sociobiological origin of transcendental ideas about religion and the human moral sentiments do not invalidate the fact that such ideas can have real effects on both the course of human being, human thinking, and on the very genetic biology that spawned them in the first place.

Unlike Briggs, Jerry Weinberger digs below the fragile and fabricated surface unity of reason and revelation in the *New Atlantis* to expose what he sees as an essential conflict between the two in Bacon's thought. For Weinberger, Bacon achieves only a superficial diachronic unity between reason and revelation through a perfunctory, merely rhetorical, tipping of the hat to that consilience. Weinberger notes that, for E. O. Wilson, Enlightenment thinkers failed to find the means to make religion consilient with scientific reasoning. They pointedly failed to come up with explanations for religious thought and feeling that were consonant with an appreciation of the kinds of emotions, desires, and needs that bring religious experience and feeling to life in human societies. Wilson has firmer hopes for an Enlightenment program informed by a sociobiology that can rationally explain religion in evolutionary terms, while still doing justice to those kinds of emotions, desires, and needs. Weinberger, however, sees Wilson as begging the difficult intellectual question of whether spiritual and rational knowledge can be unified by simply assuming that religious faith or yearning is a natural component of human beings, shaped by evolution and necessary for the survival of the species in social grouping.

Weinberger goes on to argue that Bacon "concluded both for and against the unity of [scientific and religious] knowledge" (31). Unlike the traditional scholarly view that Bacon made a clear division between the realms of faith and science (and thus that Bacon did not propose the unity of all knowledge), he argues that Bacon did believe that religious and scientific truth necessarily shared certain background assumptions involving evidence, rational understanding, and argumentation that put them on a conceptual par. For example, the New Atlantians depend upon their scientists to discern that a religious miracle is indeed a miracle and not a product of nature's common courses. In other words, rational understanding is part of our ability to believe. Moreover, human beings rationally believe in the goodness of God (why would they otherwise believe at all?) and the moral precepts that derive from that goodness. However, as Weinberger contends, Bacon realizes that many of our basic moral beliefs–such as the notion that virtue is its own reward–are in fact internally inconsistent in their logic. Consequently, there is the possibility that if one followed through on the rational conclusions that the internal inconsistencies in moral truths suggest, scientific reason would indeed undermine religious faith. Given the resistance that Bacon could have expected from believers (not to say state officials!) to such a revelation, Weinberger claims that Bacon preferred to remain obscure and secretive on this matter.

The five remaining essays in this collection demonstrate how Bacon, sometimes opportunistically, embraced synchronic unity–or even failed to do so. In dealing with Bacon's thoughts on the prolongation of life, Guido Giglioni simultaneously reveals Bacon's belief in the existence of a foundational, primitive philosophy that synchronically and phenomenologically undergirds all knowledge. Giglioni's analysis of Bacon's thoughts on medical knowledge and the prolongation of life conforms to Wilson's claim that Bacon sought to establish scientific disciplines on universal and unifying rational principles. Giglioni contends that Bacon's notions concerning the prolongation of human life were founded on a theory of vital matter, more essentially alive than fleeting animate life, because more permanent in its natural appetite for rest rather than in animate life's quest for constant, evanescent motion. As Giglioni explains, the "permanence" of "inanimate life, and not the transience of the animate lives is Bacon's most genuine meaning of life" (141). Bacon's quest is to understand the primitive, "primal and constitutive appetites of matter" (140) that he conceives as universal "physical reasons" that appertain to the entirety of the natural world, celestial and sublunar. In so doing medical scientists and natural philosophers will come to understand the deep structure of human life, prolong its extent, and heal its frailties. Ultimately, Giglioni points out, the life of man is but part of the life of matter and, consequently, according to Bacon, we may alter a man's life span if we know and control the inner workings of matter and their effects.

Daniel R. Coquillette considers the extent to which Bacon's program of legal reform still speaks to the concerns of twenty-first-century institutions of legal learning. Like Giglioni, Coquillette takes pains to show that Bacon sought to discover universal or foundational principles, here ones that subtended the operation of law in different legal systems. Bacon believes that such principles will become evident through empirical research and rational analysis of rival legal systems. Once discovered, these principles were to be compiled in the form of aphorisms and maxims for study by law students and lay persons alike. Such research into, as well as compilation and transmission of, common principles would prevent the law and legal education from succumbing to parochialism and excessive compartmentalization. Bacon, in addition, argued for the importance of a well-organized and indexed legal archive as well as for effective teaching methods that relied as much if not more upon practical experience as upon textual authority. By combining legal theory with practice, Baconian legal education institutes that consilience between principle and operation that Wilson sees as a hallmark of Enlightenment thought. As Coquillette advocates, contemporary law schools should look to Bacon to initiate badly needed reforms in current law school curriculum and pedagogy.

In writing about Bacon's reform of legal education, Coquillette demonstrates Bacon's concern for unifying and unified institutions of knowledge. In partial contrast, William Lynch considers the various ways in which members of the Royal Society pursued a Baconian methodological program in the mid to late seventeenth century. Lynch contends that the Society spawned a kind of methodological unity within diversity with which Bacon himself would have been at home. Lynch shows how Bacon's focus on "things in themselves" could be interpreted in three distinct, but often overlapping and interrelated, ways in Bacon's works and in the writings of such self-proclaimed followers as John Evelyn, Robert Hooke, and John Wilkins. "Things in themselves" could refer to knowledge in the form of 1) simple visual observation of empirical facts, 2) manual and experimental construction of material effects, or 3) "theoretical realism"–the belief that nature contains primal powers or forms that the natural philosopher could learn to assemble or "alphabetize" in order to generate or "spell" nature's complex physical processes. For Lynch, Bacon's championing of the "nature of things" versus ungrounded hypotheses provided for important collaborative and methodological unity among members of the Royal Society, even though his diverse views about how those "things" could best be accessed also led to important disagreements among those early natural philosophers. Lynch's analysis of the different ways in which members of the Royal Society accessed the empirical fact inevitably complicates Wilson's notion that empiricism in itself will lead to the unity of knowledge.

Fritz Levy explores the unifying use of Baconian method by examining the latter's notions of historical thought and the degree to which Baconian inductive method could serve historical as well as natural philosophic research. Levy examines the theory of history laid out in Bacon's *Advancement,* as well as the practice set forth in *The History of The Reign of King Henry VII*, in light of contemporary debates over historiography. These include the question of history's promotion of moral or political knowledge; advancement of private or public prudence and behavior; its relationship to fiction; and the value of new archival research and archeological remains versus traditional narrative authority. In line with his architectonic procedure of dividing all learning into coherent parts, Bacon assigns history to the part of the mind associated with memory, whereas narrative fictionalizing is associated with the imagination. In turn, he divides history writing into four categories: natural history, civil history, ecclesiastical history, and literary history. Levy's Bacon, like Wilson's, establishes consilience between natural philosophy and history by arguing for the historian's duty to first assemble a large body of historical data much as scientists must assemble natural historical data before drawing rational conclusions. Induction thus reigns supreme as the intellectual instrument of choice in both the humanities and the sciences. Bacon must therefore assign historical interpretation or "ruminated history" to the realm of civil knowledge within the domain of philosophy. However, Levy's focused analysis of Bacon's theory of history also departs from Wilson's view that Baconian history and natural science are philosophically consilient. Levy does this by suggesting that Bacon struggled with the problem of where to place the notion of "historical cause." Was it embedded in the narrative of the historical data itself and thus a component of empirical history, or was it more akin to the kind of analysis, commentary, and interpretation prevalent within "ruminated history"? Within natural philosophy, the ascertaining of "cause" was clearly the endpoint of natural philosophic analysis for which the empirical data collection was simply the means. No clear separation of "cause" from simple fact or data was actually available to Bacon in the realm of historical learning.

Finally, Timothy Reiss shows Bacon's indebtedness to the experience and discourse of early modern exploration, thus suggesting Bacon's ability to assimilate other contemporary intellectual discourses into his emerging natural philosophical perspective, if not indeed capitalize upon them. Specifically, Reiss examines the ways that Bacon redeployed contemporary cultural discourses to articulate the character of his natural philosophy. His essay establishes the profound conceptual and discursive dependency of Baconian natural philosophy upon the European project of travel to, trade with, and colonial settlement of previously little-known or unknown regions of the world. For Reiss, Bacon's notions of inductive method–experience, history-compiling, interpretation, invention, and transmission of new knowledge–is inflected by the early modern European experience of overseas voyaging, discovery, trade, and settlement. While not directly addressing Wilson's notions on the possibility of unifying human knowledge, Reiss underscores Bacon's belief that there was a '"great affinity and consent between the rules of nature, and the true rules of [human political policy)"'(224). Moreover, he points out that Bacon admired the Spanish Empire insofar as its "very organization" expressed the "primacy of 'philosophia prima,' the 'universal Science' containing these principles as 'the Maine and common way' to all knowledge"(226). While for Wilson the unity of Baconian knowledge is made possible by the universal application of the inductive method, Reiss suggests that in addition to induction the Baconian unity of knowledge also depended upon an effective use of alphabetic modeling derived from human language. This unity was dependent upon an ordering of the essences and qualities of the natural world along the lines of linguistic phonemes (what Bacon described as an "abcedarium"). These basic natural units would then be made to match up with, interpenetrate, and correct human understanding of nature. According to Reiss, the new voyages of exploration and colonization–with their dependence upon linguistic interpreters and their efforts to create such interpreters among native inhabitants–culturally converged with the new science and led Bacon to conceive of the importance of language and its structures of organization within his natural philosophic project.

A reflection on the Baconian via media between the arts and sciences

All of the essays in this collection make detailed and pointed contributions to the specific area in Bacon studies that they address, whether that be in law, medicine, rhetoric, history, religion, geo-politics, the Royal Society, or various strains of natural philosophic thought. And each in its own way contributes to an enlarged understanding of Bacon's ways of wrestling with the problem of unified knowledge. But none of our essays directly address this question: upon what fundamental grounds did Bacon seek to maintain or augment unity between the sciences and the humanities, or more largely between the study of nature and the study of a human consciousness that precipitates both products and actions?

In the remainder of this introduction, I consider this issue. For heuristic purposes I set up E. O. Wilson and William Wordsworth as my philosophical straw men to represent radically different answers to the question of how unity between the study of nature and the study of human consciousness can be achieved. Wilson, as I argue below, founds his view upon genetic materialism, Wordsworth, on the essentially complex and harmonizing character of human consciousness. By way of Shakespeare I introduce a middle way between Wilson and Wordsworth. This way acknowledges an essential dialectic between the way of matter and the way of consciousness. In positioning Bacon within this middle way, I point out the crucial Baconian concept of the unknown–whether as yet or perhaps forever–a concept that provides suppleness and flexibility to Bacon's way of thinking.

In *Consilience*, Wilson credits Bacon with the first important early modern formulation of the beneficent Enlightenment goal of rationally unifying all human knowledge and thereby viewing, as Bacon asserts, "all knowledge" as a single "province." According to Wilson, Bacon inaugurated this project by contending that inductive method should serve as the universal instrument for reforming both the sciences and the humanities. Moreover, Wilson asserts that Bacon believed in the unified structure of nature and thus of our knowledge of it. In my view, there is ample evidence in Bacon's works to support these claims. To begin with, the Lord Chancellor advocated the study of the underlying "laws" or forms of "nature" as a key to theoretical and operant understanding. And he surely believed that the various intellectual disciplines or arenas among the arts, sciences, and humanities could illuminate one another. It is not insignificant that the author of the *Novum Organum*, or new scientific method for regulating natural philosophical conduct, was also the begetter of the *Essays,* meditations on moral and civil conduct.

Michael Kiernan, the editor of the new Oxford edition of *The Advancement of Learning,* notes Bacon's belief that the various divisions of learning that he had articulated will be "fruitful and interactive" (*OFB*, IV: xxvi). As Bacon himself underscores, "we have endevoured in these our Partitions to observe a kind of perspective, that one part may cast light upon another." Elsewhere in the *Advancement* he advises that "all partitions of knowledges, be accepted rather for *lines & veines*, then for *sections and separations:* and that the continuance and entirenes of knowledge be preserved" (*OFB*, IV: 88, 93).

Bacon's belief in the unity of knowledge similarly comes to the fore in his discussion of philosophy. Bacon divided philosophy into three branches that he named, respectively, "DIVINE," "NATURAL," and "HUMANE." All of them derived from a more basic, antecedent form of philosophy that he termed "PHILOSOPHIA PRIMA, PRIMITIVE or SUMMARIE PHILOSOPHIE." Bacon explains that

> because the distributions and partitions of knowledge, are not like severall lines, that meete in one Angle, and so touch but in a point, but are like branches of a tree, that meete in a ste*m*me; which hath a dimension and quantitie of entyrnese and continuance, before it come to discontinue & break it self into Armes and boughes, there-fore, it is good, before we enter into the former distribution, to erect & constitute an universal Science by the name of PHILOSOPHIA PRIMA. (*OFB*, IV: 76)

For Bacon this foundational philosophy should "*bee a Receptacle for all such profitable observations and Axioms, as fall not within the compasse of any of the speciall parts of Philosophie, or Sciences; but are more common, and of a higher stage.*" Such would include such concerns as "*Quantitie, Similitude, Diversitie*, and the rest of those *Externe Characters of things,*" as well as other axioms derived from arithmetic, geometry, religion, music, natural philosophy, medicine, optics, physics, and civil administration, among others that are common to more than one field of knowledge (*OFB*, IV: 76–7).

As the founder of sociobiology–a wide-reaching natural science that claims genetics, animal behavior, psychology, and cultural development, among other things, as its province–Wilson, not surprisingly, sees himself as following in Bacon's path towards scientific unity and universalism. For Wilson, such unity takes shape holistically. Genetic potentialities are first realized in particular natural settings. Then, they give rise to genetically framed behaviors which in turn alter the "natural environment" and, in the higher primates, and human beings, give rise to cultural variations–including scientific study and interventions. These cultural variations also go on to shape genetic potential. In seeking knowledge about the natural world, the human place within it, and our role in adapting to and remaking that world and that place, Wilson presumes a non-Kantian unity between phenomena and noumena, between the world as we experience and alter it, and the world as it exists outside of us. Not that phenomena and noumena are necessarily the same for Wilson; just that there are empirical and rational ways to go about explaining their convergence and divergence. Ascertaining the deep structure of that relationship is in large part what science should be all about.

Thus it is with Wilson much as it was with Polixenes in Shakespeare's *The Winter's Tale*. The latter reassures Perdita that growing hybrid flowers is not a transgression of nature. For Polixenes, as for Wilson, the human experience of the world, and the arts we have developed to transform that experience, are comprehended by nature and its myriad potentialities. Hence, claims Polixenes:

> … nature is made better by no mean
> But nature makes that mean: so, over that art,
> Which you say adds to nature, is an art
> That nature makes …
> This is an art
> Which does mend nature–change it rather–but
> The art itself is nature. (4.4.89–97)[5]

[5] William Shakespeare, *The Tempest*, ed. Brian Morris (New York: Routledge, 1981).

In the course of his argument, Wilson attacks what he sees as the intellectual nihilism of postmodernist and poststructuralist thinking that undermines any effort at intellectual unity or rational universalism based on a founding faith in nature's unity. Wilson grants, however, that postmodernist nihilism has roots in the charming and disarming nineteenth-century Romantic rebellion against the prison-house of Enlightenment mechanistic rationalism that shackled both the human spirit and its freedom. Wilson sympathizes with Romanticism's fears, but finds freedom and self-determination less in metaphysical spirit than in the ever-expanding adaptability and evolution of a human consciousness constructed through the historical development of matter's potential. Of course, for rhetorical effect, Wilson simplifies the more complicated relation among Romantic thinkers and nineteenth-century science.

Yet in fact, Shelley was a great admirer of Bacon as a rebel against tradition and a liberator of human thought, and, like other Romantics, saw science as both positive and negative, not simply the enemy of metaphysical faith in the human spirit or its freedom. In the *Lyrical Ballads*, Wordsworth similarly views the poet as the materially grounded companion of the new scientist. The poet's job is to integrate new scientific knowledge into human experience and to make this new knowledge meaningful within the full range of a universal human condition that is always historically particular. Wordsworth uses the term "pleasure" to connote all the sensations, sympathies, and desires that impel our intellectual and physical interactions with the world around us–even painful ones–and make them meaningful. He is thus also a universalist trying to find a basis upon which to unify human knowledge. But unlike Wilson, he does not regard empirical or scientific rationalism as a canvas generous enough for the achievement of this goal. Certainly, Wordsworth is ideologically and historically unable to consider the sociobiological idea that human experience itself is at least initially determined by nature. For him, the most universal platform must be one based upon the human phenomenology of intellectual, emotional, and sensational experience. Human beings are human in their essence because they share a common complex consciousness. The poet achieves unity of knowledge by the ability to access that complex human consciousness that alone harmonizes perception, sensation, thought, and feeling into one. Thus, as Wordsworth explains, the poet inevitably circumnavigates the activities of the scientist:

> What then does the poet? He considers man and the objects that surround him as acting and reacting upon each other, so as to produce an infinite complexity of pain and pleasure … he considers him as looking upon this complex scene of ideas and sensations, and finding everywhere objects that immediately excite in him sympathies which, from the necessities of his nature, are accompanied by an overbalance of enjoyment… He considers man and nature as essentially adapted to each other, and the mind of man as naturally the mirror of the fairest and most interesting properties of nature… The man of science seeks truth as a remote and unknown benefactor, he cherishes and loves it in his solitude: the poet, singing a song in which all human beings join with him, rejoices in the presence of truth as our visible friend and hourly companion… the poet binds together by passion and knowledge the vast empire of human society, as it is spread over the whole earth, and over all time… If the labors of men of science should ever create any material revolution, direct or indirect, in our condition, and in the impressions which we habitually receive, the poet will sleep then

> no more than at present; he will be ready to follow the steps of the man of science, not only in those general indirect effects, but he will be at his side, carrying sensation into the midst of the objects of the science itself. The remotest discoveries of the chemist, the botanist, or mineralogist, will be as proper objects of the poet's art as any upon which it can be employed, if the time should ever come when these things shall be familiar to us, and the relations under which they are contemplated by the followers of these respective sciences shall be manifestly and palpably material to us as enjoying and suffering beings.[6]

Given these views, one could easily imagine Wordsworth sympathizing with Perdita's response to Polixenes' claim that nature simply comprehends human art and experience and that therefore she should not refuse to plant hybrid flowers in her garden. Perdita at first verbally agrees with Polixenes's argument with her response, "So it is" (4.4.97). Yet she then goes on to proclaim that she will still not plant hybrid "gillyvors" in her garden:

> No more than, were I painted, I would wish
> This youth [her beloved Florizel] should say 'twere well, and only therefore
> Desire to breed by me. (4.4.101–3)

Most critics tend to see Polixenes as the logical victor in this argument, since Perdita appears illogical in that she verbally accedes to his conclusion and yet still refuses to plant the gillyvors. I would contend, however, that Perdita is simply being polite in granting Polixenes his premises and thus his logical conclusion. In her extended response, she subtly shifts the terms of the argument away from the structural question of the relation between human art and nature, and toward the issue of her perception of the moral meaning of hybrid gillyvors as that perception is shaped by her culture, education, and personal experience. She acknowledges that, scientifically and logically, gillyvors are products of a nature that makes possible the human art of horticulture. However she, like a proto-Wordworthian poet, circumnavigates Polixenes's argument by placing it within a context of social and personal meaning, feeling, and experience that for her at this particular moment trumps his seemingly sociobiological argument. She equates hybrid gillyvors with "painted" ladies, that is with vanity, hypocrisy, and superfluousness. She finds this meaning not in "nature" per se, but in her moral environment. And for her the moral vision she has imbibed from that environment is not reducible in a meaningful way to nature or sociobiology. Her moral vision is the product of human social and spiritual practices.

Having wandered into the concerns of Wilson, Wordsworth, and *The Winter's Tale*, I must return to the question of where we should position Bacon's views on the possibility of unifying all knowledge–arts and humanities–and upon what basis. Surely, Bacon was a man who possessed, among other things, a rational, empirical, and methodical turn of mind. He revered the empirical particular. He was a list-maker, an organizer, a classifier. He was interested in examining the relation among parts and wholes, the rational, logical and empirical interconnections between the

---

[6] William Wordsworth, "Preface to the Lyrical Ballads," rpt. in *Critical Theory Since Plato*, rev. ed., ed. Hazard Adams (New York: Harcourt Brace Jovanovich, 1992), 442.

various disciplines of knowledge. If he had not been so inclined, he would not have been so apt to write *The Advancement of Learning*. And he was fascinated with the concept of fundamental natural laws or forms and the pursuit of a foundational "primitive" or primary philosophy that would lay out the essential structure of all forms of being. To that extent, Bacon does appear to conform to E. O. Wilson's vision of him as the founder of the modern Enlightenment faith in the idea that empirical research will yield the rational unity of all knowledge.

Nonetheless, Bacon was also extremely wary of any attempt at "rational unity" that would result in premature closure of the continuing pursuit of knowledge. He was suspicious of grand, overarching theories, the "Idols of the Theater" that he believed were not supported by sufficient empirical observations. One of the "Idols of the Tribe" that he particularly abhors is the tendency of the mind to find more regularity in nature that in fact exists. Consequently, the continued pursuit of knowledge must necessarily involve acknowledgment of knowledge's continued state of fragmentation–and thus of what remains unknown. As Brian Vickers notes, *The Advancement of Learning* is made up of "deficiencies and gaps; it is an 'encyclopedia' of work needed, of not yet existing knowledge, an encyclopedia of *lacunae*, as it were, which a new philosophy would fill in."[7] Moreover, by writing the *Novum Organum* (1620) in aphorisms, Bacon chose to emphasize the fragmentary state of knowledge, both then and for some time to come. In assessing Bacon's role as an early modern progenitor of the unity of knowledge, it is important to take account of his interest in the particularity or phenomenology of human vision or experience of the world. What may remain unknown or distorted, or out of our intellectual reach in the present (or in the indefinite or always deferred future) is always on Bacon's mind. Bacon's efforts to analyze and correct what he saw as those psycho-social aspects of human cognition that he felt interfered with the gathering of true and useful scientific knowledge (the various idols of the mind, encompassing, among other things, emotions, physical desires, impatience, desires for instant gratification, etc.) also bespeaks his interest in and acknowledgment of the rather Wordsworthian fact that human beings experience, know, and sense the world in particularly human ways. For Wilson that means that human knowing and its intellectual, cultural, and material achievements are the product and outcome of a lengthy period of human genetic evolution. For Bacon, however, in an era before the notion of evolution, let alone genetics, the peculiar nature of the human mind must be viewed as either a divine gift or a mysterious given. In any event, the mind was not a thing whose origins could be explained in terms of material development. Both capable and delimited, it was foundational in its own right. Moreover, while I have argued elsewhere that Bacon was interested in promoting a proto-modern form of objectivity, a form of self-distanced or disinterested cognition for the purpose of advancing natural philosophical knowledge, he did not advance what Thomas Nagel has called the "view from nowhere"–the late modern (or perhaps postmodern) notion of scientific objectivity

[7] Brian Vickers, "Francis Bacon and the Progress of Knowledge," *Journal of the History of Ideas* 53 (Jul.–Sep. 1992): 504, note 30.

that Nagel claims typifies contemporary twentieth and now twenty-first-century science.[8]

Baconian scientific disinterestedness still retains the notion that human beings must view empirical facts from an earthbound perspective–from somewhere. Universality, comprehensiveness, and conceptual unity of course are intrinsic to Bacon's intellectual goals. Nonetheless, they can only be achieved by enlarging human perspective through human efforts at ascent and development, not by eliminating that perspective. One example of Bacon's stance occurs in a memo wherein Bacon advises James I on ecclesiastical matters. In it, he argues using the analogy of surveying, saying: "it is many times seen that a man that standeth off, and somewhat removed from a plot of ground, doth better … discover it than those which are upon it, I thought it not impossible but that I, as a looker on, might cast mine eyes upon some things which the actors themselves (specially some being interested, some led and addicted, some declared and engaged) did not or would not see" (emphasis mine; *SEH*, 10: 103). Here Bacon seems to suggest that better knowing involves not the elimination of all perspective, but rather the expanding of the perspective that one does have. Similarly, in discussing "Another Errour" which plagues human knowing in *The Advancement of Learning*, Bacon asserts that "men have abandoned universalitie, or *Philosophia prima*; which cannot but cease and stoppe all progression. For no perfect discoverie can bee made uppon a flatte, or a levell. Neither is it possible to discover the more remote, and deeper parts of any Science, if you stand but upon the levell of the same Science, and ascend not to a higher Science" (*OFB*, IV: 30). For Bacon, that "higher Science" is not beyond, but very much within, human ken.

Perhaps we could position Bacon somewhere between Wilson and Wordsworth. With Wilson he advances the importance of unifying knowledge by means of empirical research and rational method. With Wordsworth he acknowledges that such unification will always take place under the aegis of an essentially human perspective that can never be reduced to material determinants. From a theological perspective, this means simply that human beings know by means of a divinely created soul that possesses certain abilities and limitations. From a non-theological, somewhat existential perspective, it means that the specificities, vagaries, and complexities of our current consciousness (at whatever time we choose to isolate it) always exceeds its material origins.

Should we then label Bacon a circumscript or moderate universalist? A practical, or pragmatic unifier of human knowledge rather than a strictly rational and Enlightenment one? Should we accept Wilson's persuasive argument that Bacon went as far with his universalism and unifying propensities as we could expect of anyone of his era? Did Bacon intentionally set out to establish a "primitive" or foundational philosophy upon which all intellectual disciplines would logically and/or empirically depend? Or did he find himself struggling to piece together–sometimes in an *ad hoc* manner–cohesive intellectual discourses of knowledge out of the complex remains of ancient traditions and newly emergent conversations? Should we

[8] See Solomon, *Objectivity in the Making*, and Thomas Nagel, *The View from Nowhere* (New York: Oxford Univ. Press, 1986).

view him as an effective progenitor of later efforts to unify science in workable public institutions or only as an imaginer of fantastic natural philosophic bureaucracies such as in the *New Atlantis*? Rather than attempt to answer these conundrums myself, I now leave to the readers of this volume the Wordsworthian task of assimilating the arguments of these essays into their own frames of reference. I would like to assume, with Bacon, that in doing so they do their best to stretch those frames into wide and generous modes of human understanding.

Chapter 2

Bacon, Atomism, and Imposture: The True and the Useful in History, Myth, and Theory

Reid Barbour

In *Cogitationes de natura rerum,* or *Thoughts on the Nature of Things,* Bacon offers his clearest and most far-reaching assessment of ancient atomism's value to his advancement of learning. "Doctrina Democriti de atomis aut vera est, aut ad demonstrationem utiliter adhibetur," he claims at the outset of a section entitled "On the Division of Bodies, Continuity and Vacuity": "the doctrine of Democritus concerning atoms is either true or useful for demonstration" (*SEH*, V: 203; X: 287). For the time being, Bacon suspends an inquiry into the relationship obtaining *between* the criteria of truth and utility. Indeed, here and elsewhere in Bacon's writings, the utility of atomism figures more prominently than any resolution in favor of its truth. With regard to the axioms of a revised natural philosophy, after all, the truth about nature ought to depend on as well as effect the changes taking place within the material circumstances of the world. In *Cogitationes de natura rerum,* the very title of which hearkens back to the great atomic poem of the first century BCE, Lucretius's *De rerum natura,* however, Bacon proceeds to explain that atomism is useful to the extent that it is rhetorically and conceptually accommodating: "For it is not easy either to grasp in thought or to express in words the genuine subtlety of nature, such as is found in things, without supposing an atom" ("nisi supponatur atomus") (*SEH*, V: 203; X: 287). For Lucretius, poetry sweetens the bitter pill of atomic truth in the face of that truth's theological and ethical implications. For Bacon, atomism facilitates and comforts the beleaguered pioneer in the wilderness of subtle particulars.

The notion that thinking and speaking about atoms might *ease* the ideation or expression of natural subtlety confronts Bacon with the possibility that particulate theories rank among those idols whose simplicity or elegance seduces the human mind. In other words, Bacon's accent on atomic accommodation concedes the possibility that particulate theories commit imposture on the very human minds whose superstitions the atomic theory is meant to explode. In short, atomism threatens the mind with an active, even aggressive falsehood in lieu of the passive lies on which the mind has come to rely. But Bacon is clearly approving atomism, even if his many efforts to use it in his musings about the nature of things reveal to him time and again that atomism in general and in its early modern context is never easily deployed.

More than any other physical theory available in Bacon's day, atomism involves its early modern students in ethical, epistemological, and theological affiliations that subvert any straightforward or confident accreditation of what is true about, useful within, or true and therefore useful regarding the subtle phenomena of nature. What is more, Bacon understands that atomism has always been so burdened, from Aristotle's hostile rejection of the theory right through 1600, when that Neoplatonic atomist, Giordano Bruno, was burnt at the stake in Rome. Paradoxically, atomism as both a physical theory and a method offers what Bacon considers his generation's very best opportunity for the advancement of learning. As the *Cogitationes de natura rerum* intimates, the separation of legitimacy from imposture in atomic theory is not simply a matter of effacing or rewriting its anti-providential theology or its ethics of pleasure so as to ensure its piety and virtue. The legitimacy of atomism also depends on keeping the theory close enough to the criterion of physical truth so as to prevent its lapse into the mental and cultural idolatry that for Bacon afflicts the contemporary field of natural philosophy.

What emerges most clearly over the course of Bacon's writings is that atomic theory is a way of seeing and, as such, epitomizes the afflictions of idolatry vexing all human ways of seeing. Thus, Bacon's complex assessment of atomism supports E. O. Wilson's argument that "It was in psychology, and particularly the nature of creativity, that Bacon cast his vision farthest ahead … He had a deep intuitive feel for the mental processes of discovery."[1] Bacon understood that ancient atomism offered its own explanation of perception–that the perceiver's atomic structures would record yet also reshape the filmy simulacra pouring from the object of perception. But in his critical histories of philosophy, Bacon knew more–that the very theory of atomic particles threw off its own ideological simulacra, and that these in turn would be received and refashioned by the distorting mirror of the observing mind.

Epicureans in the Advancement

When Bacon praises the general theory of atoms, he often invokes the name of Democritus. When he intends to vilify some feature of its legacy, he commonly takes aim at the Epicurean interlocutor of Cicero's *De natura deorum,* Velleius, one of Bacon's easiest targets. The entanglements of Bacon's critiques intimate, however, that scolding the profane "Epicure" can entail complications in Bacon's own enterprise: he must carefully separate his own fervid commitment to a concentration on nature's second causes from the abominable Epicurean rejection of providential teleology–the latter rejection always promoted side-by-side with the atomist's summons back to the alphabet of particulate second causes. One such vexed moment occurs in *The Advancement of Learning,* in which Bacon seeks to separate the philosophical truth and legitimate utility of Epicureanism from its replacement of superstition with profane impostures.

---

[1] E. O. Wilson, *Consilience: The Unity of Knowledge* (New York: Knopf, 1998), 26.

Prior to the passage in question, the *Advancement* has already criticized Velleius for epistemological reasons. In *De natura deorum,* two philosophies, the Epicurean and the Stoic, are subjected to the criticism of the Academic Cotta, whose skepticism is, like Cicero's, of the moderate and purposeful variety: "Our position is not that we hold that nothing is true, but that we assert that all true sensations are associated with false ones so closely resembling them that they contain no infallible mark to guide our judgement and assent. From this followed the corollary, that many sensations are *probable*, that is, though not amounting to a full perception they are yet possessed of a certain distinctness and clearness, and so can serve to direct the conduct of the wise man."[2] By contrast to the Academics, both the Stoics and the Epicureans are cast as dogmatists. But Velleius in particular is indicted for being helplessly addicted to the tenets of his master, whose optimism about the truth value of the senses has seduced philosophers into the most childish opinions about, for example, the size of the moon and sun.

Cicero introduces the speech of Velleius, then lambastes the Epicurean's habitually "confident manner": "nihil tam verens quam ne dubitare aliqua de re videretur, tamquam modo ex deorum concilio et ex Epicuri intermundis descendisset" ("afraid of nothing so much as lest he should appear to have doubts about anything. One would have supposed he had just come down from the assembly of the gods in the intermundane spaces of Epicurus").[3] Cicero is already introducing the theory of an otiose divinity to which Cotta soon turns with scorn, but for the moment what matters is that the Epicureans have their heads in a cloud, unable to see clearly or beyond the fog of their sectarian mind set, one in which Velleius has been raised up from a young age.

In citing this passage in the *Advancement,* Bacon rejects what he considers an overly confident or magisterial delivery of knowledge, one "for the most part Magistrall and peremptorie; and not ingenuous and faithfull, in a sort, as may be soonest beleeued; and not easilest examined" (*OFB*, IV: 31). Whether or not the Epicurean representative believes the dogma that he purveys, he certainly values affiliation over inquiry, fellowship over fidelity. His is a philosophical discourse that works in the way that Bacon says Ciceronian rhetoric can, foreshortening the pursuit of tomorrow's truths by means of an immediate gratification in the appearance of mastery that is in operation today. Bacon adds, however, that Velleius's is not the only error into which the "handling of learning" can fall, for at the other extreme is Socrates, who doubts and applies irony to everything. Regarding the attempt between Velleius and Socrates "to propound things sincerely, with more or lesse asseueration: as they stand in a mans owne iudgement, prooued more or lesse," Bacon does not yet introduce the figure of Democritus–as he will in *De sapientia veterum,* for instance[4]–as proceeding according to a legitimate method. Nor does he explain how we are to judge his own characterization thirty pages later in the *Advancement* of the

[2] Cicero, *De natura deorum,* trans. H. Rackam (Cambridge, Mass: Harvard Univ. Press, 1951), 15.

[3] Cicero, *De natura deorum,* 20–1.

[4] See Reid Barbour, *English Epicures and Stoics: Ancient Legacies in Early Stuart Culture* (Amherst: Univ. of Massachusetts Press, 1998), chapter two.

ironic Socrates as "a true and vnfained inquisitor of truth" whose "wandring manner of Inductions" is not ideal but certainly better than vain, dogmatic deduction. Even so, it is clearly the case that Velleius suffers under at least three of Bacon's idols: the cave of idiosyncratic preferences, the theater of received opinions, and the trade in sophistry called the idol of the marketplace.

When in the *Advancement* Bacon comes to analyze the general fallacies of the mind, however, he treats Velleius with comparable scorn yet also with what proves a kind of self-reflexive complexity. Referring to the human mind's corruption in general, to what elsewhere he calls the Idol of the Tribe, Bacon explains that correcting this general fallacy will do more than anything else "to rectifie IVDGEMENT" (*OFB*, IV: 116–17). The mind, he argues, is an enchanted glass that distorts and perverts the world that it should authentically reflect, and he diagnoses this mental disease as being "full of superstition and Imposture, if it bee not deliuered and reduced." The mind, in short, must be freed from the knots of superstition and returned as closely as possible to a clean, clear capacity for reflection. For an example of the kind of superstition that afflicts the mind, Bacon avers that it is far more readily influenced by affirmatives than by negatives; thus, in pagan religion with its recourse to divination, the double recurrence of an event impresses the human mind to a degree that the absence of an event simply does not.

With superstition in place as a first instance of the mind's general disease, Bacon turns to a second instance that leads him back to Velleius. Because our human spirit is essentially uniform, he maintains, we project uniformity onto nature: mathematicians, for instance, superimpose perfect patterns on the motions of the planets and strive to jettison irregularities. In consequence the human mind cannot brook the possibility of singularities in nature, of unique things operating according to their very own mode of operation. The mind invents parallels in carrying out the human conviction that man is the measure of all things, filling the study of nature with "fictions and fantasies."

At this point, Bacon draws an analogy or cites an example of man-as-measure that complicates everything he has said thus far about human error. What philosophers fantasize is "not much better, than the Heresie of the *Anthropomorphites* bredde in the Celles of grosse and solitarie Monkes, and the opinion of *Epicurus*, answearable to the same in heathenisme, who supposed the Gods to bee of humane Shape" (*OFB*, IV: 116–17). It is a challenging chain of associations–cosmic mathematicians, heretical monks, and the pagan Epicureans–even before Bacon settles on Velleius. What seems clear, however, is that the Epicureans are liable to complicity in an error more aggressive in fictionalizing the world than the largely passive, built-in superstition that they promise to expose.[5] This recoil from superstition into imposture is not theirs exclusively, for the Epicureans are Bacon's synecdoche for any promising approach to truth and utility that fails to inspect and jettison its own self-interested fabrications.

[5] For a brief history of the passive and active capacities of the mind in medieval and early modern thought, see Julie Robin Solomon, *Objectivity in the Making: Francis Bacon and the Politics of Inquiry* (Baltimore: Johns Hopkins Univ. Press, 1998), 27–37. Here and throughout this essay, I am indebted to Julie Solomon for her help with Bacon's epistemology.

The superstitious mind may well embrace the misleadingly consistent affirmatives in nature; but the imposture of atomic philosophy threatens to refashion the world as a philosopher's ethical and theological agenda dictates.

The introduction of Cicero's character Velleius sets the Epicurean atomists at cross-purposes with themselves in regard to the mind's enchanted glass. As Bacon understands Cicero, Velleius appears to be humanizing the gods, even in what Bacon calls the needless attempt to liberate the human mind from its superstitions about the gods: "And therefore *Velleius* the Epicurian needed not to haue asked, why God should haue adorned the Heauens with Starres, as if he had been an *Aedilis*: One that should haue set foorth some magnificent shewes or playes? for if that great Workemaster had beene of an Humane disposition, hee would haue caste the starres into some pleasant and beautifull workes, and orders, like the frettes in the Roofes of Houses, whereas one can scarce finde a Posture in square, or triangle, or streight line amongest such an infinite number, so differing an Harmonie, there is betweene the spirite of Man, and the spirite of Nature" (*OFB*, IV: 116–17). In *De natura deorum,* Velleius is mocking all those philosophers, especially the Stoics, who reduce divinity to the petty level of cosmic administration, an aedile whose task it was to ornament the city in preparation for festivities. From the outset of his diatribe in Book One, he has been deriding those "mere baseless figments of the imagination, such as the artisan deity and world-builder of Plato's *Timaeus*, or that old hag of a fortune-teller, the *Pronoia* (which we may render 'Providence') of the Stoics; nor yet a world endowed with a mind and senses of its own, a spherical, rotatory god of burning fire; these are the marvels and monstrosities of philosophers who do not reason but dream."[6] Insofar as Bacon's own relationship to this critique of providential thinking is intricate and even perilous, it perhaps serves his purpose to dismiss Velleius as pointless at best. But Velleius's mocks lend bravado to Bacon's own critique of the various ways in which human beings thrust their own mental habits, systems, and ideas of order on the world and, as a corollary, on the God that (they believe) created the world.

As Velleius represents them, then, the Epicureans clearly stand against superstition, but they also so virulently oppose the fictional impostures of cosmic order that they rebound from superstition into atheism and profanity. "It would be a lengthy task," Velleius complains, "to advert upon every detail of a system that is such as to seem the result of idle theorizing rather than of real research" ("optata magis quam inventa"); "but the prize example is that the thinker who represented the world not merely as having had an origin but even as almost made by hand, also declared that it will exist for ever. Can you suppose that a man can have even dipped into natural philosophy if he imagines that anything that has come into being can be eternal? What composite whole is not capable of dissolution? What thing is there that has a beginning but not an end?" Bacon too objects to the way in which theories fill the natural world with the divine mind, superimpose neat patterns and final purposes on the complexity and variety of phenomena, and substitute mental fictions for the actual and persistent experience of the world. But Velleius takes Bacon's position too

[6] Cicero, *De natura deorum*, 21.

far: if God has not created and governed the world as human beings think, then God has not created and governed the world at all. If it is wrong to replace the actualities of nature with human conceptions of the divine mind in nature, then the divine mind is irrelevant to the natural world altogether.

The Epicurean position is that the otiose gods encourage neither superstition nor imposture: they require no fear from the mortals too worried about death and damnation, and they exemplify a life without any mental activity or aggression whatever. If Velleius's point is to sublimate the gods above the compromising fray of the imperfect world, however, one implication of his critique for early modern readers is that the gods are so weak as to require a constant repose. The gods have not created and governed the world, nor could they. Velleius's argument that the gods have not created and do not regulate the world is meant to rescue divinity from demotion by the human imagination, but it also participates in what Bacon considers the Epicurean's heretical humanizing of the gods. The Epicureans, and Velleius in particular, have hounded a worthy critique of human superstition into the realm of profanity.

On the one hand, Velleius can appear to defend divinity, arguing that the gods "cannot know fatigue" or that they would never reward the foolish and unjust in the manner in which such men are rewarded in the highly flawed world. He takes aim at a host of contradictory errors and "childish fictions" in which various pagan writers have clothed the gods: they are "roving 'images'" to Democritus, the "law of nature," aether and reason to the Stoics, lusty and violent heroes to the poets.[7] On the other, Velleius also simply mocks any commitment to an active providence, joking that if the gods finally made the world beautiful, some disturbance must have woken them up at last, and their abode must have been ugly before the creation. For Velleius, the various ideas of providence denigrate god to an all too human level. But as Cotta proceeds to show in his response to Velleius, it is the Epicurean notion of gods living out their existence in the so-called intermundane spaces of an infinite universe that vainly reduces divinity to the lowly plain of human nature.

In his dismissal of Velleius, Bacon admits a basic insight in the Epicurean position: natural theology is wrong to derive the order of the universe from some neat rationalistic theory of how God would create the world in his own and in man's image. There is, Bacon says, so great a divide "betweene the spirite of Man, and the spirite of Nature," which means that human ideas about the order to things are as wrong as they are gratifying. Bacon finds himself on more stable interpretive ground when he suspends his engagement with Epicurean theology and attends instead to the atomic physics on which anti-providentialism is based. Atomism certainly figures into Cotta's assault on the Epicurean philosophy; Cotta sees the belief in indivisible particles swerving into collisions and forming worlds by chance as obviously ridiculous. But for Bacon, atomism gains a clearer credibility for its truth and utility once an historical survey of philosophy moves beyond Epicurus to Democritus. For Bacon, an orientation away from Epicurus and toward Democritus promises to jettison as much imposture as is humanly possible.

---

[7] Cicero, *De Natura deorum*, 25, 33, 37.

Democritus recovered

Bacon's works are filled with admiration for Democritus, about whom Bacon would have learned mainly from a Latin translation of Diogenes Laertius. Having sketched the life of the philosopher, Diogenes relates Hippocrates's report that Democritus was an acute observer of phenomena, then offers a synopsis of his views: the universe is made of atoms and void; the number of worlds has no limit, and each world is mortal; nothing comes from or returns to nothing. On some points, Democritus differs from his own atomic successor, Epicurus: for the older philosopher, the atoms have no limit on size, "are borne along in the whole universe in a vortex," and consequently behave according to necessity (Epicurus limits the size of the atoms–they must remain invisible–and he adds the famous "swerve" to guarantee their freedom and fortuity). All qualities are simply the constituents of atoms in motion and in combination with one another. Even though Epicurus comes to be more famously associated with the ethics and theology built on the foundation of atomism, there is a suggestion of both in Diogenes's account of Democritus, a point not missed by Bacon. "The end of action is tranquillity," Democritus is reported to have believed; if unlike Epicurus he was not prepared to identify this tranquillity as pleasure, one key corollary to the calm mind comes very close to Epicureanism, the mind free of fear from "superstition or any other emotion."[8]

According to Epicurus, one's atomic composition requires such calm especially in reaction against the fear provoked by a belief in eternal punishment and by the specter of death itself. In fact, the goal of Epicureanism is to produce a freedom from such fears, not least by way of a theology whose gods do not create, govern, or even care about the world. They too are busy being calm. The problem for Bacon is that this model of a divine nature imposing on no one and nothing scarcely conceals the Epicurean philosopher's impositions on that superstitious mind in search of any release, no matter how fraudulent. Democritus comes to the early modern reader with no such theological accouterments; the "laughing philosopher" may have his idiosyncrasies but, in Bacon's hands, he is more suitable than the Epicurean atomist for the position of cultural hero. Democritus also represents the hope of recuperating passive, empirical experience from the blindness of the superstitious persons who credulously receive that experience as the neatly patterned and direct expression of divine judgment.

In *The Advancement of Learning,* Democritus shows up early and often in Bacon's survey of natural philosophy. Dividing the study of nature into the branches of careful inquiry into causes and of the production of effects, Bacon takes Democritus as the watchword for the former: "If then it bee true that *Democritus* sayde, *That the truthe of Nature lyeth hydde in certaine deepe Mynes and Caues*," Bacon posits, it follows that some natural philosophers must commit themselves to pioneering work of the most vigilant investigation (*OFB*, IV: 80). This search for causes must be grounded in and ascend deliberately from experiments, the "digging" of the Democritean miner in the hidden veins of nature.

---

[8] Diogenes Laertius, *Lives of Eminent Philosophers*, trans. R. D. Hicks (Cambridge, Mass.: Harvard Univ. Press, 1979), 2, 455.

In the search for causes, Bacon proceeds to show, Democritus is as valuable for what he resisted as he is for what he sought. What he resisted was the search for final causes, the purposes of things rather than the matter, motion, or structure. With the Epicurean emphasis on universal fortuity, this resistance to purpose is often understood–and assailed–theologically, as a physics for a world without providence. But Bacon's Democritus is more concerned about the integrity of natural studies, "the seuere and diligent enquirie of all reall and phisicall causes." In a passage from Xenophon's *Memorabilia* that anticipates a similar critique in Lucretius, Bacon finds examples of the way in which the flight to final causes impedes the understanding of nature as it truly is: "For to say *that the haires of the Eye-lidds are for a quic-sette and fence about the Sight: Or, That the firmenesse of the Skinnes and Hides of liuing creatures is to defend them from the extremities of heate or cold: Or, That the bones are for the columnes or beames, whereupon the Frames of the bodies of liuing creatures are built; Or, That the leaues of trees are for protecting of the Fruite; Or, That the cloudes are for watering of the Earth; Or, That the solidnesse of the Earth is for the station and Mansion of liuing creatures:* and the like, is well inquired & collected in METAPHISICKE, but in PHISICKE they are impertinent" (*OFB*, IV: 86). Bacon concedes that final causes are worthy targets of study if kept within their own province, but they have become usurpers on material, efficient, and (properly understood) formal causes. For Bacon, the obsession with final causes is comparable to the remora, that fish with the power out of all proportion with its size to prevent the forward movement of the largest ship.

Bacon's enlistment of Democritus as the hero of "*reall and phisicall causes*" packs all the attractions and dangers of atomism, including an Epicurean atomism, into one elaborate sentence: "And therefore the Natural Philosophie of *Democritus*, and some others who did not suppose a *Minde* or *Reason* in the frame of things, but attributed the *form thereof able to maintaine it self to infinite essaies or proofes of Nature*, which they tearme *fortune*; seemeth to mee (as farre as I can iudge by the recitall and fragments which remaine vnto vs) in particularities of Phisicall causes more reall and better enquired then that of *Aristotle* and *Plato*, whereof both intermingled *final causes*, the one as a part of *Theologie*, and the other as a part of *Logicke*, which were the *fauorite studies* respectiuely of both those persons" (*OFB*, IV: 86). Here is that atomic legacy in all of its anti-teleological brass. Nature is denuded of divine agency and runs its own operations. Those operations are governed by "fortune," which might point to some kind of Democritean determinism but in all likelihood is meant to capture the Epicurean stress on chance: hence, worlds are able to be formed in the void because the infinite atoms are making unlimited (if always hapless) assays at collision and combination. Eventually those atoms hook up and form a world, but no thanks to divine creation or guidance, and by no means in accordance with human logic. Above all, Democritus stands for the careful inquiry into the particularity of natural phenomena, with atoms the ultimate emblem for such a particulate focus.

Crucially, however, Bacon's Democritean philosophy requires what Graham Rees has called a "[t]heological respectability" (*OFB*, VI: 1). The serious and protracted search for this warrant risks imposture as surely as the failure to find any such warrant collapses into profanity. Far from ignoring the Epicurean theological superstructure reared on top of atomic physics, Bacon immediately argues that the Democritean approach to natural particularity in fact saves divinity from vulgarity. This rescue of the divine from the mundane is (as Bacon sees it) an Epicurean argument missed or abused for centuries by polemicists interested only in caricaturing the Epicureans as atheistic gourmands. In *De natura deorum,* however, even Velleius makes the claim that Epicurus plays an extraordinary role in shoring up the basis for human piety. "For he alone," Cicero's interlocutor boasts, "perceived, first, that the gods exist, because nature herself has imprinted a conception of them on the minds of all mankind ... Such notions Epicurus designates by the word *prolepsis*, that is, a sort of preconceived mental picture of a thing, without which nothing can be understood or investigated or discussed." Not only is there a universal and preconceived consensus on the existence of the gods; more than this, "nature ... also engraved on our minds the belief that they are eternal and blessed," leading Epicurus to the conviction that the gods "can neither know trouble itself nor cause trouble to another, and accordingly cannot feel either anger or favour, since all such things belong only to the weak."[9] In the *Essays,* Bacon revisits this argument that the supremacy of the gods requires their dispassion. In "Of Atheism," he remarks that "*Epicurus* is charged, that he did but dissemble, for his credits sake, when he affirmed; There were *Blessed Natures*, but such as enjoyed themselves, without having respect to the Government of the World." Bacon is prepared to defend Epicurus, not just from the charge of closet atheism but from the more commonplace accusation of an ignoble (because anti-providential) theology: "But certainly, he is traduced; For his Words are Noble and Divine: *Non Deos vulgi negare profanum; sed vulgi Opiniones Diis applicare profanum. Plato* could have said no more" (*OFB*, XV: 51–2). In this paraphrase of a Latin translation of Diogenes Laertius, profanity is derived from vulgar providentialism, that is, from the kind of base theology that implicates the gods in all manner of lowly human devices and desires.

Published for the first time in 1612, "Of Atheism" takes a bolder route, however, in the theological defense of Democritean atomism. For Bacon argues that atomic physics simply could never work without divine assistance: "Nay even that *Schoole*, which is most accused of *Atheisme*, doth most demonstrate *Religion*; That is, the *Schoole* of *Leucippus*, and *Democritus*, and *Epicurus*. For it is a thousand times more Credible, that foure Mutable Elements, and one Immutable Fift Essence, duly and Eternally placed, need no God; then that an Army, of Infinite small Portions, or Seedes unplaced, should have produced this Order, and Beauty, without a Divine Marshall" (*OFB*, XV: 51). The benefit of the attack on vulgar opinion–on the commonplace assumption that the Epicureans are wretched atheists–is that it offers a version of the sublime that rivals Plato's. The problem with arguing that the atoms (but not the elements) require a "Divine Marshall" is that it threatens to install the divine mind too invasively in things themselves. In short, it risks an imposture capable of duping the superstitious in much the same way as those theological fictions at which the Epicureans have always taken aim.

---

[9] Cicero, *De natura deorum*, 45–7.

Bacon begins this portrait of atomism with the conviction that "I had rather beleeve all the Fables in the Legend, and the Talmud, and the Alcoran, then that this universall Frame, is without a Minde" (*OFB*, XV: 51). He goes on to warn that a little natural inquiry will yield an atheistic obsession with dispersed physical causes, and that only a profound commitment to the study of nature will reveal the providential linkage in phenomena. But the argument connecting Democritean physics to a celebration of divine providence in the *Advancement* neither pictures the divine at rest nor busies the divine mind in the minute administration of natural affairs. Instead, the divine is made a political genius, with nature his hapless instrument. God is comparable to "the greater and deeper pollitique, that can make other men the Instruments of his will and endes, and yet neuer acquaint them with his purpose: So as they shall doe it, and yet not knowe what they doe" (*OFB*, IV: 87). By politicizing the divine agent, Bacon evades the dangerous traps of profanity–the human mind supposes that God is not at all involved in the creation or governance of the world– and imposture–the human mind neatly and self-flatteringly fabricates divine involvement in the world and deciphers the world accordingly.

In Bacon's version, then, atomism's assault on any putative evidence for the agency of providence in nature converts into an argument for the oblique and obscure instrumentality of the politic divine. As a rewriting of the Calvinist God, whose obscurity derives from the unutterably high prerogative yet nitty-gritty micromanagement attributable to the divine will, Bacon's construct of providence makes surprising orthodoxy of that early modern caricature that conflates two figures of vice, the shameless "Machiavel" with the feckless "Epicure." As a reshaping of Democritean atomism, Bacon's God underwrites not the supine divinity of *De rerum natura,* not the Calvinist God of incalculable will, and not the transparently rational creator and purposeful governor beloved by the natural theologian who gazes at nature as a hieroglyph. This God's mystery is itself a form of rationality, but one more like prudence, craft, and theatrical effect than it is like pure will. For Bacon, the most significant corollary of this construct is that whereas such a God might be accused of dissimulation, the divine dodge bemuses human speculation without allowing human idolatry to have its way with innocent nature. In nature, human beings should recognize the power of this politic God but look to Scripture for a clearer understanding of any divine plans for them.

Thus Bacon comes to terms with the ways of reconciling atomism with providence. But he must also reckon with an obvious trajectory in the legacy of ancient atomism: that as the physical theory passes from Democritus to Epicurus and then onward again to Lucretius and the Epicureans, its ethical, theological, and political accretions loom increasingly large as the very purpose of that physics. Bacon knows, moreover, that this development is already present in the records about Democritus and also that the intellectual allegiance between Democritus and Epicurus is much too great simply to divorce a regard for the former from the bogey of the latter. Over the course of his works, therefore, Bacon mentions that Epicurus borrowed from Democritus his notions of color, infinitude, "their doctrine of atoms," and even the subjection of natural philosophy to an ethics of pleasure: "But Epicurus, accommodating and subjecting his natural to his moral philosophy … would not willingly admit any opinion that depressed or hurt the mind, and troubled or disturbed that *Euthumia* [tranquillity] of his, which he had adopted from Democritus."[10] Far from simply the truth-seeking father of atomic physics, Democritus is responsible for his children's subordination of physical truth to psychological exigency and moral purpose. The result (Bacon concludes) is a philosophy prepared to lie about nature if nature can be used to protect the philosopher from a hostile world. Such imposture is altruistic in intent, for it assumes that the ethical passivity of otiose tranquillity is far healthier for human life than the epistemological passivity liable to anxieties of divine punishment. Above all, Bacon concedes, the heirs of Democritus are unprepared to face the truth about the natural world if the confrontation deprives them of philosophical accommodation.

Crucially, "accommodating" can mean that the philosopher endeavors to translate hard natural truths into sweeter, more digestible terms, whether it is Lucretius's poetic sweetening of harsh ideas in *De rerum natura* or Bacon's own recourse to myth in *De sapientia veterum*–the very passage about Democritus under review appears in Bacon's interpretation of the myth of Pan in *De augmentis*. But Bacon intends another meaning for "accommodating," the sense suggested by Shakespeare's "unaccommodated man" in *King Lear*. Accommodated man turns to the accouterments of human culture–to shelter and clothing but also to theology and philosophy–for comfort in the face of painful truth in what amounts to a version of Bacon's idol of the theater: logics, systems, or schools of thought in which the mind reeling from complexity can take refuge.

That atomic theory might play an integral role in the accommodating impostures invented by the human mind appears all the more likely when Bacon finds a place for the theory in the *Advancement*'s explanation of the so-called prime philosophy. Whereas the prime philosophy seems to rely on the obsessive analogical thinking that energizes contemporary poets such as Donne, the atom's position in Bacon's version of that thinking is emphatically not the exclusively metaphorical effect to which the atom is put in contemporary verse, according to which the atom is comparable to everything from dreams and fairies to lover's souls and farts. Bacon's notion of the prime philosophy posits that all the branches of learning can be traced to the same trunk by means of analogy, and "that meane and small things discouer great, better then great can discouer the small." As "*Aristotle* noteth well" in the *Politics,* the uncovering of the very commonwealth begins with the smallest social unit, the family: "*the nature of euery thing is best seene in his smallest portions*, and for that cause hee enquireth the nature of a Common-wealth, first in a Family, and the Simple Coniugations of Man and Wife; Parent, and Child, Maister and Seruant, which are in euery Cottage" (*OFB*, IV: 65). The same can be said for the natural world, the cosmic city, for one must begin "in meane concordances, and small portions." In the 1605 version, the helpful minutiae involve the discovery of magnetic effects "in needels of Iron, not in barres of Iron" (*OFB*, IV: 65). In the 1623 version, atoms have replaced magnetized iron needles as the minimum analogous to the family unit.

[10] *De Augmentis,* in John M. Robertson, ed., *Philosophical Works of Francis Bacon* (New York: E. P. Dutton, 1905), 86.

On the one hand, then, atomic physics is apt to help Bacon refute analogical thinking in an effort to evacuate natural studies of all cultural and epistemological contaminants. Atoms offer the natural philosopher a potentially true and useful way out of the magician's bogus world of signatures or the natural theologian's impositions on nature. In revising his *Advancement,* however, Bacon finds atoms especially useful for showing how analogical thinking might be reasserted legitimately. Epistemologically, Bacon finds, atomism can support the ambition of those dogmatic impostors whose sense of subtlety is premised on a network of correspondences and analogies. But it can also serve skeptically to subvert such imposture as well as awaken the credulous who are eager to believe in lies. Indeed, Bacon knows that at times the Epicureans are foolishly confident in their knowledge-yielding faculties, especially in the senses whose gauge of, say, the size of the moon the Epicurean philosopher is prepared to accept without question. But Bacon knows that at other times the Epicureans support a moderate and instrumental skepticism: in their recognition that the most significant phenomena are invisible; in their concession that explanations of celestial phenomena will often fracture into competing hypotheses; and in Lucretius's acknowledgments that in practice, the circumstances of perception can mislead the unwary.

At times, then, Bacon is scornful of the hubris afflicting Epicurean confidence in the philosophy's methods. At other times, he admires their attempts to steer clear of extreme dogmatism and hapless skepticism. The very sophistication of Epicurean epistemology is at its greatest not with the arguments from the bare senses about lunar size or the arguments from analogy about atomic properties and motions, but rather with the ways in which the sect accounts for the basis of all perception in touch and, as a subset of that problem, explains the conceptual access that human beings have to the gods. For Bacon understands that, far from not caring about the gods, the Epicureans premise their ethical desideratum on their construction of the gods, who are said not to care about human or even cosmic affairs. Their essential notion–that atomically constituted human beings can be happy only if they live simply and calmly–is an ethical superstructure about which Bacon has very much to say. The question for Bacon is always, however, whether tranquillity motivates imposture or (as the Epicureans themselves claim) liberates the human soul to accept the truth about the natural world in which that soul lives.

Bacon's *Essays* epitomize his more sympathetic response to the Epicurean ethics of pleasure. In "Of Superstition," he notes that atheists cause no trouble in states and that they do much to help bring about civil times. In *De Augmentis,* the Epicureans are named as these atheists and distinguished from the superstitious Stoics whose convictions propel them to resistance and rebellion (Charles I was so provoked by this thought that he noted in the margin of the English translation of *De Augmentis* his own preference for the superstitious subject[11]). In the essay "Of Truth," Bacon offers a backhanded compliment to Lucretius in translating the famous opening to Book Two of *De rerum natura*: "The Poet, that beautified the Sect, that was otherwise inferiour to the rest, saith yet excellently well: It *is a pleasure to stand upon the*

[11] See Barbour, *English Epicures and Stoics,* 145–6.

shore, and to see ships tost upon the Sea: A pleasure to stand in the window of a Castle, and to see a Battaile, and the Adventures thereof, below: But no pleasure is comparable, to the standing, upon the vantage ground of Truth." At this point in his English rendering of Lucretius's Latin, Bacon interrupts the sentiment to insert the proviso that truth's "hill [is] not be commanded, and where the Ayre is alwaies cleare and serene" (*OFB*, XV: 8). There is no question that Lucretius values the true knowledge of causes and so that clarity of mind as against the knots and mists of superstition has a premium for him; nor is there any question about the fact that the Roman poet must constantly urge his readers, and especially his beloved Memmius, to pay rigorous attention to his difficult arguments. But Bacon's interruption might also be said to modify the tendency of Epicureanism: whereas Lucretius values truth in its service to tranquillity, Bacon values distance from the turbulence of life in its service to true knowledge about the nature of things. Lucretius envisions the sage on the hill; Bacon reminds the reader that the hill of truth, like Donne's in "Satire Three," is steep and crooked, "not to be commanded." Even the word "serene" signifies "clear" rather than "peaceful" for Bacon. As he continues the translation–"*And to see the Errours, and Wandrings, and Mists, and Tempests, in the vale below*"–Bacon stresses misprision (three terms) over perturbation ("Tempests"), then he adds a proviso to the whole passage very much like those found in the writings of many seventeenth-century students of Lucretius: "So alwaies, that this prospect, be with Pitty, and not with Swelling, or Pride. Certainly, it is Heaven upon Earth, to have a Mans Minde Move in Charitie, Rest in Providence, and Turne upon the Poles of *Truth*" (*OFB*, XV: 8). Thus Lucretius requires three modifications: his pleasure in experiencing distance from the tumult of ambitious, passionate human life must be wary of lapsing into misanthropy or cruelty; tranquillity should never be divorced from a strong belief in the guardianship of divine providence (hence the citations of Genesis and the Gospels in the essay "Of Truth"); and Epicurean tranquillity should not rob the investigator of a constant orbit around the axis of truth.

When the *Advancement* offers an extended meditation on the Epicurean ethics of tranquillity, Bacon's assessment is at first condemnatory. Listing the sect along with the Cyrenaics in contrast to the followers of Zeno and Socrates, he compares the prizing of pleasure over virtue to the switches in identity that one encounters in comedies of error, in which the maid and the mistress exchange their customary garments. Unlike Montaigne, who claims that the normal preference of Stoicism over Epicureanism is unfair to the genuine views of the latter,[12] Bacon makes it clear that widespread European approval of virtue over pleasure as the ultimate goal of living is legitimate. Even "the reformed schoole of the Epicure*a*ns, which placed [happiness] in serenity of mind and freedome from perturbation," prove foolish, for "they woulde haue deposed *Iupiter* againe, and restored *Saturne*, and the first age, when there was no summer nor winter, spring nor Autumne, but al after one ayre and season" (*OFB*, IV: 137–8). The reformed Epicureans long for the golden age, without mental disputation, without fixing in place notions of good and evil, but wholly with concern for a

[12] Michel Montaigne, *The Complete Essays,* trans. Donald M. Frame (Stanford: Stanford Univ. Press, 1958), 307.

mind evacuated of disturbance but not replenished with virtue or truth. Such longing to live without longing is also ungodly, Bacon argues, comparing the Epicurean pursuit of tranquillity to the Anabaptist gauge of all matters "according to the motions of the spirit, and the constancy or wauering of beleefe, all which are manifest to tend to priuate repose & co*n*tentment, and not to poynt of society." Bacon stresses that God expects far more from a moral agent than simply private contentment or the avoidance of perturbation; God expects "Ciuile businesse" or vocational involvement in human affairs (*OFB*, IV: 138–9).

Bacon's assessment of the value of tranquillity is a little less clear in the *Advancement* when he sets Socrates against the Sophists, the former seeking "an equall and constant peace of mind," the latter hankering after "much desiring, and much enjoying." Bacon notes the overlap between Socrates and the Epicureans, conceding that the latter understand virtue as a major constituent of felicity. The Sophist desire always to want something and then to receive it over and over again amounts to a kinesthetic pleasure quite at odds with the Epicurean steady state; as Bacon says, Socrates compares that former and inferior pleasure-in-motion to scratching an itch. Ironically, however, Bacon calls the same process "Advancement": the Sophists believe "that *good of Aduancement*, is greater than *good of simple Preseruation*: because, euery obtayning a desire, hath a shew of aduancement, as mocion though in a Circle, hath a shew of progression" (*OFB*, IV: 141–2). The preservation approach to living aims at minimizing what is ill; the advancement strategy aims at maximizing what is good. One is sincere and evenhanded; the other is vigorous and changing. But what about Bacon's own "Advancement"?

Bacon raises the possibility that man might preserve with the Epicureans and advance with the Sophists, but he is daunted by the difficulty of negotiating that balance. There are men, he says, who delight in their pleasure even more than other men, and yet are not bothered when they have to set that pleasure aside. But which pleasure? The kinetic scratching pursued by the Sophists or the serenity sought by the Epicureans and Socrates? Bacon loosely cites a claim in Petrarch that only weak and fearful people refuse to enjoy a stimulant for concern that they will long for it and then feel dread about it. Bacon adds "that most of the doctrines of the Philosophers are more fearefull and cautionary then the Nature of things requireth. For, when they would haue a mans whole life, to be but a discipline or a preparation to dye: they must needes make men thinke, that it is a terrible Enemy, against whom there is no end of preparing." But is it the Sophist who has no fear about the scratching, or the Epicurean who is undaunted about questing for the golden age? Bacon seems to come down against the Epicureans when he proceeds to argue that it is folly to be so protective of the mind and to encase the mind (as much as can be) in harmony and uniformity, apart from "co*n*trary Motions" and the applications of civil life. Even so, a marriage between serenity and motion would be best (*OFB*, IV: 141–2).

In order to clarify and conclude his meditation on the marriage between advancement and tranquillity, Bacon introduces two analogies: the lutist and the jeweler. With the first, he accentuates the difference between the philosophical and civil lives as two approaches to music, for one harps on the same line or phrase, sweet and subtly varied but far more staid than the other with "such strange and hard stoppes and passages." With the image of the jeweler, though, Bacon means to show how one might carefully reconcile the two penchants, "to Imitate the wisedome of Iewellers, who, if there be a graine, or a cloude, or an ise which may be ground forth, without taking to much of the stone, they help it: but, if it should lessen and abate the stone to much they will not meddle with it." Bacon draws the conclusion that Epicurean ethics can be badly or productively used, for "So ought men, so to procure Serenity, as they destroy not magnanimity" (*OFB*, IV: 141–2). If rightly approached, the articulation of tranquillity with a particulate physics can result, he knows, in rendering natural truth and human utility mutually supportive of one another. Bacon also understands that Democritus's dogged pursuit of particulate nature can devolve into a fabricated pretext for and legitimization of a human agency rendered utterly useless if no longer fearful.

Subsequent to the *Advancement,* Bacon approaches the problem of how atomic physics has been extended into civil use from two vantages. In *De sapientia veterum* he seeks to modernize the cultural currency in which an atomic physics can best be traded. However, in revisiting those myths in *De Principiis Atque Originibus Secundum Fabulas Cupidinis & Coeli,* Bacon begins to ponder whether the truth of atomic physics–as against the truth of pneumatic and plenistic theories of the fundamental constituents of nature– might prove a more valuable question to raise than the political, ethical and rhetorical query into use value. Although Bacon does not expressly state as much, it begins to seem possible that atomism belongs in the fifth section of the Great Instauration, the one in which non-rigorously derived hypotheses provide a respite from the difficulties of a laborious inductive method. As such, atomism offers access to an ethics of pleasure as well as a physical pretext for that ethics. Perhaps too it continues to provide useful ways of conceiving natural subtlety. But as a human fabrication, it requires the skeptic's suspension of judgment that, for Bacon, must last only temporarily on the way to truth. If the proper use of atomic theory is especially hard for Bacon to separate from imposture, then a relocation of atomism in the fifth section of the Great Instauration might relieve the anxious philosopher of having to succeed at that division. All at once, atomism is a singularly hard case of mediating the true and the useful in the advancement of learning, as well as the epitome of just how difficult that mediation always must be for the Great Instauration.

Bruno and the atomic myth

Bacon revived and assessed atomic theory in the wake of Giordano Bruno's influential philosophical writings, the tendency of which was to celebrate the mythic, cosmic, spiritual, and ethical registers of the atom. In short, Bruno's cultural presence challenged early seventeenth-century atomists to decide whether the reclamation of the atom with its anti-providential legacy needed the intervention or removal of human invention in both its metaphorical and systematizing capacities. English writers such as Nicholas Hill, Robert Burton, and Thomas Carew, could read Bruno as infusing the material and impious atom with profound significance at long last, or as muddling an honest natural philosophy with heavy doses of pneumatics, poetics, and heroics.

In *The Philosophy of Francis Bacon: An Essay on Its Development from 1603 to 1609*, Benjamin Farrington argues that Bruno was Bacon's true "philosophical precursor," not least in the view that knowledge is power. Just as Bacon was beginning to conceive his six-part project in Elizabethan England, Bruno brought to England in person and in print "a new breath, a new vision, an enlarged horizon" for the study of nature; and though Bacon rarely mentions Bruno, "certain it is that Bacon … had become convinced that a new order of events was at hand, the specific quality of which would consist no longer in a mere *imitation* of nature but in her *domination* by man."[13] More generally, Steven Menn has argued that philosophy in the lifetime of Bacon confronted new expectations–for the defeat of skepticism, for the comprehension of nature, for the overhaul of human life, and for a heightened theological legitimacy framing all these endeavors.[14] A large part of Bacon's manuscript labors in gauging these expectations involved his investigating modernity's relationship to antiquity. As with Bruno, so with Bacon: the ancient atomists figured centrally in this investigation. At the crux of Bacon's position on antiquity was his conviction that "science is to be sought from the light of nature, not from the darkness of antiquity."[15] But, Farrington shows, his studies of history led Bacon to admire the Presocratic philosophers as a group and Democritus in particular: "He sensed in their fragments a more direct contact with nature than in later Greek schools and, in his enthusiasm, made a pioneer study of their scattered remains," with "a preference for the materialism of Democritus over the idealism of Plato." Hence Bacon's oft-repeated notion that river time submerged the weightiest matter; his claim that "Democritus attributed to nature immense variety and infinite succession"; and his belief that serious students of nature will inevitably engage Democritus's severe method, physical premises, and particulate emphasis.[16]

Bacon often resists the tendency to situate atomism within the frame of a mythological poetics that fuses the infinite maximum and the particulate minimum by means of divine spirit and love. When he mentions Bruno in *Historia naturalis et experimentalis* (1622), the "Nolan" keeps company with Telesio, Gilbert, and Patricius, whose imaginary and stage-play philosophies have (he believes) largely missed an audience. Bruno has become part of the vexing ideological baggage of which a legitimate atomism would largely have to be rid. Accordingly, in *De Sapientia Veterum,* essays on Pan, Proteus, and Prometheus attempt to lighten or at the very least to measure the heavy ideological freight of atomic physics. In these instances, myth is useful as a way to provoke readers to consider whether an atomism ultimately stripped of myth might help explain physical phenomena and liberate natural philosophy from theology and logic.

But Bacon clearly reflects Bruno's impact on the Elizabethan generation of intellectuals when, among the same mythological essays in *De Sapientia,* he attempts to

13 Benjamin Farrington, *The Philosophy of Francis Bacon* (Liverpool: Liverpool Univ. Press, 1964), 28,

14 Stephen Menn, "The Intellectual Setting," in *The Cambridge History of Seventeenth-Century Philosophy,* eds. Daniel Garber and Michael Ayers (New York: Cambridge Univ. Press, 1998), 1, 33–58.

15 Farrington, 48.

16 Farrington, 48.

recover atomism from the depths of history precisely by embodying its tenets in the features and desires of Cupid. As Julie Robin Solomon has shown, Bacon's debt to Bruno in the atomic myth of Cupid is laden with a language of seals, imprints, and numismatics that stresses ideologies of concealment, royal prerogative, mercantile interests, and pneumatic theology. In refining Bruno's mythical physics, Bacon purveys far more than just a culturally current vehicle of instruction; rather, he participates in the complex ideological impostures into which atomic theory can be fashioned.[17]

Bruno's time in Bacon's England (1583–85) had produced an outpouring of six philosophical dialogues in Italian, including *De gli eroici furori,* in which the philosopher's desire for truth and wisdom was retrained in the direction of the Epicurean infinite. Philosophical desire doubled, on the one hand, as liberation from Aristotelianism and, on the other, as the heroic skepticism prohibiting the complete comprehension of what is true. In other Italian dialogues, then again in the triplet of Latin Lucretian poems published in Frankfurt in 1591, Bruno followed Nicholas of Cusa into an emphasis on the meeting of contraries that conflated maximum with minimum, infinite with atom and integrated the universe into a unity made possible by the spirit of divinity. As Paul Henri Michel has put it, Bruno "has a deep-seated feeling, a taste for unity, which constrains him to place Unity at the heart of existence, if not even to confuse one with the other." Atoms, then, are inextricable from this oneness and exist to be penetrated as seals enclosing the divine, as surely as the infinite universe and innumerable worlds unfold the divine without any boundaries at all. Thus, Michel argues, "The eternal identifies itself with the indivisible, with the atom, with the 'minimum.'" God, soul, and atom are essentially interchangeable in the Brunonian universe; but they are also "distinguished not only by analysis, but common experience." Hence the atomic constituents of the world stress motion and discontinuity, part of that imperfection that provokes the philosopher into a constant longing for the infinite and divine.[18]

Bruno's animated atomism, his marriage of Cusa and Epicurus, had two sorts of influence. There was something of a cult following in writers such as Nicholas Hill, whose *Philosophia Epicurea* (1601, 1619) produced its own band of followers. Bruno and Hill were at the center of the atomic revival until Gassendi, Charleton, Evelyn, and Stanley took over at mid-century. In *The Anatomy of Melancholy,* Robert Burton depicted Bruno and Hill as the heirs of ancient atomists in their emphasis on infinitude, atoms, innumerable worlds, and heliocentrism. Burton understood also that Bruno and Hill had crucially evolved a new theological framework and legitimacy for atomic physics, however unorthodox or at least pagan that theology might be. But it was precisely this point–the recovery of a cultural currency for atomism itself, rather than the infusion of atoms with divinity in a mythopoetics–that formed a second kind of Brunonian influence. From this vantage, the challenge undertaken by Bruno was to liberate the intellect from the customary limitations on philosophical argument and to refocus and expand the philosopher's vision so that divinity itself, the infinite universe and all the particulate matter within it might be genuinely seen and heroically pursued.

---

17 See Solomon, *Objectivity in the Making,* 201–24.

18 Paul Henri Michel, *The Cosmology of Giordano Bruno,* trans. R. E. W. Maddison (Ithaca: Cornell Univ. Press, 1973), 77, 80–1, 98.

Writing philosophy in Italian dialogues and dressing it in mythological guises were Bruno's humanist strategies for converting a difficult philosophical vision into cultural currency. But Bruno had a penchant for alienating would-be patrons and audiences with his contentiousness. In turning to Latin poetry with prose commentary in the 1591 triplet, he confronted what for him was the distasteful problem of philosophical rhetoric. Far from being encouraged to pursue the expansion of intellectual vision into the infinite or the concentration of that vision into a boldly delineated minimum, Bruno notes how a speaker without eloquence was bound to be "hissed out of the forum." The philosopher's impropriety in language served as the excuse for audiences to reject his thought as both ignorant and unwieldy. Bruno ventures so far as to decry a stress on verbal decorum as a form of sacrilege, for a rhetor (he concludes) "profanes the sacred images of the gods with purple robes, gold inlay and threads of fine ornament" when those gods–Saturn, Mars, Neptune, and the others–"want to be depicted in their true naked majesty … with the simplest weapon."[19] As Dorothy Waley Singer has argued, Bruno's choice of Lucretian verse "as the vehicle of his thought" was an attempt at grandiloquence,[20] for Lucretius offered the Nolan an epic language and versification far more heroic than the vernacular legacy of commentaries and sonnets on love.

In *De sapientia veterum,* however, Bacon tries to have it both ways: to dress up atomic philosophy in myth so as to make it more decorous, palatable, and useful, yet to illuminate the naked majesty of the atom as an explanatory device. That nakedness entails that while the ideological freight of atomism might be usefully repackaged, the theological and ontological superstructure of Bruno's philosophy must be jettisoned. The process of sorting the former from the latter is so difficult that the fabrications of myth become all the more consoling as the means of lulling a delighted reader into not noticing the slippage between securing a viable myth and rejecting aggressive impostures.

Bacon's atomic allegory of Cupid aims at simplicity in analysis. Even though there is more than one Cupid, Bacon begins, one can sort out the differences. The love in question is more ancient than all the gods except Chaos, with whom Cupid "begot all things, the gods included." Cupid's four main attributes–he is an infant and an archer, blind and naked–pertain "to the cradle and infancy of nature, and pierces deep."[21] Like Bruno, Bacon differs from the Epicureans, who posit that atoms are devoid of vitality or sentience; the myth of Cupid points rather to the idea that the atom is invested with appetite or instinct, though Bacon glosses such desire as motion. The theological monster lurking around atomism raises its head when Bacon turns to Cupid's lack of parents, which he glosses as the atom's claim as a causeless point of origin for all things. On this point, Bacon simply exempts God, who is the atom's only efficient cause. Like Bruno or Nicholas Hill, he shrouds the atom in an aura of mystery: one can learn "the method and process of it" perhaps, but the atom,

---

19 See *Iordani Bruni Nolani Opera Latine Conscripta,* eds. F. Tocco and H. Vitelli, 1, part 3 (1889, reprinted 1962), 138 (my translation).

20 *Giordano Bruno, His Life and Thought* (New York: Henry Schuman, 1950), 156.

21 Sidney Warhaft, ed., *Francis Bacon: A Selection of His Works* (New York: Macmillan, 1982), 293.

like God, is unapproachable, the seal and the secret cause of all causes, the physicist's reverential retort to the Platonist's ideal forms. In connecting the causeless atom to its also secretive divine creator, Bacon quotes Solomon's characterization of the beauty, propriety, and elusiveness of created things, all of which (nonetheless) God has offered to man's investigation.

At times, like Bruno, then, Bacon mythologizes the atom not just decoratively but ontologically, even to the point of fusing the minimum with the maximum. "For the summary law of nature," he argues, "that impulse of desire impressed by God upon the primary particles of matter which makes them come together, and which by repetition and multiplication produces all the variety of nature, is a thing which mortal thought may glance at, but can hardly take in."[22] At the same time, however, Bacon's essay on Cupid scarcely reads like Bruno's theosophy. Turning to the ancient Greeks, Bacon accentuates how "careful and acute" their study of matter was, even if they were hapless on the question of motion. Bacon chastises the hasty thrust of ancient studies toward theologizing explanations of motion: despite his further discussions of a prime philosophy uniting all the branches of learning in a summary law of nature, he applauds the perceptive and patient work of Democritus on the physics of the atom. Democritus and Epicurus alike made mistakes about the atom–the former in his views on motion toward a center, the latter with his theory of the atomic swerve–but if on the one hand these mistakes darken the cloak of mystery around the atom, they also support Bacon's clarion call for a wider theory based extensively on "particulars."

Bacon's atomic Cupid plays three roles then: the myth translates a hard and unpopular but promising physical theory into standard, even royalist,[23] cultural currency; it makes gestures toward rescuing the theological warrant of that theory; and it simply instructs the reader about how the atom (a naked child) is uncompounded, how it manages to operate at a distance (archery), and how (in keeping with Lucretius's own lexicon) the atom is "blind," in simplest terms, "invisible." Bacon takes the attribute of blindness as another opportunity to mine an untapped support for providentialism in atomic accounts of creation. After all, the atom itself "has very little providence, but directs his course, like a blind man groping, by whatever he finds nearest." The Epicurean argument for fortuity is then converted into an extraordinary boost for providence, for the blindness of atoms "makes the supreme divine Providence all the more to be admired, as that which contrives out of subjects peculiarly empty and destitute of providence, and as it were blind, to educe by a fatal and necessary law all the order and beauty of the universe."[24] The threat of anti-providentialism, Bacon knows, might be the especial burden of ancient atomism. But it will also prove troublesome for any philosopher whose boldest stroke is to reject the Brunonian theologizing of natural phenomena and whose project is to focus investigative energy on matter and motion with only a contingent relationship to God's will and not a necessary and poetic relationship to God's reason.

22 Warhaft, ed., 293–4.

23 Solomon, *Objectivity in the Making*, 217–19.

24 Warhaft, ed., 295.

Elsewhere in *De Sapientia Veterum,* in the allegory of Prometheus as providence, Bacon revisits the vexing question of whether a Democritean natural philosophy simply subverts or surprisingly supports a marriage between theology and induction. The mythological essays are useful not just for selling but also for crafting Bacon's potentially most improper merchandise: a godly yet particulate physics. To make matters more difficult for the merchant, Bacon is increasingly unhappy with the inevitable compromises that merchandising philosophy is bound to produce. All along he has conceded that his allegories might not so much be "found" in the ancient myths as put there, the marriage between Cupid and the atom more forced than affective. But in *Thoughts and Conclusions on the Interpretation of Nature,* he appears to undercut the credibility of his mythological essays, at least those with a physical bent. "Eager to miss no possible path" in reviewing the history of philosophy, Bacon "cast his eyes towards that region of cloud and darkness, the secret recesses of remote antiquity." He admits, however, that in seeking among ancient myths those critical physical notions that had fallen from favor over time, he acted without complete sincerity in the interest of marshaling greater dignity for ideas such as atomism. So too, he notes, newly rich men invent ancient genealogies for their families.[25] Ironically reflecting on his own tendency to render atomic properties in terms of royal prerogative, Bacon concedes that the metaphors and myths in which he, like Bruno, has enshrouded the atom might prove the most embarrassingly obvious imposture.

The alternative is of course to focus on the physical evidence supporting or subverting atomic theory, to "avoid any sort of imposture," to ignore as irrelevant the factor of antiquity or modernity, and to seek truth (rather than eloquence) in "the light of nature, not recovered from the darkness of antiquity."[26] Whether atomism can be made acceptable and decorous, and whether it is useful in thinking about and expressing the subtlety of things, the genuine philosopher must eventually ask whether atomism is true.

The truth about atoms

In *Cogitationes de natura rerum,* Bacon not only theorizes that atomism is either true or useful. In the first section of that work, entitled "On the Division of Bodies, Continuity and Vacuity," he offers some notes toward a determination of whether the utility of atoms might result from, rather than corrupt and conceal, their truth. Bacon proceeds toward this resolution with caution as well as eagerness. Dividing "atom" into two key definitions–as the final term guaranteeing a limit on divisibility and as the body without vacuum–he proceeds to conceive the probable role of that "ultimate term" in the spread of sensation. As he compares the views of Hero and Democritus on collective and interspersed vacuities, Bacon once again exhorts his reader not to be daunted by "the exceeding subtlety of nature, but to think that both the units and

25 Farrington, 87.
26 Farrington, 87.

the sums of things are equally subject to calculation." Atomism seems mainly useful at this point, a form of encouragement for philosophers struggling against the subtlety of particulate nature. Indeed, Bacon urges his reader not to relegate such thinking to abstruseness, for atomism should be put to "work and use" by those philosophers who labor "diligently in experience and particulars, and cut nature as it were to the quick," even if what they produce is incomplete and infelicitous.[27]

As he moves on to "the equality and inequality of atoms or seeds," Bacon clarifies that there are really rival atomic theories. He also italicizes a deficiency in all the rivals: they have dealt ineptly with motion and change. At the same time, he suggests that there are different ways in which a theory can be useful: it can stir up hope, soothe the intellect with its allure, or it can offer practical solutions to problems. But the "most useful of all" will act as "the true moderator of hope and works." As Parts Three and Four direct their focus on the insufficient ways in which atomists and alchemists alike have handled motion, Bacon lumps atomism together with the vulgar, commonplace, and fashionable theories of his day, a far cry from the submerged and provocative theory needing to be recovered from the ravages of time and the biases of Plato and Aristotle. "For if we consider for a while the philosophies in fashion," he asks, "will the principle of stimulus of matter by privation, of the shaping of matter according to an idea, of the aggregation of similar particles, of the fortuitous agitation of atoms in a vacuum, of strife and friendship, of reciprocal impressions of heaven and earth, of alliance of the elements by symbolising qualities, of the influence of celestial bodies, of sympathies and antipathies, of secret and specific virtues and properties, of fate, fortune, necessity–will, I say, such generalities as these, which are nothing but spectres and appearances that float and play on the surface of things, as on water, enrich mankind or increase their possessions?" In this dismissal of a horde of natural philosophies, atomism neither requires a mythic dress nor sends vulgar theories packing: rather, atomism joins theories of the elements, of correspondence, of Platonic forms, and of the occult qualities in constituting those myths of natural order that emerge from the inflated, deluded, and idolatrous human intellect. Such myths cannot even lay claim to elitism, for they belong in "popular discourse." They are not true and, because of this, not useful, for they neither understand "the principles, fountains, causes, and forms of motion" nor control these "appetites and passions of every kind of matter." For the point of natural study is not simply to study motion, but also "to acquire the power of exciting, restraining, increasing, remitting, multiplying, and calming and stopping any motion whatever in a matter susceptible of it; and thereby to preserve, change, and transform bodies."[28] Bacon makes use of a Lucretian metaphor–the constituents of nature are comparable to letters in the alphabet–but that comparison is divorced from its specifically atomic context.

As Graham Rees has emphasized, Bacon's attempt to determine the truth of a theory such as atomism was not simply an empirical affair. Even though an emphasis on particulate evidence was the foundation of Bacon's inductive method, an

[27] Farrington, 290–1.
[28] Farrington, 292, 294–5, 297.

explanatory pyramid was to be erected on top of that foundation. The burden of a methodized induction was the discovery of middle axioms from the gathered evidence, but with this discovery in progress, the Baconian philosopher was permitted respites in theory that might illuminate a wider view of natural causes. What is more, even if theology were not allowed to dictate to natural philosophy, the evidence and the axioms would, nonetheless, have to pass the test of theological legitimacy (*OFB*, VI: xlviii–li). The more Bacon came to dismiss the legacies of Democritus and Epicurus, the more he turned to pneuma or spirit and to basic qualities such as heat and cold in the manner of a systematic natural philosopher with whom Bacon once lumped Bruno: Telesio. Bacon's hope in this reorientation was that a natural philosophy geared toward pneuma would carry greater theological warrant; truthfully describe and explain natural philosophy over the course of the manifold "histories" of phenomena; and translate into ethical and social utility in the preservation of human life and happiness. But all these goals would come at a price: if Democritean and Epicurean atomism proved to be imposture despite its claims of rigor and its critique of superstition, a philosophy of pneuma would have to prove rigorous and critical despite what for Bacon was its all too obvious legacy of fraud.

In *Descriptio Globi Intellectualis* (around 1612), Bacon continued to take aim at the atomists, for instance, at Democritus for erecting "a huge and dazzling speculative edifice on some slight and silly mistake." Democritus is still very much a brilliant force with which to reckon on so large a question as the structure, fullness, and integrity of the world, but Bacon is apt to underscore his ineptitude: thus, Democritus "was a good dissector of the world, but in matters concerning its structure he was even worse than ordinary philosophers" (*OFB*, VI: 115, 117). But in the manuscripts of the second decade of the seventeenth century, and in the natural histories of the third, Bacon's attention has turned to "pneumatic stuff"; such spirit has "frequently played the roles ascribed to the void in Democritus' philosophy," while "in the received philosophy its operations are generally assigned to qualities" (*OFB*, XIII: 187).

The clearest evidence of Bacon's turn to Telesio appears when he revisits the myths of Cupid and Coelum in *De Principiis Atque Originibus Secundum Fabulas Cupidinis & Coeli. Sive Parmenidis et Telesii & praecipue Democriti Philosophia tractata in fabula De Cupidine.* First, Bacon stresses that the distinctive characteristic of Cupid is his efficacy "in uniting bodies," an integrative tendency not pinpointed in *De sapientia veterum.* Then, Bacon turns to Democritus himself, whose exceptional severity in the study of nature's principles and origins did not prevent his losing control of his own contemplation. Indeed, Democritus's ideas are divorced from experience in the way of empty fables, for both derive "from the intellect left to itself and not firmly supported by experience advancing step by step." As he proceeds to decipher the myth of Cupid, Bacon avoids any mention of atoms at all. He lingers on Cupid's lack of parents in an effort to chastise those philosophers who have run aground in pursuit of a fraudulent genealogy for original matter, and he speaks of Cupid's "mode of being" as a force or essence secretly "impressed by God on the primary particles." Thus God empowers the primary particles before they enter into combinations, and (Bacon adds) it is impossible for the human mind to understand this primary force with any certainty. When he returns to Democritus and his legatees, it is to stress how dark and clandestine the primary seeds of nature essentially are, even to the point of using Lucretius's language against the Roman poet's conviction that visible things can lend us analogical access to the invisible. Bacon sums up the significance of this darkness: "Thus atoms are not like fiery sparks, drops of water, bubbles of air, specks of dust, nor tiny amounts of spirit or ether. Nor is their power and form something heavy or light, hot or cold, dense or rare, hard or soft, such as are found in larger bodies, since these virtues and others of the kind are products of composition and combination" (*OFB*, VI: 199–203). Removed from the range of our analogies, atoms jettison their ideological accretions and pretenses as the "seals" appropriate to an oblique and politic divine maker.[29] But these denuded atoms also run the risk of evaporating into thin air, like those other "fleeting things"–Platonic ideas, fairies, and fanciful dreams–to which poets compared them in the seventeenth century.[30]

At the very point that Democritus appears to triumph–and Bacon features him in a way that he does not in *De Sapientia Veterum*–Bacon contends that in fact atomic theory falls far short of the parable of Cupid in its explanatory power. For "in this very matter, namely the atom's motion compared with that of larger bodies, the philosophy of the parable seems to differ from the philosophy of *Democritus*. For we find that *Democritus* is not only quite at odds with the parable, but also at odds and virtually in contradiction with himself in the other things he says on the matter" (*OFB*, VI: 203). The Greek philosopher failed to assign heterogeneity to atomic motion as well as atomic bodies. But Bacon's reprisal of the Cupid myth seeks to accomplish something else that Democritus clearly did not: in returning to God's primacy and to a comparison between divine mystery and the obscurity of primal causes and origins, Bacon reminds the reader that the mythologizing of natural philosophy is responsible for theological warrant as well as for evidentiary accuracy and explanatory prowess.

Even as ancient atomic theories fall from favor in his meditations on cause and origin, Bacon retains the heroic stature of Democritus for the dangers and difficulties that he incurred in his severe analysis of the nature of things. Democritus understood the way of the world: the deeper one's explorations penetrated into nature, the more antagonistically one would be treated by the philosophical herd. In antiquity it was acknowledged that Democritus was the philosopher's philosopher, an intellectual athlete with encyclopedic knowledge. Among the genuine thinkers of antiquity, Bacon argues, Democritus was considered a magus, by which was meant a sage "who immersed [himself] more profoundly in silent and demanding kinds of contemplation." If Bacon still finds Democritus useful, then, that status has as much to do with the heroic way of a true philosopher's life as it does with the specifics of atomism. Indeed, Bacon defers his discussion of Cupid's "infancy, wings, arrows and so forth" for an expanded discussion of the theories of primary matter among the pre-Socratics in general (*OFB*, VI: 205, 207).

---

[29] Solomon, *Objectivity in the Making*, 215–20.

[30] See Gerald Hammond, *Fleeting Things: English Poets and Poems, 1616–1660* (Cambridge, Mass.: Harvard Univ. Press, 1990) and Barbour, *English Epicures and Stoics*, chapter one.

As *De Principiis* unfolds, Bacon is clearer on the physical theories that he would reject and on the general criteria for a legitimate theory than he is on the specific theory that he would embrace. Because Cupid is born from chaos, we are told, he is a person with "some form." His nakedness is now a chief indicator of how other theories have erred in clothing primary matter in qualities or characteristics that it simply does not have. This insight leads Bacon to a taxonomy of pre-Socratic schools, into the second of which atomism clearly fits without being singled out: "those who hold that the principle of things is one in substance, a principle fixed and invariable, but who deduce the diversity of entities from the different magnitudes, shapes and positions of that same principle" (*OFB*, VI: 209, 211). But Bacon never comes around to a careful discussion of atomism. In attacking those Presocratics who postulate one kind of primary matter, he calls on both Telesio (the stress on contraries such as cold and heat) and Lucretius (the critique of theories with mortal first principles). Then, having set aside the fourth view, essentially the theory of homoeomeria held by Anaxagoras,[31] Bacon makes a critical move. He moves to the third generic theory, the one premised on multiple but not infinite types of primary matter. In this category, Bacon turns to Parmenides (who stressed two, "fire and earth, or heaven and earth") and to the modern philosopher who has done the most to recuperate that philosopher: Telesio. What Bacon has to say about Telesio will be mixed: that philosopher is very good at destroying Peripatetic views but not as skillful at constructing his own. But Telesio's basic views consume Bacon's attention to the end of the fragment, namely, that "the primary forms and the primary active entities, and therefore the primary substances, are heat and cold," that all natural phenomena involve the interaction of active and passive matter; and that the calculus of nature's economy expresses the permutations derivable from two opposite quaternions (heat-light-tenuity-mobility vs. cold-dark-dense-static) (*OFB*, VI: 225).

There is no question that Telesio's system has more integrative explanatory power and clearer theological warrant than atomism, though staunch seventeenth-century atomists contend that pneumatic theories tend to require divinity less and to make greater claims for the divinity of nature than a focus on particulate matter. But Bacon is so restless with Telesio that he continues to countervail his theory with the views of Democritus. At times, he even suggests that a true and useful natural philosophy–one devoid of imposture as much as humanly possible–will emerge only if Democritus and Telesio are made to converse with and correct one another.

Indeed, Telesio's improbable tenets have an almost Epicurean haze about them, inasmuch as "his seems a kind of pastoral philosophy which contemplates the world calmly and as if in idleness. For he talks not badly about the system of the world, but very ignorantly about principles." Setting forth a system without regard for either chaos or macrocosmic change, Telesio touts a philosophy that "seems rather a frivolous product of the limitations of the human mind. For a person who philosophizes according to the sense alone may assert the eternity of matter but deny the eternity of the world as we see it; and this was the opinion both of the most ancient wisdom, and

[31] Anaxagoras's theory postulates that any material comprises little particles of that same material; thus, bone is made of little "bony" particles, and so forth.

of the man who comes nearest to it, *Democritus*" (*OFB*, VI: 251–3). Suddenly, Democritus appears to have evidence, theory, and even theology on his side: "Holy Writ bears witness to the same thing, but with the crucial difference that whereas Holy Writ holds that matter comes from God, the ancient philosophers represent it as something original to itself." Democritus both is and is not on the side of divine warrant, close to it but so far away from the three spiritual premises that Bacon says faith always teaches: that "matter was created from nothing … that the elaboration of system was dictated by the word of the Almighty, and not that matter elaborated itself out of Chaos into the present schematism; and thirdly, that this schematism was (before the Fall) the best of those which matter (as it had been created) could support." Bacon ticks off the fundamentally impious triad of the atomists in return: "For they abhor creation out of nothing, and suppose that this schematism was fashioned after many circuitous processes and efforts of matter; and they do not worry about it being the best possible schematism, since they claim that it is perishable and variable" (*OFB*, VI: 251–3). Far from being simply true or useful, atomism is the natural philosophy that Bacon cannot live either with or without. Pneumatic theory offers an attractive way to reconstitute and to criticize atomism; but as the addition of atoms to his treatment of the prime philosophy in the 1623 *De Augmentis* illuminates, Bacon never quite jettisons atomism from the fundamental cargo of his Great Instauration.

Atomic ways of seeing

In the first decades of the seventeenth century, Bacon tried and failed to secure theological legitimacy for atoms. He tried and failed to separate its theory from all the stage-play fantasies of idle and arrogant deduction. He admired the severity with which Democritus had studied particulate nature, protesting toward the end of *De Principiis* that the atom avoided the pitfalls of abstraction (which is not materially real) and mortal principles (which are all too real). Indeed, the atom's success in securing the material world was so impressive "that a clearly irresistible necessity drives men's thoughts (if they want to be consistent) to the atom, which is a true entity, having matter, form, dimension, place, resistance, appetite, motion and emanation" (*OFB*, VI: 251–3). Once again, Bacon revisited the question of how Christian philosophers might baptize atomism or at least inoculate it against its infamous ideological contaminants. But atomic properties had a way of eluding induction. Hence, at the end of *De Principiis,* Bacon returned to Telesio, largely unknown and yet "a lover of truth, a man useful to the sciences, a corrector of certain doctrines, and the first of the new philosophers" (*OFB*, VI: 259). Much of Bacon's own natural histories in the 1620s would be devoted to a severe investigation of the evidentiary foundation of Telesio's theories.

By the 1650s, some heirs of Baconian philosophy would claim that the Great Instauration had at long last driven the occult and the scientific apart. But in his last years, Bacon struggled to separate a genuine natural philosophy from superstition by means of mediating between atoms and pneuma in an intellectual milieu that very often fused the terms and the postulates of the two primary materials–one Epicurean, the other Stoic–into hybrid constructs. One thinks of Robert Fludd in this capacity.[32] Bacon's natural histories not only contributed *in fact* to the emulation–that is, the interaction and rivalry–between spirit and particles. Just as clearly, Bacon was *perceived* by contemporaries to play a critical role in measuring the theological, theoretical, and evidentiary legitimacy of the contending physics. For instance, in Nathaniel Torporley's *A Synopsis of the Controversie of Atoms,* the ancient quarrel between Aristotle and Democritus is rehabilitated as a debate between Torporley himself and Thomas Harriot. Harriot himself, we now know, was essentially an atomist, though he approached that ancient theory in the mode of skeptical inquiry and fresh experimentation. In the so-called "second squadron" is a brief discussion of the "rarefaction of water into ayre," into which matter Torporley inserts a reference to "Verulams experiment of spirit of wine and ayre."[33] For Torporley, Bacon's contribution to the debate between atoms and spirit was practical and experimental rather than theoretical and theological. New manuscript evidence has taught us that Bacon's intervention in the search for the principles and causes of matter operated at all three levels.

In the seventeenth century, atomism appealed to the poets and to natural philosophers alike, for what early modern intellectuals and dreamers had come to understand was that atomism was an extraordinary way of thinking and seeing the subtlety of the world. Bacon's point about truth and utility was that if early modern culture was to see the world aright, it was going to have to start afresh from the vantage of the atomist. In a letter, Harriot made the same point to Kepler. In 1606, the two men were exchanging ideas about refraction, Kepler with his concern that his previous work on optics was far too theological and not sufficiently physical. Having offered a rigorously atomic explanation for how refraction operates, he concluded with a metaphor or allegory for what he and Kepler had seen and still might see: "I have led you to the doors of nature's mansion, where her secrets are hidden. If you cannot enter on account of their narrowness, abstract yourself mathematically, and contract yourself into an atom, and you will enter easily. And after you have come out, you will tell me what wonders you have seen."[34] Clearly Harriot meant to say that quantification and particulate matter were the keys to nature's clandestine ways. But the contraction into an atom recommended by Harriot demanded the analogical ways of a Lucretian poet, and even then there was no guarantee that one would enter the tiny and elusive vessel.

Nor, given the ethical and theological affiliations of atomism, was there a guarantee that the philosopher contracted into an atom would ever make it back out into the light of God's grace or into the uncloistered duty of citizenship. Montaigne once told

---

[32] See Reid Barbour, *Literature and Religious Culture in Seventeenth-Century England* (New York: Cambridge Univ. Press, 2002), 181–95; 234–5; Brian Vickers, ed., *Occult and Scientific Mentalities in the Renaissance* (New York: Cambridge Univ. Press, 1984), 95–155.

[33] See Jean Jacquot, "Thomas Harriot's Reputation for Impiety," *Notes and Records of the Royal Society,* 9 (1952); rpt. in *A Source Book for the Study of Thomas Harriot,* ed. John W. Shirley (New York: Arno Press, 1981), separately paginated; Torporley's synopsis appears as an appendix; the reference to Bacon is on 185.

[34] Jacquot, 181.

a joke that Bacon would have understood well. The question of conversion between Stoics and Epicureans was like the matter of cooking a chicken. Just as a chicken could be cooked but not uncooked, so a Stoic could be turned into an Epicurean, but that Stoic so proselytized would never be returning. On occasion, Bacon recognized that sober natural philosophers were almost fatalistically driven toward atomism.[35] He also recognized that the garden of the Epicureans beckoned as the bower of bliss waiting as the ethical reward at the end of Democritus's severe investigations. Bacon's turn to spirit, to Parmenides and Telesio, ensured that his researches were godly and holistic. But this pneumatic way had its own problems: the threat of pantheism with which occultists such as Fludd were charged, the betrayal of a rigorously inductive method, and the ethical specter of determinism no less socially worrisome than the tranquil hedonism of the Epicureans.

For Bacon, the need for a natural philosophy with a minimum of imposture was heightened by the fact that nature was itself subtle, for as the *OED* makes plain, contemporary associations between subtlety and imposture were commonplace. The subtlety of nature was heightened further still by Bacon's emphasis on its contingency, part and parcel of his treatment of God as politic in his providential designs. And if the truth about nature was subtle, the uses to which nature might be put were fraught with impostures of many sorts, ranging from political pageantry to private pleasure. In the *OED,* pageant makers, quacks, empirics, papists, actors, and prostitutes are all found committing imposture on others, but Bacon understood that people might waste their time fooling themselves as surely as they might duping others.[36]

In the *Sylva Sylvarum,* Bacon dismisses a certain experiment in ventriloquism as of "no great use … but for imposture" (Century III, #241).[37] But his lifelong struggle to assess the value of atomic ways of seeing constantly reminded him that in pursuing the subtlety of nature under the pressure of the exigencies of human life, a philosopher could neither isolate nor dismiss imposture. Very early in his career, in the *Religious Meditations,* Bacon does his best to anatomize imposture, with its "three formes of speaking" comprising the divisive "webbes and clatterings of the Schoolemen"; the "tales, stories, and examples" of "Church poets" and "ancient heretikes"; and the again heretical enchantments offered in "misteries, high parables, Allegories, and illusions."[38] But in his lifelong engagement with atomism, Bacon came to understand that imposture inflects the ways in which nature operates on its own, the ways in which human beings operate on nature, and the ways in which divine providence has set up and continues to preside over all natural and human operations.

---

[35] Montaigne, *The Complete Essays,* 307.

[36] *OED* references accessed on-line.

[37] *Sylva Sylvarum* (Montana: Kessinger Publishing Co., n.d.), 40.

[38] "Of the seuerall kinds of Imposture," in *Essaies. Religious Meditations. Places of Perswasion and Disswasion* (London, 1598), separately paginated, 22–3.

Chapter 3

The New Science and the *Via Negativa*: A Mystical Source for Baconian Empiricism

Michael McCanles

Instruments and obstacles

Francis Bacon thought and wrote in a period of western intellectual history that post-dated that crucial cusp where technical questions about how human beings know were beginning to be replaced by questions as to whether human beings know anything at all with certitude. That cusp consisted of a group of fourteenth-century scholastic philosophers–pre-eminently the Franciscan philosophers Duns Scotus and William Ockham–who were engaged in an endgame struggle with the fundamental logical contradiction that shadowed all thinking about thinking from Boethius and the fourth century on: the contradiction entailed by the "fact" that an immaterial intellect can have no contact through the material senses with a material world.[1]

Bacon grasps very early on, in the writings that precede the 1605 *Advancement of Learning*, that the fundamental instrument of his new science, namely, the combined human power of material observation and intellectual conception, is an obstacle as well. This is why the whole tenor as well as the title of the 1605 exposition is primarily propadeutic: a new science requires first of all a radical revision of both the "how" and the "whether" of human learning in general. Bacon treats questions of whether human beings have the capacity to know anything with precision and certainty in the *De Augmentis Scientiarum*, while "how" questions–matters of definition, procedure

[1] Marilyn McCord Adams begins her two-volume study of Ockham "at the beginning," so to speak, with Boethius setting the question of human knowledge as grounded necessarily in universal concepts, a form of the question that was to be set aside only when Ockham disposed of all the difficulties deriving from Boethius' position by throwing out the question itself. Adams paraphrases Boethius' argument: "If there are no extramental genus-things or species-things, our genus-concepts and species-concepts will not represent extramental things exactly as they are and will thus be 'vain and false.' This consequence seems the more pernicious when viewed against the common assumption of Plato and Aristotle that having such genus- and species-concepts is our means of having knowledge properly speaking. For they hold that knowledge properly speaking and definition are of what is universal *and* real; if nothing real is universal, scepticism is the result" [Marilyn McCord Adams, *William Ockham*, 2 vols. (Notre Dame: Indiana Univ. of Notre Dame Press, 1987), I:8].

and falsification–are taken up in the various sections collected under the title *Novum Organum*. Bacon conceives the first as matters of a broadly ethical nature, while the latter are technical and analytic.

The question of "how" humans know as Bacon inherited it focused on the mid-point of the process: the transformation of material data into conceptual information. The locus of this middle point in medieval philosophizing was the so-called "species intelligibilis"– literally, a (conceptually) intelligible (material) representation–a theoretical entity that solved the problem of being a bridge between material sense experience and immaterial conception by being somehow both. Not surprisingly, the "somehow" was the problem.[2] In similarly framing the question of how humans think as a matter of immaterial knowledge about a material world, the epistemological component of Bacon's new science became entangled in the same instability that not only characterizes the history of western epistemology before the fourteenth century and nominalism's supposed "solution," but reappears after it as well.

This instability is reflected in the oscillating values Bacon gives human conceptionalization in the *De Augmentis Scientiarum* and the *Novum Organum* respectively. The *De Augmentis* subordinates mental conception to sense experience in calling for a purge of the debilities of human fantasizing that have for too long (Bacon holds) usurped the place of rigorous thought. Bacon's privileging sense experience over mental conception in the *De Augmentis* in turn reflects Ockhamist nominalism's insistence that universal concepts are mentally-constructed systems of signs that in no way constitute a "representation" of that world. On the other hand, the *Novum Organum*, particularly in the theoretically fertile prefatory material Bacon called "The New Instauration," gives us a more sophisticated take on this subject. Here, sense experience is tested by a method of investigation that accepts the same nominalist distrust of human concepts expressed in the *De Augmentis,* but which by the same token enables construction of conceptual models for talking about the material world the truth of which is a matter of pragmatic testing rather than Aristotelian claims of concept-object correspondence. Bacon, in short, treats human conceptualization at different times and places as at once an obstacle to, and the instrument of, experimental "objectivity." Some of the sources, both historical and theoretical, that constrained and guided this formulation of the problem are my subject here.

The history of the problem begins with the Ockhamist revolution, which Gordon Leff sums up in a perspicuous epigram:

---

[2] For Leen Spruit, "The functional need for intelligible species can be described roughly as follows: to represent the formal properties of sensible reality in a way that is 'proportionate' to the human mind" [*Species Intelligibilis: From Perception to Knowledge; Volume One, Classical Roots and Medieval Discussions* (Leiden: E. J. Brill, 1994), 10]. Spruit expands on this "need": "The need for a cognitive principle that is unknown prior to reflection, and that makes knowledge possible, is envisaged in medieval philosophical discussions on the problem of how an immaterial mind may grasp material reality in its intelligible kernel. Although direct evidence for this principle is not available, one may still introduce it as a theoretical postulate, needed for its extensive explanatory power. Abstracted by the mind and with a content grounded in sensory information, the intelligible species seems apt to bridge the gap between mind and sensible reality, and thus to account for the objectivity of mental concepts"(4).

> So far as the process of intellection itself is concerned, then, the crucial distinction is between images as the *principle* of cognition and as the product of cognition; Ockham rejects the first and accepts the second … . [emphasis mine][3]

For Aristotle as mediated to early medieval philosophers by Boethius and later through the Arabic Aristotelians, the human mind's assurance that what it knows of the material world is objective depends on an epistemic model whereby the mental images the mind derives from that world are the cause the mind's concepts of it. When Ockham argued that mental concepts are not the cause (i.e., "principle") of knowledge, as in the species theory, but rather the *ex post facto* product of conceptualization working from and on (i.e., responding to) direct sense and intellectual intuition, he instituted an important trade-off. The sole ascertainable "fact" that intuition delivers is the object of intuition as it exists at the moment in which the intuition takes place.[4] A central consequence of this cognitive model is that the content of intuition is not and cannot be articulated, because articulation–the break-up of the perception into discrete conceptual parts–only occurs when the intellect invents concepts, and the propositions and arguments built out of them.[5]

In the *De Augmentis*, then, we find Bacon giving an Ockhamist account of unmediated intuition of material things–what he calls "judgment by induction"–that contrasts with the judgments characteristic of the deductive syllogism:

> For here the same action of the mind which discovers the thing in question judges it; and the operation is not performed by any middle term, but directly, almost as in the same manner as by the sense. For the sense in its primary objects at once apprehends the appearance of the

---

[3] Gordon Leff, *William of Ockham: The Metamorphosis of Scholastic Discourse* (Manchester: Manchester Univ. Press, 1975), 43.

[4] Ockham's definition is the following: "Intuitive cognition of the thing is cognition that enables us to know whether the thing does or does not exist, in such a way that, if the thing exists, then the intellect immediately judges that it exists … ." Quoted in Armand Maurer, *The Philosophy of William of Ockham in the Light of its Principles* (Toronto: Pontifical Institute of Medieval Studies, 1999), 477. Adams I, 502: "According to Ockham an intuitive cognition of a thing is that by virtue of which one can have evident knowledge of whether or not the thing exists, or more broadly, of whether or not a contingent proposition about the present is true."

[5] "Since the concept which the mind abstracts from the object of the senses would be absolutely simple in content, no proposition asserting its instantiation could be contradictory," Adams II, 943. This is because such a proposition would have the same concept as both subject and predicate, e.g., "This existing object exists," a tautological predication that by definition lacks affirmative power. As a formidable indicator of Ockhamist thinking's dominance on this point in the history of western philosophy, Hegel in *The Phenomenology of Mind* gives a straight Ockhamist analysis of sense experience without once mentioning the fourteenth-century Franciscan. Hegel makes explicit the import of Ockham's sundering the cognitive domain of intuited contact with the material world from that of human thought and language: "The I does not contain or imply a manifold of ideas, the I here does not *think*; nor does the thing mean what has a multiplicity of qualities. Rather, the thing, the fact, *is*; and it *is* merely because *it is. It is*–that is the essential point for sense-knowledge, and that bare fact of *being*, that simply immediately, constitutes its truth," G. W. F. Hegel, *The Phenomenology of Mind*, trans. J. B. Baillie (New York: Harper Torchbooks, 1967), 150.

> object, and consents to the truth thereof. In the syllogism it is otherwise, for there the proof is not immediate, but by mean [i.e., by the mediation of the minor premise].[6]

The implications of this conclusion for Bacon's evolving scientific method are difficult to overestimate. He distrusts "Methodical delivery" of reported scientific findings because it entails "a kind of demonstration in a circle, one part illuminating another," a closed system of tautologies or analytic propositions where connections are void of any reference or even relevance to the non-conceptual material world (*De Augmentis*, *SEH*, IV: 450). This distinction follows from the restriction Ockham puts on the content of sense intuition, namely that it refers to an object existing only at the moment of perception. This restriction entails the conclusion that one cannot assume a priori any temporal links to the object backward and forward in time, just as Ockham's reduction of all existing material things to integral individual objects joined by no ontological "universals" isolates the object in space as well. Establishing such connections, then, will have to depend wholly on continual empirical observation of specific moments in the time/space continuum, a method that of necessity can make no assumptions about what the content of these observations will be before the fact.

Bacon makes a related argument when he envisions the human mind as a mirror cleansed of human thoughts and emotions and thus made ready to receive the true light of nature, but with a significant addition. In an early treatise preceding *The Advancement* he declares that the intellect is

> not so scoured and clean as to admit the genuine native images of things without some colouring from the imagination. Speaking generally, the human mind is a mirror so uneven as to distort the rays which fall upon it by its angularities. It is not a smooth flat surface.[7]

Bacon speaks here of mental concepts as the products of the imagination distorted by the emotions. And it is at this point that a nominalist concern for technical questions of how the human mind knows passes for Bacon into ethical questions of human bias, the answers to which determine for him whether the mind can know any extra-mental reality truthfully at all. The *De Augmentis* includes a number of statements scattered throughout that collectively identify two distinct emotional blocks to human knowledge of the material world: human desires and the lust for power. Both are matters of human pride which Bacon would humble in submission

[6] Bacon seems to be conflating–or confusing–three different epistemological and logical concerns: (1) the (semiotic) mental/verbal proposition which consists of a predication and is the subject of judgment as to whether it is either (logically) correct, or (ontologically) true; (2) the middle term in a (logical) syllogism, which mediates between the generalizing major premise and the syllogism's conclusion. Both of these are posed against (3) the notion of an (epistemological) act of sense perception which simultaneously–as for Ockham–"apprehends the appearance of the object" and "consents to the truth thereof." Whatever of these alternatives Bacon intended, it is apparent that the intuition/conception opposition is the assumption governing his argument. See "Translation of the 'De Augmentis,'" *SEH*, IV: 428.

[7] *Thoughts and Conclusions* [*Cogitata et Visa*], in Benjamin Farrington, *The Philosophy of Francis Bacon* (Chicago: Univ. of Chicago Press, 1966), 88.

to the truth of *res ipsas*. The moral and indeed religious origins of these ideas suggest that Bacon is not merely talking about "mistakes" in thinking and logical fallacies in epistemic model-building–the sole concern of the strictly technical nominalist critique–but rather a matter as much of the will as of the intellect. What one might call this "voluntaristic" component of Bacon's vision of the new science requires likewise the creation of a "new scientist," one characterized by a commitment not only to an experimental praxis, but also to an ethic, of truth-seeking.

Consequently Bacon in the *De Augmentis* repeatedly calls attention to conditions of the mind that require a discipline that goes beyond merely changing one's epistemological theory. One of these obstacles is pride and vanity which characterize "those that are puffed up with their own knowledge, and rebellious against God" (*SEH*, IV: 362). Desire to see the world as we would wish it rather than as it is is a constant theme in Bacon's critique. He distrusts poetic fictions specifically because they accommodate "the shows of things to the desires of the mind, not (like reason and history) buckling and bowing down the mind to the nature of things" (*SEH*, IV: 316). He finds the myth of Bacchus an allegory of "the nature of Desire, or the passions and perturbations of the mind" (*SEH*, IV: 333), desires that drive those who pursue natural magic, who "are carried away by insane and uncontrollable passion after things which they only fancy they see through the clouds and vapours of imagination" (*SEH*, IV: 367). Bacon wittily assimilates philosophers' contempt for particulars to the snobbery of the proud "[who are] too dainty and lofty to pay attention to particulars, and especially to dwell any time upon them. For they used examples or particular instances but as serjeants or whifflers to drive back the crowd and make way for their opinions" (*SEH*, IV: 410–11).

Ethically-oriented passages like those above mingle in the *Novum Organum* with technical "how" questions as if the mentally-induced obstacles to the empiricism of the new science had their origins indifferently in either. The first two idols of the mind–those of the Cave and the Tribe–are epistemic dispositions peculiar to individuals and to humankind respectively, while the second two, the Idols of the Marketplace and the Theater, are matters of culture, what might be called "social constructions," that is, culturally-created fictions invented according to human desires. These latter are obstacles peculiar to the prejudices induced by the different education and experiences of each individual, and assume an ethical critique of mentally-constructed entities: "The human understanding is no dry light, but receives an infusion from the will and affections; whence proceed sciences which may be called 'sciences as one would.' For what a man had rather were true he more readily believes" (*SEH*, IV: 52).

Bacon applies the same critique specifically to the personal idols of the Cave and to the social idols of the Theater, which though different in their origins differ not at all in their results: both are forms of human self-delusion imposed by taking what the undisciplined human mind desires for truths of the non-human world: "Whatever his mind seizes and dwells upon with peculiar satisfaction is to be held in suspicion" Bacon says of the Idols of the Cave (*SEH*, IV: 60). And of the Idols of the Theater he says they are "more compact and elegant, and more as one would wish them to be, than true stories out of history" (*SEH*, IV: 63).

And Bacon strikes a technical note in his description of the Idols of the Market-Place, which are semiotic systems too imprecise to allow accurate notation of the data derived from experiments or articulation of generalizations based on these experiments. This push toward expanded lexicons of more precisely differentiated lexemes, for instance, also appears in Ockham's central philosophic work, the *Summa totius logicae*, which deals with how the data given in intuition may be rendered in propositional form, and is particularly concerned with disambiguating the references of specific terms in specific types of propositions. As will appear, this emphasis on human-made sign systems is a characteristic of nominalist models of human knowing wherever they are found. In the twentieth century various logical positivisms have pushed toward the invention of logical languages, the components of whose lexicons are strictly defined and thereby not subject to the vagaries of the general "marketplace" of human intercourse.

Fourteenth-century Nominalism and the *Via Negativa*

Bacon's treatment of scientific experimentation as a matter of ethical askesis, i.e., programmatic self-discipline, however, is not something he inherited explicitly from Ockhamist nominalism. That human beings were subject to error by reason of emotional and habitual biases was a fact hardly unexamined in either classical or medieval philosophy. However, such ethical issues were treated as subjects independent of epistemological model-building. On the other hand, that Bacon considers both technical and ethical issues as obstacles to the new science is not an historical anomaly. It derives from the history, partially submerged in fourteenth-century expositions of the nominalist revolution, of the concept of *intuitio,* a concept which grounds the all-important distinction between intuition and abstraction underlying the purely technical dimension of Bacon's proposed empirical method.

This concept appears at an historically problematic break between the fourteenth-century Franciscan philosophers, particularly John Duns Scotus and William Ockham himself, and the whole history of medieval epistemology that precedes them. Scotus was the first philosopher to advance the notion of *intuitio* in order to account for the mind's knowledge through the senses of particular material objects. Ockham's own notion of intuition derives from his critique of his older Franciscan colleague on the matter of the intelligible species. Where Scotus yielded thus far to the traditional demand for a mediating entity between mind and matter, Ockham contended that concepts were mental constructions that did not produce cognition but were instead its result. Consequently, the so-called "intelligible species" was not only unnecessary, it was also necessarily false. All attempts to "break up" the material world into collocations of universal concepts was merely a projection of mentally constructed concepts onto a world that did not contain them. However, having departed from Scotus on the need for an intelligible "species" and thereby cut the problem out of the relation between mind and the material world, Ockham's vaunted razor only managed thereby to relocate it as an analogous split within the human mind between intellectual intuition of particulars and mentally constructed concepts and semiotic languages.

The radicalness of Ockham's critique of Scotus can be grasped by recognizing that nothing in the history of western philosophy of human knowledge after Aristotle prepared Ockham for such a rejection, and that is because the source of both Scotus' and Ockham's notions of intuition was in fact not philosophical at all in any usual sense of that term. One recent account finds *intuitio* first used in the writings on optics by a still earlier Franciscan, Roger Bacon, as a borrowing from a treatise on the subject by the Arab philosopher Alhazen, where "intuitio" was lifted from Augustine in order to translate an Arabic word into Latin. However, as that account acknowledges, it was left to Scotus and Ockham to bring the notion of *intuitio* to its place of centrality in epistemic model-building.[8]

There was however another source also within the Franciscan tradition, but one concerned with a wholly different matter: the trans-verbal knowledge of the divine pursued by Franciscan mysticism, a pursuit given its definitive form and direction in the writings on contemplative practice by the Franciscan philosopher Bonaventure in the thirteenth century, but developed throughout the early history of the Franciscan order out of the mystical experiences of the order's founder, Francis of Assisi. Here, the mind/reality and word/thing confrontation is fully acknowledged, and develops in specific arguments that with remarkably little change reappear in Ockham's reconfiguring the domain of human thinking about the material world. Present in the former, however, is the ethical dimension that is otherwise absent from Ockham's epistemological theorizing but lies at the forefront of Francis Bacon's speculations. One may be a sinner and still be an epistemologist, but mystical experience is another thing altogether. And it is here, where transcendence of human thought and language as a necessary prelude to mystical union with the divine, fundamental to Christian mystical theology initiated by the writings of the pseudonymous sixth-century "Dionysius the Areopagite," that the moral perfection of the knower becomes necessarily entailed by the rigors of the knowledge aimed at.

Particularly relevant are specific parallels between the ways in which the two arguments are laid out. The object of knowledge in both cases–God on the one hand, the discrete objects of the material world on the other–is a reality that is integral, simple and beyond grasp through the segmentations wrought by conceptual thinking. The obstacle to human cognition is the same, namely, the necessary human reliance on conceptual thinking, an obstacle that must be consciously set aside as the price of human knowledge of this reality. The result in both cases is that the knower must impose upon himself a rigorous denial of normal human modes of conception and verbalization in order to achieve a consciousness of a different order. In both cases, what is thinkable and sayable is not true, if "true" means the capacity to construct in the mind concepts representing mimetically the reality in question.

Because discourse about mystical experiences of the divine produced by mystics themselves is often a second-order semiotic discussion of the complex relation

[8] See Katherine H. Tachau, *Vision and Certitude in the Age of Ockham: Optics, Epistemology and the Foundations of Semantics 1250–1345* (Leiden: E. J. Brill, 1988), 69–70. Also Sebastian J. Day, *Intuitive Cognition: A Key to the Significance of the Later Scholastics* (St. Bonaventure, N.Y., The Franciscan Institute, 1947).

between the experience itself and human ways of conceiving and communicating it, modern studies of medieval mystical theology treat it as in part a semiotic phenomenon. In typical nominalist fashion, the nonverbal intuited experience itself is communicated by verbal discourse that is in effect a self-referential semiotic analysis focusing on just this incommensurability between human verbal discourse (i.e., human conception) and that intuited experience. Thus the discourse that seeks to communicate mystical experience is to a significant degree already a nominalist theory of knowledge, and this is how Dionysian treatise *On the Divine Names* treats it, namely, as a matter of the humanly constructed "names" of God.[9]

The specific points that constitute the nominalist articulation of epistemological and mystical experience respectively can be laid out and enumerated, and in fact, such points are characteristic of nominalisms wherever they occur, beginning in the early modern period with that advanced by Bacon's secretary, Thomas Hobbes, in the opening chapters of *Leviathan*; also Jeremy Bentham's *Theory of Fictions* in the eighteenth century; and proceeding through the twentieth-century structuralism of Ferdinand de Saussure, and the varying semiotics that developed out of the structuralist model–Umberto Eco's *Theory of Semiotics* is a significant example–and not excluding the historically independent semiotics of Charles Peirce. As Ockham can be seen to have foretold, his *Summa totius logicae*, with its regrounding human epistemology within the structures of human conception and by extension conventional languages, completes at one stroke the radical relocation of epistemology's center of gravity from the mind-object relation to a concern with how the human mind invents codes of signs and the rules that govern them.[10]

[9] Michel De Certeau produces a history of what he calls "mystics" (by analogy to the study of "physics") in the later Renaissance by explicitly subjecting mysticism to a Foucauldian analysis where its structuralist, semiotic underpinnings are excavated from the written and printed texts dealing with the subject. See De Certeau, *The Mystic Fable, Volume One: The Sixteenth and Seventeenth Centuries*, trans. Michael B. Smith (Chicago: Univ. of Chicago Press, 1992).

[10] By the time Hobbes comes to publish *Leviathan* (1651) he can make the Ockhamist assertion that there is "nothing in the world universal but names; for the things named are every one of them individual and singular" with little fear that anyone (save those like Puritans, i.e., the deluded) can contradict him; Thomas Hobbes, *Leviathan: On the Matter, Forme and Power of a Commonwealth Ecclesiastical and Civil* (New York: Collier Books, 1962), 35. Hobbes' distinction becomes in Jeremy Bentham's hands a distinction–eminently Baconian–between "entities" which are "real" and those which are "fictitious," yielding the semiotic division "of noun-substantive into names of real entities, and names of fictitious entities," C. K. Odgen, *Bentham's Theory of Fictions* (Paterson: Littlefield, Adams & Co., 1959), xxvii. In the twentieth century C. K. Odgen and I. A. Richards, *The Meaning of Meaning* (New York: Harcourt, Brace, 1947) places the referential function of human language at the center of how "meaning" occurs, where meaning is a matter not of its intentional, semantic component, but of its extension only. Umberto Eco, *A Theory of Semiotics* (Bloomington and London: Indiana Univ. Press, 1976); also, an earlier attempt at such a theory, *La struttura assente: Introduzione alla ricerca semiologica* (Milano: Bompiani, 1968). For a discussion of Eco's nominalist attack on "natural," i.e., mimetic or otherwise "motivated" relations between sense data and language, see Michael McCanles, "Conventions of the Natural and the Naturalness of Conventions: Umberto Eco's A Theory of Semiotics," *Diacritics* 10, No. 1 (1980): 77–87. A tell-tale illustration in modern times of a typical nominalist view of this problem occurs in De Saussure's *Course in General Linguistics*, which illustrates how far epistemology has shrunk as a necessary consideration once

The argument I'm advancing here is that the epistemology which Bacon succeeds in outlining in *The Advancement of Learning* and its continuation the *De Augmentis Scientiarum* is indebted to late-medieval nominalism, and through that debt linked back to a still earlier origin, namely one of the founding group of writings underpinning medieval Christian mysticism, the texts by a sixth-century writer who identified himself as "Dionysius the Areopagite" whom St. Paul converted in Athens to Christianity. Thus similar theoretical moments are crucial to all of the following works: Dionysius's *Mystical Theology* and *On the Divine Names*; the Franciscan philosopher Bonaventure's treatise on mystical experience, *Itinerarium Mentis in Deum* derived from Dionysius; Ockham's appropriation of Dionysian nominalism when applying his own epistemological theory in his discussion of the "attributes" which the human mind assigns to the divine; and Bacon's call for an askesis that radically tests human concept- and language-making powers as the price of the capacity to use these powers to discover the "laws" of the physical universe.

These moments include: (1) Insistence on the radical structural incommensurability between intuited experience and conceptual thought. (2) The consequent limitation of the connections between these two dimensions to matters of "extension" or reference, as distinct from "intension" or meaning, the latter being a construction wholly interior to mental conceptual cognition. (3) A compensatory emphasis on the concept- and sign-forming powers of human thinking, and consequently on the semiotic and logical systems that state the rules for articulating such mental and culturally-conditioned entities.

The Dionysian influence was present in Franciscan mysticism from the founding of the order, but comes to its definitive expression in the brief treatise by Bonaventure, the *Itinerarium Mentis in Deum* ("The Mind's Journey to God"), which outlines the progressive steps by which in meditation the mind reaches an *excessus mentis*, the moment of grasping the divine beyond human conception as expressed by paradoxical affirmations that violate human thinking governed by the law of noncontradiction.[11] Bernard McGinn summarizes Bonaventure's conclusion:

---

the determining factors of human knowing are conceived as matters of concept and sign generation. In order to illustrate graphically his notion that the semantic component of any language constitutes a way of "cutting up" the flow of sense data, thereby making the differences nameable, Saussure produces an illustration of a river-like flow, the diachronic course of which is intersected at intervals by vertical demarcations. This illustration constitutes a "picture" of the external world prior to mental articulation according to the paradigmatic oppositions and syntagmatic linkages that together constitute the semantic components of human "natural languages," a world which is prior to language being nothing other than a fluid stream of unstructured and meaningless sense experience. On the basis of this extreme, and totally unexamined, empiricist view of the mind's view of the external world, it is reasonable to infer that De Saussure did not believe the epistemological question needed to be dealt with at all. Ferdinand de Saussure, *Course in General Linguistics*, eds. Charles Bally and Albert Sechehaye in collaboration with Albert Riedlinger, trans. Wade Baskin (New York; McGraw-Hill, 1966), 112. Typical of structuralist semantics indebted to Saussurean linguistics are A.-J. Greimas, *Sémantique structurale: Récherche de methode* (Paris: Librairie Larousse, 1966), and John Lyons, *Semantics*, 2 vols. (Cambridge: Cambridge Univ. Press, 1977).

[11] Bernard McGinn, *The Flowering of Mysticism: Men and Women in the New Mysticism (1200–1350)*, vol. III of *The Presence of God: A History of Western Christian Mysticism* (New York: Crossroad Pub., 1998), 78, 108–10.

> In order to know any truth in secure fashion, Bonaventure insists, the human mind needs to have a direct, if not necessarily clear, contact with Truth itself, the *Divina Ars*. In other words, we need to have an implicit "contuition" of the Uncreated Word in any act of affirming the truth. This innate presence of the Word can be made explicit and conscious through the practice of the "contemplative contuition" sketched out in the *Itinerarium*.[12]

What Bonaventure means by "contuition" is subject to debate. Whereas some commentators have held that Bonaventure meant by it direct intuition of the divine by the mind of the mediator while "in this life," it would appear that he is referring to a process of moving both by means of and through, i.e., past, the stages of the *itinerarium* which consist of the material world of creation and then of human language and thought, toward an understanding of the divine that transcends both.[13] Ewart H. Cousins' consideration of the meaning of *contuitus* points implicitly to Scotus and Ockham's later choice of *intuitio* when faced with similar epistemological problems presented by the mediating species assumed by previous medieval philosophers:

> It is true that Bonaventure presents this contuition as a contemplative path and not as an ecstatic experience. But there is solid reason to include such an intuitive contemplative consciousness within the sphere of mysticism … The [meditative] journey begins in sense knowledge, and proceeds to the soul, then to God, and finally to ecstasy. The third stage contains as an essential element the mysticism of language. From a psychological point of view, however, it is possible to see all seven stages as moments in a dynamic process that culminates in ecstasy.[14]

"Ecstasy" is a quasi-technical term that connotes a range of meanings for different mystics and mystical theologians. In the Franciscan mystical tradition the word comes to mean "standing outside of" normal human means of cognition, and is thus defined as the opposite, or even negation, of human concepts and languages.

The influence of Franciscan mysticism on purely philosophical considerations makes its appearance in Ockham's treatment of what became a standard topic of philosophical discussion following upon the revival of interest in Dionysian mystical theology in the thirteenth century, namely the problems arising from attempts to reason back from creation to Creator in order to assign positive attributes to the divine. The similarity between Ockham's analysis and cognate points occurring in the Dionysian texts have been noted in the major Ockham literature. Armand Maurer mentions this connection in passing, while Adams incorporates the element of negative theology in her extensive discussion of Ockham's critiques of Aquinas and Scotus on the question of divine attributes, and acknowledges the historical influence

---

[12] McGinn, 108; he translates "*Contuitus*" as "concomitant gaze, insight, or grasp," 374.

[13] Dunstan Dobbins, *Franciscan Mysticism: A Critical Examination of the Mystical Theology of the Seraphic Doctor, with special references to the sources of his doctrines*, Franciscan Studies No. 6 (New York: Joseph F. Wagner, 1927), 146–7.

[14] Ewart H. Cousins, "Bonaventure's Mysticism of Language," in *Mysticism and Language*, ed. Steven T. Katz (New York: Oxford Univ. Press), 252.

of Dionysian thinking on fourteenth-century interest in this matter.[15] She goes so far as to indicate that Ockham's discussion of divine attributes is founded on the same reasoning that produces Ockham's radical revision of Scotus' intuition/abstraction split. However, since Ockham is engaged in a strictly philosophical analysis of this topic, she no more adverts to the mystical element latent in this influence than does Ockham himself. Like Maurer and Leff,[16] Adams is content to emphasize that Ockham's response to attempts at justifying application of concepts drawn from the created world to the divine creator falls under the same unrelenting nominalist strictures that Ockham imposes on human epistemology in general.[17]

Ockham's treatment of the divine attributes is in effect a late example of the same genre of commentary to which the Dionysian *Divine Names* belongs. In fact, historically the Dionysian texts generated two different lines of development in the thirteenth and fourteenth centuries, one as mystical theology and the other as natural (i.e., philosophical) theology. The former is typified by Bonaventure's *Itinerarium*, which emphasizes the affective dimension of the mystical experience, and which in turn recapitulates the thinking of what is now recognized as a crucial moment in the assimilation of Dionysian mysticism into Franciscan thinking, the writings of the late thirteenth-century Victorine monk, Thomas Gallus.[18] The other is a line of strictly philosophical and theoretical analysis that includes Aquinas' teacher Albertus Magnus, Aquinas himself as well as Scotus and Ockham.[19] Where the first tradition is directed to developing the inner spiritual life of mystical contemplation, the second deals with what knowledge of the divine human reason unaided by revelation and church doctrine can reach to. Consequently the points where Ockham's analysis of the divine attributes–an exercise in "natural theology"–exhibits indebtedness to the "mystical theology" of the Dionysian texts, can be easily missed if the fundamentally different directions in which both kinds of exposition are pointed remain the focus of attention. If, however, only epistemological considerations guide the analysis, the parallels consistent with the three central theoretical points characteristic of nominalist theorizing in general noted earlier become evident and determinative.

The radical incommensurability between intuited experience and conceptual thought is the fundamental assumption which both Dionysian treatises on mystical experience, *Mystical Theology* and *The Divine Names*, take as their point of departure. Consequently human thinking and expression are limited to "performative" rather than "expressive" functions, whereby verbal signs are discussed as performatively referencing, i.e., pointing to the transcendent divine reality, by way of denying their capacity to express it conceptually.

This sundering of the linguistic and the logical in order to distinguish and develop the strictly extensional, referential function of the former is replicated in Ockham's

[15] Maurer, 201; Adams II, 908 ff.

[16] Leff, 410.

[17] Adams II, 935.

[18] McGinn, 78ff.

[19] Paul Rorem, *Pseudo-Dionysius: A Commentary on the Texts and an Introduction to Their Influence* (New York: Oxford Univ. Press, 1993), 214–24.

hands in the original uses to which he puts the traditional medieval theory of supposition, discussed below. It is in any case on this basis that Dionysian mystical theology can be labeled specifically a "negative theology," namely because it consists of specifically verbal statements that alternately affirm attributes of the divine, and references such statements themselves in order to negate them.[20]

And it is here that the third theoretical moment of nominalist theorizing makes its appearance. What in Ockham's philosophizing becomes the central work, namely a logic that specifies and catalogs the different extensional, referential relations among different kinds of concepts and conventional signs for articulating the data given in intuition, was already in Dionysian negative theology a similar second-order classification of the semiotic instruments by which signs invented by the human mind can do nothing to express their meaning except by denying their capacity to do so. In other words, this third moment recapitulates and goes beyond the second moment by making explicit the notion that the capacity of human expression for extensionally referencing the divine comes at the expense of denying its intensional, semantic content. This trade-off is developed in the Dionysian texts as an interplay between "cataphatic" and "apophatic" uses of human language, and as such gestures toward a second-order theory of human semiosis that Ockham will open out to encompass a fully developed, exhaustive theory of all suppositional, referential functions.

As Denys Turner points out, "*apophasis* is a Greek neologism for the breakdown of speech, which, in face of the unknowability of God, falls infinitely short of the mark."[21] The term plus its opposite *cataphasis* occur "in the vocabulary of late Platonist Christian theology,"[22] and the use of the two terms together are standard in modern discussions of negative theology in general and Dionysian mystical theology in particular:

> "Apophaticism" is the name of that theology which is done against the background of human ignorance of the nature of God. It is the doing of theology in the light of the statement of Thomas Aquinas in the thirteenth century, that "we do not know what kind of being God is." It is the conception of theology not as a naive *pre*-critical ignorance of God, but as a kind of acquired ignorance, a *docta ignorantia* as Nicholas of Cues called it in the fifteenth century. It is the conception of theology as a strategy and practice of unknowing, as the fourteenth-century English mystic called it [the anonymous writer of *The Cloud of Unknowing*], who, we might say, invented the transitive verb-form "to unknow" in order to describe theological knowledge, in this its deconstructive mode. Finally, "apophaticism" is the same as what the Latin tradition of Christianity called the *via negativa*, "the negative way."[23]

---

20 Ninian Smart, "What Would Buddhaghosa Have Made of *The Cloud of Unknowing*?" in Katz, ed., 120. Smart makes the speech-act distinction between "performative" and "expressive" statements regarding the explicitly Dionysian rhetoric of the anonymous *Cloud of Unknowing*, but the appropriateness of the distinction is clear as applied to the Dionysian texts themselves.

21 Denys Turner, *The Darkness of God: Negativity in Christian Mysticism* (Cambridge: Cambridge Univ. Press, 1995), 20.

22 Turner, 19.

23 Turner, 19.

A "cataphatic" use of language is straightforward propositional assertion by which a predicate is affirmed of a subject. Assigning attributes to the divine is quite simply a matter of cataphatic statements–affirmations about God–and the difficulties that attend doing so is the crux of the whole discussion of this topic as a standard subject of commentary in medieval philosophizing.

"Apophaticism" refers to a second-order analysis of language about a reality that this language cannot directly grasp, reflect, and represent: the *via negativa* is thus a theology that is "about" the divine by way of being about language about the divine. Denys Turner finds implicit in Dionysius's *Mystical Theology* a distinction between first-order language, which is full of violations of semantic and syntagmatic well-formedness, i.e., paradoxes; and a second-order language that takes as its concern the first's "logical and epistemic foundations":

> For just as the first-order language is revealed in its inadequacy by the necessity of our constructing simultaneously mutually cancelling images of God, so the second-order logic of that first-order language is imposed by the necessities which derive from our having to say mutually cancelling things about the foundations of theological language as such … .[24]

Ockham's theory of supposition similarly distinguishes between a first-order reference of words to things (personal and simple suppositions: mentally constructed concepts referring to the data given in the senses, and universal concepts respectively); and a second-order use of language (material supposition) that references the first order of language.[25] This distinction is important for enabling recognition that when universal concepts such as "man," "beauty," "red" are used, it is not directly to existents in the material world–for nominalist thinking the fundamental mistake of epistemic "realism" is the belief that universals have an existence in the extra-mental world–but only to entities constructed by the mind and verbalized in conventional languages.

Illustrative of the above points of comparison is the fact that for both Dionysius and Ockham the obstacle to assigning attributes to the divine lies in the transcendent divine simplicity. The *locus classicus* of this point in the Dionysian texts is the second chapter of *The Mystical Theology*:

> [T]he writings [i.e., biblical scriptures] always celebrate the appropriate divine names not partially but of the whole, all-complete, wholly complete, and full divinity. Thus, all of these are to be attributed absolutely, wholly, and unreservedly to the entire wholeness of the wholly complete and entire divinity. As we have been reminded in the *Outlines of Theology* [of the neoplatonist Proclus, a major Dionysian source], if someone declares that this celebration is not about the whole divinity, then in a boldness contrary to the divine law, he blasphemes and cleaves asunder the one beyond unity.[26]

---

24 Turner, 33.

25 Adams I, 329.

26 John D. Jones, trans., *Pseudo-Dionysius Areopagite, The Divine Names and The Mystical Theology Translated from the Greek With an Introductory Study* (Milwaukee: Marquette Univ. Press, 1999), 27. C. E. Rolt provides a less literal transition: "all the Names proper to God are always applied in Scripture not partially but to the whole, entire, full, complete Godhead, and that they all refer indivisibly, absolutely,

Jones in his commentary on this passage indicates its purport, namely that the "simplicity and unity" of the divine is "beyond nature," i.e., beyond human comprehension:

> The divine names are to be attributed in this way so that we do not cleave asunder the unity beyond unity … For Pseudo-Dionysius the divinity is named unity … "on account of the simplicity and unity of its partlessness beyond nature." The term "whole" refers to the divinity beyond unity, to the unity and simplicity of the divine partlessness or simplicity. Here, "whole" means "partless"; that is, "whole" is to be understood without reference to parts.[27]

Jones's summary statement that "the opening section of *Divine Names II* is largely concerned with showing all *logia* [i.e., "cataphatic" affirmations] about the divinity are said of the whole divinity"[28] assumes that the Dionysian text functions as a second-order commentary that has as its reference the necessarily limited referentiality of first-order statements about the divine. This latter form of reference corresponds to the second of the three theoretical moments in nominalist epistemologies, where mental concepts and semiotic systems (e.g., natural languages) are understood to relate to an extra-mental reality in a purely extensional manner, i.e., they "point to" the divine but cannot adequately represent its nature. Though such cataphatic statements may themselves articulate the divine into distinct parts, that articulation is not to be taken as a mimetic representation of the divine reality referenced, but only a manifestation of the system of concepts/signs by which human thinking makes comprehensible to itself the reality of a divine entity whose integral wholeness is beyond its comprehension.

As indicated above, second-order meta-linguistic statements in the exposition of medieval mystical theology about specific predications of divine attributes, stand to the latter as "apophatic" statements stand to "cataphatic" statements. What the latter says "is," the former says "is not," if "is" is taken to mean "is structured as human conceptual/semiotic means of signifying are structured," that is, a projection–as a nominalist would put it–onto the real world of human modes of thinking and expression.

I shall take up later the question of "objectivity" implicit in the above discussion when considering the evacuation of personal subjective perspective as Franciscan mysticism deals with it, because it is a major topic in Bacon's discussion of scientific method. But it should be clear by now that mystical theology's split between conceptual/semiotic affirmations of "truth," and a second-order system of analysis that reverses that affirmation, is a significant historical source of how questions of objectivity are put in the early modern period.

---

unreservedly, and wholly to all the wholeness of the whole and entire Godhead," *Dionysius The Areopagite, On The Divine Names and the Mystical Theology* (London: Society For Promoting Christian Knowledge, 1940), 65. See also Pseudo-Dionysius, *The Complete Works*, trans., Colm Luibheid, with Paul Rorem (New York: Paulist Press, 1987), 58.

[27] Jones, 27.

[28] Jones, 36.

The surd of knowledge of the divine that is left over once the cataphatic/apophatic dialectic has done its work is typified in the Dionysian text of *Divine Names VII*:

> Thus the writings call the completely lit light "invisible"; they call that which is greatly celebrated and many named "ineffable and nameless". They call that which is present in all and discoverable from all "incomprehensible and inscrutable". For this reason it is even now said that the divine Apostle celebrates the foolishness of God by calling it that which appears contrary to reason and absurd in itself but which leads us to the ineffable truth before all logos.[29]

The *logia* referred to in this passage are paradoxes in which contradictions are predicated of the divine, such as "unknown" and "all intelligible," "all named" and "nameless." However, contradictories "are unified in the divine unity; thus, they are freed from difference (and from sameness). To be sure, we 'see' this only in the culmination of negative theology: the abandonment to unknowing."[30]

Whatever "unknowing" means or refers to here, it is clear that, consistent with a nominalist map of human cognition, true contact with the divine–like true contact with the objects of the material world given in intuition–requires the negation not only of human conceptual and verbal signs, but in addition the negation of the basic principle of their "well-formedness," namely the principle of noncontradiction, a consequence born out in the experience of early twentieth-century physicists regarding the behavior of sub-atomic matter. In other words, the intuited experience of the material world in the Ockhamist epistemic model corresponds structurally to what in the Dionysian model of mystical theology constitutes the human soul's union with the divine.

That this is so is manifest in Ockham's own argument regarding the topic of "divine attributes." In this connection, Adams links Ockham's thinking on "conceptual" vs. "real" distinctions to the same assumption that produced the *via negativa* of mystical theology: the assumption transmitted ultimately from Plato's *Parmenides* to Plotinus of what she calls the assumption of "concept-object correspondence": that the propositional semantic distinction between a subject and a predicate implicitly asserts a real, ontological distinction between the two. Conversely, if what is to be asserted is the integral simplicity of the subject of a proposition, then no distinct predicates are possible. As Adams states:

> Plotinus seems to have adopted this analysis … For starting from the claim that the One is simple … he concludes that it would be incorrect to say even that the One is, since that would be to posit a duality of property-bearer and property in the One. And under the influence of Neoplatonism, theologians took over these metaphysical and semantic postulates, and inferred that no positive, non-relational properties can be truly predicated of God.[31]

---

29 Jones, 175.

30 Jones, 39.

31 Adams II, 909–10.

Ockham's position is built atop his critiques of Aquinas and Scotus, both of whom argue that while human thinking requires the distinguishing of various attributes of the divine, propositions asserting such distinctions do not contradict divine simplicity. But if one assumes, as Ockham does, that all ontological propositions–those understood as having "simple" supposition–must be understood to assert concept-object correspondence, his response is, as Adams notes, going to be precisely what one would expect given his nominalist analysis of human cognition:

> Ockham argues that claims of this form–such as that divine wisdom and divine goodness are really the same but distinct *rationes* [Aquinas's concepts]–are at best misleading and at worst commit a category mistake. Taken literally, they imply that what is really one and the same can be made somehow distinct by a thought process in which it is conceived of in distinct thought objects.[32]

Consequently Ockham's conclusion is identical with that of Dionysius and of the *via negativa* in general, when he says that " … divine wisdom is the same as the divine essence in every way in which the divine essence is the same as itself. Similarly, for divine goodness and justice. Nor is there any distinction at all, or even any nonidentity there in the nature of the thing … ."[33]

However, his treatment of the verbal means of expressing this divine simplicity involves not at all the use of what Turner calls the "self-subverting utterance" of paradoxical rhetoric cultivated in medieval mystical theology.[34] This is because Ockham has at hand a conceptual apparatus for expressing this apophatic moment unavailable to the sixth-century writer of *The Divine Names* and *Mystical Theology*, namely the logical classifications of reference that constitute the scholastic theory of supposition. What is only implicit throughout those two treatises, namely that they are in effect treatises on a second-order meta-language for talking about talking about the divine, becomes in Ockham's case deliberate and overt. The barrier of concept-object correspondence is leaped at one bound by the nominalist countering assertion of concept-object disparity, matched with the compensatory notion that human concepts

---

[32] Adams II, 922. Significantly, Adams finds it necessary to advert to her earlier discussion of Ockham's nominalist epistemology in order explain his position on the divine attributes: "Recall … that Ockham rejects any ontology that posits a plurality of formally distinct entities in one and the same real thing, on the ground that none can provide a criterion of distinction for real things. For contradiction is 'the most powerful way' of proving such distinction. But if a and b are formally distinct entities in the same real thing x, it follows that contradictories–viz., 'a is formally the same as a' and 'b is not formally the same as a'–are simultaneously true about what is or is in the same real thing," II, 935. Thus Ockham's natural theology, which submits "contradictory" predications to a logical test and finds them therefore unacceptable, remains at the level of the "cataphatic"; whereas the mystical theology of a Bonaventure (or a Dionysius) would explore how the apophatic failure of cataphatic affirmation–signaled by just the contradictions that Ockham rejects–prepares the way for an "intuition" of the divine that transcends human conceptual and verbal articulation. And it is at the privileging of intuition over conceptualization that Ockham and the mystical theologians, following different, even mutually-exclusive paths, converge and meet.

[33] Quoted in Adams II, 942.

[34] Turner, 21.

and semiotic expressions of human understanding of the divine have only extensional, referential force. They are, in short, "about" the divine, not knowledge of it.

"Knowledge of" the divine is for Ockham's natural theology exactly what "knowledge of" is for his epistemology: an intuitive cognition of the object of cognition:

> there are not many attributal [*sic*] perfections, but there is only one perfection there, indistinct in reality and reason, which properly and literally should not be said to be in God or in the divine essence, but in every way is the divine essence … .[35]

This issue of knowing the divine attributes under the constraints imposed by this conception of divine simplicity lies at the point where his argument intersects most explicitly with Dionysian mysticism: "God cannot be cognized in Himself in such a way that the divine essence itself is the immediate term of the act of understanding [i.e., of intuitive cognition] … ."[36] Intuitive cognition of God, like intuitive cognition of the objects of the material world, yields the evident (i.e., non-propositional) assent to divine existence, a state which Ockham says is reserved for the Blessed who have intuitive knowledge of the divine essence. As Adams, citing Ockham, puts the difficulty:

> Given that we cannot naturally have an intuitive or abstractive cognition of which God Himself is the immediate object, Ockham's conceptual empiricism dictates that "… He cannot be conceived by us in this life in a simple concept proper to Him": at least by our unaided natural powers, because "nothing can be cognized by us by our purely natural powers in a simple concept proper to it, unless it is first cognized in itself."[37]

Ockham's theory of supposition resolves the problem in the same way as does the discourse of mystical theology, namely by restricting the uses of human logic and semiosis in this case to reference only: the disparate articulations of both are "true" of the divine to the degree that in propositions the subject and predicates suppose the same referent.

However, for the Blessed, as Ockham projects the case, the matter is somewhat different. They alone, granted immediate intuition of the divine essence, can formulate propositions–if formulating propositions be their pleasure, Ockham seems to imply–in which the disparate subjects and predicates of human propositions not only supposit for the same referent, but intentionally, i.e., in their meanings "mean" the same thing.[38] In other words, as for Dionysian mysticism, so for Ockhamist nominalism: talking about the divine means talking that violates the rules of logical/linguistic well-formedness. And similarly in both cases, the truth that the combined cataphatic and apophatic means of expression points to is one given only in pre- and non-discursive intuitive cognition: respectively intuition through the

---

35 Quoted in Adams II, 942.

36 Adams, II, 942.

37 Adams, II, 944.

38 Adams, II, 963–4.

senses of the objects of the material world, and mystical union with the divine in this life and intuition of the divine essence for the Blessed in the after-life.

The point of this comparison is that the structural parity between the epistemological options available both to theological speculation and to the epistemic workings of human cognition, marks the point where Dionysian/Franciscan mystical theology supplies an epistemic scaffolding for a nominalist theory of human knowledge of the material world as well.

Intuitive cognition and empiricism

As such, however, this connection raises an even more challenging possibility in the pre-history of early modern empiricism. Does this intuitive knowledge of the "Blessed" constitute Ockham's norm of human knowledge in general, to the extent that all other forms of knowledge by comparison with the cognitive value of that intuition fall short as merely human inventions and therefore subject to negative critique? That conclusion would be consistent with the goal of Franciscan mysticism such as Bonaventure detailed it in the *Itinerarium Mentis*, just as it would be consistent with Ockham's nominalist critique of philosophic realism, and not least, with Baconian and post-Baconian critiques of human knowledge not based on empirical intuition. A crucial point in this comparison of what are, after all, quite different cognitive programs, is that all three assume that intuition is normative–a kind of epistemic gold standard–in that it is the only form of cognition that is epistemologically complete and perfect. All others–reflected in the human constructions of mental conceptualization and the semiotic codes that mediate this conceptualization to the self and between human selves–would therefore appear to that degree at best incomplete, if not downright erroneous.

And it is at this point–the epistemic norming of the beatific vision transmitted from Franciscan mysticism to Franciscan nominalism as the norming of intuitive cognition–that the parallel technical and ethical issues in Bacon's exposition already discussed come back into focus. In one direction, if intuitive cognition (read: empirical investigation) is the privileged access to the truth of the material world, then Bacon finds himself (like medieval mystical theologians) necessarily concerned with the ethical matter of bracketing and eliding the personal fantasies and emotions of the self that would inhibit that access. Again, the consequent relegation of mental concepts and human semiotic systems to the status of non-correspondence with this material world raises the technical issue of assigning to them the merely extensional function of "pointing to" this reality consistent with the third moment of nominalist theorizing already noted, thereby rendering them inadequate to express this truth. Both issues are in their different ways the result of establishing non-conceptual, unmediated intuition as the norm of true knowledge. I close this discussion with some indications of how Bacon resolves the aporia created by these two fundamental components of his projected new science.

In a statement that applies felicitously to Bacon's project, Denys Turner remarks about the place of the human self in mystical experience that "'Detachment' … is the ascetical practice of the apophatic."[39] For such fourteen-century mystics as Meister Eckhart and the author of *The Cloud of Unknowing* as well as the sixteenth-century Spanish mystic St. John of the Cross, there is what he calls an

> "apophatic anthropology" as radical as their apophatic theology, the one intimately connected with the other. All three in *some* sense deny that I am "a self"; or at least, they appear to say that whatever may be the proper description of the fullest union of the human self with God, there is no distinction which we are able to make *between* that "self" and the God it is one with.[40]

Turner's discussion of detachment from the self as Meister Eckhart develops it is particularly relevant to Bacon's argument. In a sermon on the subject Eckhart says that "detachment" "approaches so closely to nothingness that there can be nothing between perfect detachment and nothingness."[41] More specifically, Turner argues that Eckhart espouses an "'apophaticism of desire' which parallels his apophaticism of intellect. As intellect is a 'nothing' and is nameless, so the will must be a 'nothing' and motiveless."[42]

Turner's argument is particularly suggestive here in connection with Bacon's continually expressed desire to enforce a psychological detachment from the biases of the human mind characterized by desires to control, possess, or otherwise project onto the physical world an ego-driven agenda:

> What, then, is the practice of detachment? I think that as we can say that detachment is the solution, so the problem is not human desire itself, but its possessiveness. The strategy of detachment is the strategy of dispossessing desire of its desire to possess its objects, and so to destroy them … Detachment, for Eckhart, is not the severing of desire's relation with its object, but the restoration of desire to a proper relation of objectivity; as we might say, of reverence for its object.[43]

Detachment from the self's desires on the one hand, and bracketing conceptual and semiotic inventions on the other, are thus similar apophatic moments that engage the will and the intellect respectively. If the latter become projections of mental concepts through inadvertence–inviting thereby the admonitions of the nominalist critique–the faults of the first instance are rather a matter of human desire, its weaknesses and aggressions. It is therefore reasonable to distinguish technical and ethical forms of apophaticism, the one concerned with the technology of human cognition,

[39] Turner, 179.

[40] Turner, 5–6.

[41] Turner, 172. "Perfectly disinterested, a man has no regard for anything, no inclination to be above this or below that, no desire to be over or under; he remains what he is, neither loving nor hating, and desiring neither likeness to this or unlikeness to that. He desires only to be one and the same; for to want to be this or that is to want something; and the disinterested person wants nothing. Thus everything remains unaffected as far as he is concerned," "About Disinterest," *Meister Eckhart: A Modern Translation*, trans. Raymond Bernard Blakney (New York: Harper & Row, 1941), 83.

[42] Turner, 180.

[43] Turner, 183.

and particularly its limitations, the other with matters of a willed commitment to the effort to understand these limitations and find ways to circumvent them. Bacon admonishes against both not only in his discussion of the Idols of the Mind in the *Novum Organum* but throughout the *Advancement* and the *De Augmentis*. If the technical apophaticism of Bacon's new science entails his nominalist privileging of an empirically based experimentation, the objective results of this apophaticism in turn depend upon an ethical apophaticism which guards against taking the desires and aggressions of the self as guides in reasoning and articulating the results.[44]

But there is more to a scientific askesis in Bacon's discussion than the apophatic moment of the *via negativa*. At the beginning of this discussion I noted an apparent anomaly in Bacon's discussion of what he calls "sense" experience, which he sometimes privileges over against intellectual conceptualization, and at others insists on subordinating to the conceptual system of an intellectual procedure which he projects as identical with the "new science" that he envisions. This latter tack is particularly evident in the prefatory material attached to the *Novum Organum*. Clearly Bacon's theoretical speculations, directed by his nominalist, apophatic strictures, leave him convinced that negation of conceptualization is only part of the process, and a propadeutic one at that. The same could be said of the positive residue of Ockham's reduction of natural theology to the conceptual and semiotic constructions of the human intellect not yet gifted with the intuition of the divine essence granted to the Blessed. In both cases, apophatic critique has a purpose that looks beyond itself. For Ockham, natural theology proceeds apace but now understood to be a matter wholly confined to extensional reference within the limits of human semiosis. What Turner calls the third moment of mystical theology, a "negation of the negation" in the interest of reaffirming cataphatic statements about the divine now purged of the delusions of concept-object correspondence,[45] becomes in Bacon, remarkably enough, a discernible if necessarily undeveloped foreshadowing of modern, post-Newtonian experimental practices.

---

44 The points made here about the relevance of self-detachment to Bacon's thinking are further developed and opened out into their socially-defined configurations by Julie Robin Solomon, *Objectivity in the Making: Francis Bacon and the Politics of Inquiry* (Baltimore: Johns Hopkins Univ. Press, 1998). She rightly notes (43ff.) that for Bacon the "restraint of the passions" was to work for facilitating not merely virtue but cognition, a perspective made possible by an "early modern nominalist episteme that stressed the mind's productive role." The apophatic agenda I discuss here, however, is complicated by the fact that, as Solomon observes: "Bacon necessarily confronts historical, social, and political contingencies that never formed part of the problematic of Scholastic epistemology … [W]hen philosophers see the essences of things as even the partial product of a productive mind, then the natural philosopher must face the more complex task of unifying a socially heterogenous group of minds that actively produce and pursue knowledge"(46). That this situation seems to make true self-detachment at once more elusive and therefore more desirable, is one inference that arises from Solomon's argument, as well as from other discussions of the same issue, such as Lorraine Daston, "Objectivity and the Escape from Perspective" *Social Studies of Science* 22 (1992): 597–618; and Peter Dear, "From Truth to Disinterestedness," *Social Studies of Science* 22 (1992): 619–31. Pursuing this trajectory, it is not surprising that one finds apophatic paradox making its reappearance in the title and substance of Thomas Nagel's discussion of "objectivity," *The View from Nowhere* (New York: Oxford Univ. Press, 1986).

45 Turner, 22.

And it is here, where critique turns to procedure-building, that there occurs an intriguing moment in Bacon's development of the consequences of his own nominalist critique. Bacon's shifting evaluations in both the *De Augmentis* and the *Novum Organum* of mental conceptualizing mark an ambiguity which he inherited along with the nominalist critique. If the *species intelligibilis* was for Ockham at last consigned to the dust heap of history, the purpose which it had existed to serve was not.[46] That purpose was to guarantee that the contents of the mind was indeed an accurate reflection of the mind's experience of the material world. And implicit in that guarantee was the notion that the *species intelligibilis* always performed some sort of mimetic function: thoughts were "true" because they were effects of external causes (i.e., like causes like) and/or because they reproduced in some way copies of the outside world. And the price of foregoing such a theoretical entity was steep indeed: if you don't have concept-object correspondence, then you're left with concept-object split, and that means skepticism, the conclusion that there is no such thing as accurate contact between mind and matter.[47] On that basis, for Ockham the only solution was to posit no species whatsoever. He gives an essentially Johnsonian kick to the edifice of medieval epistemology and its troubled liaison with the mediating species: "Thus I refute the chimera of concept-object correspondence." There is no such thing as a mediating species to explain, and consequently no such thing as "epistemology"–indeed, there was nothing at all to explain in the first place: the mind sees things through the senses, and that is that.

---

[46] Leen Spruit concludes to the same effect: "Ockham's position ultimately begged the question of the origin of knowledge, for his postulated mental capacity is a huge black box that is presumed to execute just the complex function it was rang in to explain", *Species Intelligibilis: From Perception to Knowledge, Volume Two: Renaissance Controversies, Later Scholasticism, and the Elimination of the Intelligible Species in Modern Philosophy* (Leiden: E. J. Brill, 1995), 389. For a discussion of how such circular reasoning can work to subvert the very thing defined, see my article on Jacques Derrida's deconstructive critique of nominalist privileging the intuitive moment as something allegedly escaping human concepts. What Derrida attacks overtly is Husserlian phenomenology and its privileging pre-conceptual intuition, but it is an attack that reaches across the centuries to the fourteenth-century nominalist revolution. Derrida's *coup de main* is his undeniable assertion that perception is a concept: "Perception is precisely a concept, a concept of an intuition or of a given originating from the thing itself, present itself in its meaning, independently from language, from the system of reference. And I believe that perception is interdependent with the concept of origin and of center and consequently whatever strikes at the metaphysics of which I have spoken strikes also at the very concept of perception. I don't believe that there is any perception," 29. Michael McCanles, "From Derrida to Bacon and Beyond," *Francis Bacon's Legacy of Texts: "The Art of Discovery Grows with Discovery,"* ed., W. A. Sessions (New York: AMS Press, 1990), 25–46. It's as if Derrida had informed members of the Sierra Club that "nature" is a human concept, therefore a human construct, and therefore either "nature" is not "natural" at all, or human constructs (including urban development) are all "natural."

[47] Robert Pasnau cites the case of Walter Crathorn, a Dominican contemporary of Ockham's, whose account of the *species intelligibilis* as a "likeness" of the object perceived leads to the *reductio ad absurdum* of copy theory epistemology. Pasnau states: "According to Crathorn, cognitive representations of the external world must have the very qualities of the things they represent. A species of redness must actually be red, and the same goes for other sensible qualities" (89). If "red" ideas of red objects become the norm of ontological epistemic reference, it is clear that failure to meet such a requirement can only lead to just the sundering of concept and intuited object that Ockham proposed; *Theories of Cognition in the Later Middle Ages* (Cambridge: Cambridge Univ. Press, 1997), 89ff.

But of course, not quite. And that is what Bacon's equivocations on the subject act out for us. If mental concepts are human constructs, then they don't obey the regulative principle of concept-object correspondence. Alternatively, if the theoretical skepticism consequent to concept-object split is not to be the inevitable result, then a new way of understanding the relation of mental concepts as *ficta* (i.e., "something made") to the extra-mental world is necessary.[48] I close this discussion with a brief exposition of Bacon's solution to the problem–at least the beginning of such a solution–one that slips past the dilemma created by making a copy-theory epistemology the norm and test of true knowledge of the material world. In both cases, we discover something closer to a hermeneutic process of testing, which takes for granted that the model is not and can never be "true" in a copy-theory sense of the term.

Relevant here is Antonio Pérez-Ramos' assertion that Bacon had, in effect, already foreseen a form of pragmatic testing of hypotheses that is normal, for instance, in modern physics' reliance on what is in effect a hermeneutically circular process of generating tests that in turn modify original hypotheses, leading in turn to further tests and the refining or changing of hypotheses.

Pérez-Ramos' immediate concern is the meaning of Bacon's "simple natures" in Book Two of the *Novum Organum*. These constitute the underlying causes of the empirically sensible qualities of heat, which is Bacon's example. As Pérez-Ramos says, "Bacon presents his Form as a rule of operation," and quotes Bacon's recipe for creating "heat" by reciting a procedure in which the distinguishable material characteristics of heat–which include "a dilating or expanding motion," so "repress[ing] this motion and turn[ing] it back on itself" that dilation proceeds in a non-uniform manner, one part counteracting another part–are produced, with the result that "you will undoubtedly generate heat."[49] Pérez-Ramos concludes:

> What Bacon has called "Form" is at once a statement purporting to unveil the quiddity or essence of a phenomenon by redescribing it in a lawlike manner *and a recipe for a successful operation or reinstantiation of it on the basis of the pragmatic effectiveness of that law.*[50]

The truth of the theoretical recipe lies in the resultant material pudding. What the theory predicts, the experiment may (or may not) confirm. If you can reproduce the natural object using your scientific description of its make-up as the basis of reproducing it, that fact becomes proof of the truth of the scientific description. Thus we have the standard test for new scientific formulations: can the same results be reproduced by other scientists using the propounder's "recipe" for these results? In brief, reproducibility establishes "truth." And what has been neatly sidetracked is the need for a copy-theory epistemology as a guarantee of the truth of the hypothesis' ontological

[48] The issue of mental concepts as *ficta* (i.e., "something made" by the mind, not to be confused with *figmenta*, e.g., a concept of a fictional mythical being, Maurer, 505) was central in fourteenth-century debate over the epistemic status of universal concepts.

[49] Antonio Pérez-Ramos, "Bacon's forms and the maker's knowledge tradition," *The Cambridge Companion to Bacon*, ed. Markku Peltonen (Cambridge: Cambridge Univ. Press), 109.

[50] Perez-Ramos, 109

claim, burdened as it is by the objective/subjective conflict consequent on the nominalist reduction of mental conceptualization to a purely extensional function.[51] An extensional moment remains but radically altered in its material content and therefore in the theoretical test for its credibility: what is now "pointed to" is whether or not the test succeeds in reproducing the results predicted by the original hypothesis.

Fritjof Capra locates a similar moment when comparing the nominalist epistemological theorizing articulated by eastern mysticism, and the post-Einstein world of modern physics. Like Franciscan nominalism of the fourteenth century in the west, Buddhist logic pursuing a similar nominalism derived from Indian logic at roughly the same time argues an apophatic split between human concepts and language and the data of the material world given in the senses.[52] Capra, who seems unaware of similar intellectual forces operating to similar consequences in the history of western thinking about scientific procedure, finds the same rupture between human conceptual articulation (read: scientific hypothesizing) of sense data and the data itself. Citing Einstein's aphorism "As far as the laws of mathematics refer to reality, they

---

[51] Cf. Richard J. Bernstein's argument that the only way to escape from the binary opposition, ultimately derivable from Descartes' sundering of subject and object, between the dialectically related methodological alternatives of "objectivism" and "relativism," is through a Gadamerian hermeneutic. As in Pérez-Ramos' insightful interpretation of Bacon's passage on prediction from a constellation of discovered forms, the hermeneutic resolution Bernstein poses allows "mistaking" fictional mental concepts for realities in order to generate a discovery procedure that can test this mistake; Richard J. Bernstein, *Beyond Objectivism and Relativism: Science, Hermeneutics, and Praxis* (Philadelphia: Univ. of Pennsylvania Press, 1988), 131 ff., on the "hermeneutic circle."

[52] Buddhist meditational askesis parallels and reflects a corresponding nominalism in Buddhist epistemology, a correspondence that also invokes some of the same ethical exhortations, and encounters some of the same epistemological and procedural conundrums that Bacon explores at length in his two major works of the 1620s. Tom J. F. Tillemans in his study of the relation between Buddhist epistemology and Buddhist religious scriptures comments on "the well-known principle in Buddhist logic that perception … sees only particulars, undifferentiated into substances, qualities, actions, etc., and in reality it is only the undifferentiated particular … entity which exists. The other side of the coin for the Buddhist is that these fundamental distinctions between separate substances, universals, etc. are only invented by the conceptual mind and hence are not mirrored by the facts," Tom J. F. Tillemans, *Scripture, Logic, Language: Essays on Dharmakirti and his Tibetan Successors* (Boston: Wisdom Pub., 1999), 8. Grounding Buddhism's nominalist analysis of mental constructions is the central text of Buddhist psychology, the *Abhidamma,* a highly formalized phenomenology of human consciousness deriving from texts purportedly as old as Indian Buddhism itself, i.e., 2500 years BCE, including various commentaries later incorporated into the canon; Nyanaponika Thera, *Abhidamma Studies: Buddhist Explorations of Consciousness and Time,* 4th ed., ed. Bhikku Bodhi (Boston: Wisdom Pub., 1998). As in Bacon's similar arguments, these commentaries are intended to clear the mind of psychological obstacles to achieving "immediate experience" of the world around us. So far from being a "given" of human consciousness, such experience can only be achieved by a deliberate discipline, the purpose of which is "to open man's eyes to that which is not speculatively arrived at by the logical method of hypothesis and deductive verification, but which can be immediately apprehended … ." Herbert V. Guenther, *Philosophy and Psychology in the Abhidharma* (Berkeley: Shambhala Pub., 1976), 3. What Bacon calls "Idols of the Mind" the *Abhidharma* more systematically distinguishes as variously constellated moments of consciousness which, coming and going with daunting rapidity, collectively commit all human beings to enslavement to continual delusion.

are not certain; and as far as they are certain, they do not refer to reality"[53] as evidence of this rupture, he goes on to show how inadequate hypothesizing produces crises of well-formedness, i.e., paradoxes, in the "results":

> Every time the [early twentieth-century] physicists asked nature a question in an atomic experiment, nature answered with a paradox, and the more they tried to clarify the situation, the sharper the paradoxes became. It took them a long time to accept the fact that these paradoxes belong to the intrinsic structure of atomic physics, and to realize that they arise whenever one attempts to describe atomic event in the traditional terms of physics.[54]

Like the violations of noncontradictory well-formedness in Dionysian language anent the divine, a major discovery in post-Newtonian physics is what one might call the apophatic character of traditional hypotheses: what they predict about the behavior of sub-atomic particles assumes a model which is rendered self-contradictory by the results of testing.

> This state of affairs is bound to remain paradoxical as long as we adopt the static view of composite "objects" consisting of "basic building blocks". Only when the dynamic, relativistic view is adopted does the paradox disappear. The particles are then seen as dynamic patterns, or processes, which involve a certain amount of energy, appearing to us as their mass. In a collision process, the energy of the two colliding particles is redistributed to form a new pattern, and if it has been increased by a sufficient amount of kinetic energy, this new pattern may involve additional particles.[55]

The nominalist reduction did not resolve the dilemmas posed by a copy-theory epistemology by rendering human thought wholly independent of unmediated intuition, and vice versa. And Bacon's equivocations on the subject, not to mention the history of the "objectivism"/"relativism" debate that dogs post-Renaissance epistemology in general testify to the truth of that observation. But as Pérez-Ramos' reading suggests, the problem eventually becomes a clue to its own resolution. What Ockham did was for the first time set western thinking free of its role as dutiful daughter of sense experience, and thereby enabled it to develop her own potentialities as semiotic systems whose independent creativeness makes possible human understanding of the material universe in a detail and at a depth impossible as long as western thinking remained in tutelage to an Aristotelian copy-theory epistemology. If Bacon indeed unwittingly prophesied a hermeneutic theory of experimental procedure, one which has by now in molecular and astral physics become simply business as usual, it is only because he accepted the gambit thrown him by his own tutelage under a nominalist epistemological regime.

---

[53] Fritjof Capra, *The Tao of Physics: An Exploration of the Parallels Between Modern Physics and Eastern Mysticism*, 2nd ed. (Boulder, Co.: Shambhala Pub., 1983), 41.

[54] Capra, 66.

[55] Capra, 78.

Chapter 4

The Feminine Birth of the Mind: Regendering the Empirical Subject in Bacon and His Followers

Catherine Gimelli Martin

> The human mind in studying nature becomes big under the impact of things, and brings forth a teeming brood of errors.
>
> Francis Bacon, *The Masculine Birth of Time*

> Man, being the servant and interpreter of Nature, can do and understand so much and so much only as he has observed in fact or in thought of the course of nature: beyond this he neither knows anything nor can do anything.
>
> Francis Bacon, *Novum Organum*

To speak of the birth of a "feminine" mind in either Francis Bacon or his followers initially seems to be a contradiction in terms. If Baconians invoke the feminine mind at all, they seemingly do so only to banish her insidious "brood of errors," fallen methods that not only forestall "the golden future of the human race" but "sacrifice its promise of dominion by turning aside to attack transitory shadows."[1] This apparent antifeminism has caused the vast majority of recent critics to accuse Bacon of demeaning, despoiling and even raping female "Nature," a charge that very few have sought to challenge.[2] Bacon's *Masculine Birth of Time* particularly leaves itself liable to this accusation by insisting that the "brute" Amazonian forces of the natural world must be tamed through a "chaste, holy wedlock" with "*things themselves.*" At once imitating and overgoing Theseus's conquest of Hippolyta, the Amazonian queen in *A Midsummer Night's Dream*, his "son[s] of science" thereby secure "an increase

---

[1] See "The Masculine Birth of Time" in *The Philosophy of Francis Bacon,* Benjamin Farrington (Chicago: Univ. of Chicago Press, 1964), 70. Hereafter references to *The Masculine Birth of Time* will be cited parenthetically as *PFB.*

[2] Ever since Carolyn Merchant's study of *The Death of Nature: Women, Ecology, and the Scientific Revolution* (San Francisco: Harper and Row, 1980), a host of feminist scholars have "tried" Bacon's philosophy of nature and found it guilty. Merchant influentially identified the victim with the holistic *natura* of the pre-scientific hermeticists, chemists, and natural magicians, which Baconian empiricism destroys and/or "deadens." Only a few scholars have defended Bacon's gender mythology; Peter Pesec reminds us that his favorite personification of "interrogated" *natura* is actually a male sea god, Proteus, and that Bacon never actually admitted legal evidence based on torture. See Peter Pesec, "Wrestling With

beyond all hopes and prayers of ordinary marriages, to wit, a blessed race of Heroes and Supermen who will overcome the immeasurable helplessness and poverty of the human race."After exterminating fallen Mother Nature's race of "giants, monsters, or tyrants" (*PFB,* 72), they enter the new Promised Land that Thomas Sprat and Abraham Cowley later declare "opened" by Bacon.

The problem with this monological reading of Bacon's program is that like his "things themselves," his "Nature" is never securely masculine or feminine. Like his shape-shifting Proteus, natural realities are on the one hand passive matter awaiting conquest, and on the other supremely active gods or goddesses scheming to outwit the passive human mind. As a result, Bacon never securely represents the mind needed to "marry" things themselves as either passive or active, feminine or masculine. As in the passages cited above, a truly companionate marriage between man and fallen or "tyrant" Nature typically restores their original harmony, but given the failures of the past, the sons of science must first arm for battle with an Amazonian spouse who will never become a true "help-meet" without active intervention. This situation alternately reminds Bacon of the problems associated with the shape-shifting Proteus, the sea or nature god who delivers false prophecies unless he is physically forced to assume his true form. The solution is thus to become a similar type of shape-shifter. To gain *and* become a truly Christ-like "bride" perfectly receptive or passive before her "Lord" (in this case, protean Nature), the human mind must become perfectly active in cleansing its inner chambers or "lamps" of all traces of idolatry. These physical labors demand a type of zealous Christian warfare, but they also require the patiently submissive seeking of the wise virgins in the biblical parable. Unlike the foolish virgins, these "brides" imitate the vigilance and even the virility of the bridegrooms who will conduct them to their inner chambers. The result will be a true marriage free from the endless rivalry between masters and foolish or treacherous concubines: a union blessed both by marital reciprocity and peaceful reproduction of blessed "fruit" or works.

At this point in Bacon's complex dialectic, the masculine impulse of conquest gives way to the birth of a receptive feminine mind in both partners, as the "true" sons

Proteus: Francis Bacon and the 'Torture' of Nature," *Isis* 90.1 (1999): 81–94 and I.D.D.D. Landau, "Feminist criticisms of metaphors in Bacon's philosophy of science," *Philosophy* 73 (1998): 47–61; Kate Aughterson, "'Strange things so probably told': gender, sexual difference, and knowledge in Bacon's *New Atlantis*," in *Francis Bacon's New Atlantis: New Interdisciplinary Essays,* ed. Bronwen Price (Manchester: Manchester Univ. Press, 2002), 156–79; and Catherine Gimelli Martin, "The Ahistoricism of the New Historicism: Knowledge as Power versus Power as Knowledge in Bacon's *New Atlantis*," in *Faultlines in the Field,* eds. Claude Summers and Ted-Larry Pebworth (Columbia: Univ. of Missouri Press, 2002), 22–49. Besides Merchant, important Bacon critics include Brian Easlea, *Science and Sexual Oppression: Patriarchy's Confrontation with Women and Nature* (London: Widenfeld and Nicholson, 1981); Londa Schiebinger, *The Mind Has No Sex?: Women in the Origins of Modern Science* (Cambridge, Mass.: Harvard Univ. Press, 1989), 119–59; Maria Mies and Vandana Shiva, *Ecofeminism* (London: Zed Books, 1993); Valerie Plumwood, *Feminisms and the Mastery of Nature in Western Philosophy* (Oxford: Oxford Univ. Press, 1994); Evelyn Fox Keller, *Feminism and Science* (Oxford: Oxford Univ. Press, 1996); and Mark Breitenberg, *Anxious Masculinity in Early Modern England* (Cambridge: Cambridge Univ. Press, 1996). For additional bibliography, see Pesic.

of science guide an ambiguously gendered "nature" by assuming ambiguously gendered minds.[3] Of course, this "feminine mind" retains a strong residue of traditionally masculine qualities: it must continually search, test, and probe into unknown, alien, or Amazonian territory. These qualities explain why Bacon and his followers consistently disparage both traditional femininity and male effeminacy as passive, garrulous, vain, indolent, and ineffectual. Yet because Bacon typically associated these faults with the scholastics and their false "concubines" or methods rather than with real women, his followers were encouraged to contrast their meretricious arts with the more truly industrious, pragmatic, and charitable crafts of the good wife. Traditionally identified as domestic and/or feminine, her "fruits" rank only a little lower than the masculine "light" that unveils them–and some critics would argue, even higher. Yet this utilitarian bias is offset by Bacon's firm belief that Greek science was derailed by focusing on the easy "fruit" promised by purely "verbal" methods. Since the same methods led them to "boyishly" premature conclusions rather than real "male" maturity, Bacon and his successors in fact employ a rhetoric consistently haunted by a fear of effeminacy. Nevertheless, this rhetoric should not obscure the very real birth of his feminine mind, an event that has never gone completely unnoticed in the history of ideas. As Lord Macaulay famously observed, "To make men perfect was no part of Bacon's plan; his humble aim was to make imperfect men [and women] comfortable." Thus while he pitched his "foundations" low, he framed them to support the solid and, to use his own metaphor, marital and domestic pursuits that will restore the harmonious balance between man and nature, male and female.[4]

However, there is little doubt that many of Bacon's immediate followers understood his masculinist rhetoric as monologically as his many modern critics. Sprat and Cowley's Royal Society rhetoric is steeped in an ideology that endlessly reiterates the claim that only a clear, strong grasp of words and things will foster "firme … Masculine virtues." Although these virtues may be "defective in … gentleness" and "pliableness," Sprat's 1667 *History of the Royal Society* likens them to the toughness of "the strongest, and the Noblest" metals, materials best fitted for rapid conquest and exploration. As shown by "all the late discoveries of this Age in the reall knowledge of mankind, and nature," these "manly" tools allow the fellows of the Royal Society to abandon the effeminate "meadows of security" for the wild and uncertain plain of doubt. Because these charming meadows only fill "the head with the matter and artifice of discours" which breaks "the force of things by the subtilty of words," their flowery "embroidery" is worse than useless. Like scholastic method, subtle arguing, questioning, and debating "weakens mens arms, and slackens all the sinews

[3] Even strong feminist critics like Evelyn Fox Keller concede that Bacon's science of nature is described as "an hermaphroditical birth"; see her "Baconian science: an hermaphroditical birth," *Philosophical Forum* 11 (1980): 299–308.

[4] From Thomas Babington Macaulay's "Lord Bacon," cited in Robert K. Faulkner, *Francis Bacon and the Project of Progress* (London: Rowman & Littlefield, 1993), 13. Faulkner is typical of critics who see Bacon as a Machiavellian materialist without fully crediting his appeals to Christian charity and companionate marriage. But his emphasis on Bacon's pragmatism is accurate.

of action." At once over-refining and corrupting "mens wits," "feminine" rhetoric spins gossamer threads "found to be more useless, than those which are homespun, and gross."[5] Sprat typically reinforces his masculinist program with the Protestant rhetoric of Christian liberty: here "real men" prefer the Abrahamic freedom of the wilderness to the soft, enslaving luxuries of Egyptian domesticity, and the freeborn Sarah (whose name means "princess") to concubines like her "slavish" handmaid, Hagar. Yet as in its source, this female/female comparison illustrates just how and why the strong male/female oppositions of Baconian rhetoric typically break down. Unlike most Calvinists who used the Hagar/Sarah comparison to symbolize the binary opposition between "free" biblical interpretation and the "Babylonian captivity" of the Roman Church, Bacon saw startling continuities between man's unfallen or free and fallen or "slavish" condition. These continuities explain why he not only portrays nature as both masculine and feminine but both Eve and Edenic nature as alternately innocent and experienced: the human lapse was neither predestined, irreversible, nor the sole result of "female" fault.

His novel interpretation of Genesis allowed Bacon's 1605 *Advancement of Learning* to imagine both the original and future Eden as an ideal synthesis of wilderness and domesticity, nature and culture reflected in Adam and Eve's originally dialogical marriage. Under this ideal dispensation, even the oriental untamed other can fill a positive feminine role. In fact, he believed that Abraham's promised land of milk and honey will only be fully realized once a new Eden overspreads the earth and a redeemed Mother Nature or new Sarah takes on his concubines fertility. Sarah's freedom is thus further associated with boundless "oriental" fertility via its typological counterpart in the ideal marriage between the Queen of Sheba and Abraham's most advanced heir, King Solomon, Bacon's archetypal "son of science." Their masculine/feminine partnership is fulfilled in many different forms, the best known of which occurs in Bacon's utopian romance, the *New Atlantis.* There the seers of Salomon's House perform the same benign paternal function for nature and society that its Tirsans perform for their human families. Although the maternal role in this fiction is much more ambiguous, with many critics claiming that the mother is completely displaced or "exiled" in the ritual Feast of the Family, John Archer convincingly shows that her hidden position–seeing but not seen–marks the very essence of both Bensalemite and feminine power.[6] Here as elsewhere, Bacon's unstable binary oppositions never simply align "oriental," civilized, and soft on one side of the spectrum, and "western," wild, male, and strong on the other. Not only are his archetypal mothers, spouses, and handmaidens only adequately understood in relation to an ambivalent

---

[5] Thomas Sprat, *Observations on Monsier de Sorbier's Voyage into England* (London: John Martyn and James Allestry, 1665), 290–1 and the *History of the Royal Society*, eds. Jackson I. Cope and Harold W. Jackson (London: Routledge & Kegan Paul, 1958), 32–33, 332, 326. Cf. 129, and see John Evelyn, *Public Employment and an Active Life Prefer'd to Solitude, and All Its Appanages* (London: H. Herringman, 1667).

[6] See John Michael Archer, "Surveillance and Enlightenment in Bacon's *New Atlantis*," *Assays* 6 (1991): 112–13, 116 and my analysis of the romance's New Historicist-Foucauldian misinterpretation in "The Ahistoricism of the New Historicism."

range of fathers, husbands, and suitors who alternately encourage the benign creativity of the mythic female, deliberately thwart her most positive achievements, or hypocritically "father" her brood of errors; when these Baconian males fail to forge successful "weddings" of art and nature, the fault is usually no one's but their own.

The *locus classicus* of this unstable gender dialectic occurs in Bacon's seminal retelling of the myth of Atalanta and her suitor, Hippomenes. By identifying Atalanta as "art" and Hippomenes as "nature," Bacon shows that even his "sons of science" are at times identified as female and their opponents as masculinized forces of nature. As a swifter and more civilized "art," Bacon's Atalanta seems to be pitted against slow, sluggish "nature" in order to demonstrate not just two very different Natures but two very different masculine minds. One mind adopts a positive androgyny while the other lapses into the "effeminate" indolence condemned by Sprat. Full only of itself and its own infertile ideas, the effeminate male mind spawns the sterile "brood of errors" that Bacon accuses of making no real contact with natural "things" or with the seminal ideas (*rationes seminales*) residually informing nature. Puffed up in a kind of false pregnancy with its shallow preconceptions, its mental "womb" is as grotesque as the nether parts of Bacon's Scylla, the deformed water-deity perverted by false method, pride, and presumption. For Bacon, this hapless creature has fallen so far from the fair virgin symbolically wedded to God and first given to Adam that she has instead become like Spenser's Errour, the chthonic, parthenogenic "mother" whose divorce from divine truth makes all of her offspring abortive. Both Scylla's perverters and Atalanta's crude conquerors are thus identified with the fallen masculine mind first born in Adam's false dreams of godlike power, later inherited by the Ancients, and perpetrated by the scholastics. By turning the true "womb" of fruitful ideas into a nest of "barking monsters, from which she [Scylla] could not be delivered" (*SEH*, IV: 14), this mind creates both the first and "second disease of learning." The first "distemper" (the word slyly implying a canine affliction) is spread by men who "study words and not matter" (*SEH*, III: 287), and the second when they fall in love with the wordy "women" they have divorced from both Nature and natural fruition.

The cure for both diseases is not only to avoid the Scyllas lurking in the ungrounded male mind–the brood of errors "conceived" in excessively mental or narcissistic attempts at "insemination"–but also the fate of Hippomenes. Although nominally a winner, Hippomenes has been emasculated by the opposite but equally fatal error of relying too little on the mind to improve his methods. Inheriting all the defects of the original race of Titans spawned by *physis* or lawless nature, he is bulky and strong, not swift nor imaginative like his conquered bride. Hippomenes thus represents the fundamental error of Greek science, which by idealizing Jove and his *nomos,* foolishly silenced his spouse rather than harnessing, husbanding, or nurturing its feminine "other," Art. This unequal marriage enslaves both husband and wife to arbitrary, poorly understood, yet ironclad conventions and stereotypes. Like Jove's cruel conquest of his first wife, Metis, the cunning female who would have borne the son that deposed him had she not been swallowed by her spouse, Hippomenes's unjust victory over Atalanta produces a male-dominated marriage that commands nature without first obeying her, as Bacon's *Novum Organum* recommends (*SEH*, IV:

47). His *Wisdom of the Ancients* goes even further in teaching Atalanta to regain her natural advantage by learning to despise the false or trivial artifacts that her slow-moving suitor, Hippomenes, throws in her path–the golden balls or premature "fruits" that cause her to lose the race. Symbolically freed from the distractions that waylaid both the Greek philosophers and their heirs, the alchemists and natural magicians, Art thus acquires a disciplined form of feminine intuition divorced from idle curiosity long associated with Eve or Pandora. In regaining her natural reciprocity with her spouse by ignoring the petty baubles that soon turn into golden chains, Atalanta will also bear him vastly superior children. Together, Art and Nature thus join hands and "win the race" against tradition and superstition.

Once again, this myth envisions a nature ruled neither by a Titanic Gaia nor an Olympian Jove but by an essentially Edenic marriage that will end mankind's long exile from its garden. It will also overcome man's "second fall" as the human race rethinks its idolization of the lost paradise. The pre-Christian Ancients invented this idol when they made nature "an image of God and … man an extract or compendious image of the world" (*SEH*, III: 350). This teaching led to a misplaced reverence for nature's "magical" divinity (*SEH*, III: 349–59) even though fallen nature had long since become more like Hippomenes than Eden and humans had never been made in the image of the created world but rather in the image of God. Returning to Eden thus depends upon recognizing that *both* Adam and Eve fell not by desiring God's knowledge of nature but by transgressing his true command to "dress and keep" the garden. By failing to understand it as a spouse or fellow creature, not as an untouchable mystery, they failed to "know" nature in both the abstract or contemplative and in the active or sexual sense in which Adam was to "know" Eve: not by brutally subjugating her, but by fulfilling her needs and desires. Although not entirely divorced from Atalanta's error, this "fall" into hegemonic Nature is chiefly caused by the male fault of listening only to themselves. By indulging this tendency, the early natural philosophers either produced a science "characteristic … of boys; it can talk, but it cannot generate; for it is fruitful of controversies but barren of works" (*SEH*, IV: 14); or fell into the Egyptian error embodied in Hippomenes's victory over Atalanta. Forsaking method, they were slavishly "beholden to a wild goat for surgery, to a nightingale for music, to the ibis for clysters, to the pot lid that flew open for artillery, and in a word to chance, or anything else, rather than to Logic" (*SEH*, IV: 409). This "lame" or sluggish empirical method errs in relying only on "simple experience; which, if taken as it comes, is called accident," and which must be replaced by real "experiment" understood as knowledge actively and deliberately "sought for" (*SEH*, IV: 81). Once found, the new experimental science will thus fill its temples of science with human statues, not the images of animals.

Bacon's *New Atlantis* anticipates this change by featuring statues of human discoverers not only "as monuments to the success of the high art of the place," but also as reminders "that method, not the chance or accident characteristic of man's sluggish history is here the driving force of all inventiveness." Thus unlike the "timid ancients," Bacon's Bensalemites can freely invent "voluptuary" feminine arts because they possess the truly masculine fortitude, discipline, and courage to reproduce earthquakes, lightening, or other "marvels of nature" once considered impossible or impious to imitate (*SEH*, IV: 99). They are blessed rather than cursed for this artful imitation of God's six-day creation because unlike the secrets of his holy rest, these works belong to God's natural dispensation.[7] Yet as we have seen, their achievements are conditional on following an inductive method conceived as a Atalanta-like craft. Without the feminine mind to assist it, the masculine mind is much too inclined to indulge his own "puffed up" narcissism, an analysis ultimately borrowed from the Greek association between "effeminate" but effective cunning and officially despised but successful prudence. As Marcel Etienne and Jean-Pierre Vernant have shown, this prudential alliance is chiefly associated with Odysseus, the client of Athena, as opposed to the male chauvinism of Achilles, the self-assertive hero favored by the Greek male gods.[8] Julie Solomon further shows that Bacon needed the archetypal skills of Odysseus to achieve the kind of self-distancing required by his new science. Although Odysseus could literally become "Nobody" in the cave of the Cyclops, he could also become a kind of "everybody." Mobile, shifting, and polymorphic in relation to emergent circumstances, his strategy of being everywhere and nowhere explains his capacity for "dialogical and … empathetic 'introjection,'" which is not (*pace* Plato) the strategy of the weak, but the hidden "mother" of wise Athena, Metis herself. Like her daughter, Metis thus stands for the dialectical wisdom "born" in an amorphous space "shared by at least two minds engaged in self-estrangement." In Roman times, Publilius Syrus significantly heralded the rehabilitation of this master/slave reversal by teaching that the wife who submits to her husband will command him, much as Bacon's new philosopher commands nature.[9]

According to Solomon, Bacon heralded the modern rehabilitation of the Metic or "triangulated" model for "conceiving political and intellectual authority" as an "interactive dialogue." By combining ideas which no longer discretely belong to either party, Metis's daughter Minerva again "emerges armed to the teeth from Jupiter's pate." His "wound" also creates a "mental break" with the past whereby both monarchy and, to a more limited extent, patriarchy are at once legitimated and made obsolete as hierarchical and/or natural superiority gives way to the disciplined observation of the empirically real.[10] Bacon's fundamental interest in ending, not mending the mutual subjugation of master and slave, king and counselor, husband and wife also explains the paradoxical "masculinism" of the *New Atlantis*. Through a series of inversions that critique the "Western imperial and gendered hierarchies" of previous travel and utopian literature, the salvific "wilderness" of the New Atlantis is

---

[7] J. Peter Zetterburg, "Echoes of Nature in Salomon's House," *Journal of the History of Ideas* 43 (1982): 179–93, cited 187, 184. Zetterburg usefully corrects Paolo Rossi's overly schematic assertion that Bacon completely rejected rather than strenuously reformulates Aristotle's conception of mimesis (179).

[8] Marcel Detienne and Jean-Pierre Vernant, *Cunning Intelligence in Greek Culture and Society,* trans. Janet Lloyd (Chicago: Univ. of Chicago Press, 1974); see especially 108.

[9] Julie Robin Solomon, *Objectivity in the Making: Francis Bacon and the Politics of Inquiry* (Baltimore: Johns Hopkins Univ. Press, 1998), 109, 117–19, 137. Bacon could of course also have found similar teaching in Paul, whom he often cites on the virtues of humility and charity.

[10] Solomon, *Objectivity in the Making,* 154–7, 160.

not described as a virgin land passively awaiting a conquering male culture but, as Kate Aughterson shows, as an "other" that is neither foreign, dangerous, nor feminized. Moreover, by banishing Europe's sexual double standard and rearticulating "masculinity as chaste husbandry in an equitable marriage," its relatively equal marriages closely parallel the new reciprocity between king and counselor. Agreeing with Solomon that Bacon at once "advocates a Jacobean patriarchal model and supersedes it," Aughterson finds its land and people, nature and culture, men and women all ideally becoming "nursing" fathers or fruitful mothers who epistemologically realign "nature *with* technology."[11]

Mythical, real, and rhetorical handmaidens

Ironically, although many of Bacon's followers failed to grasp the radical implications of these ideas, many of his enemies did. Often anticipating his modern critics, not just the Spanish Inquisition (which banned his books) but English Aristotelians like Henry Stubbe considered his interpretation of both the natural and marital contract sacrilegious. As a Galenic physician, Stubbe was especially outraged that Bacon dared to consider "queenly" mother nature a bride *or* a handmaiden, and the true "sons of science" as ultra-flexible negotiators like Odysseus rather than inflexible but dutiful servants like Hippocrates. Stubbe instead heroizes Hippocrates as one who made "himself absolutely the *slave* of *Nature,* attended on *her motions,* sometimes gently *leading* her on, as it were a *Gentleman-Usher;* sometimes *following* her, as a *Page,* never pretending to *command her* by his *Medicines*: So cautious he was, that he would not adventure to *do good unseasonably,* and at other houres then his *teacher* [nature] directed him, lest he should *do harm.*" Yet he places feminine nature on this exalted pedestal for less than flattering reasons: her queenly arbitrariness and deeply veiled mysteries are so impenetrable that it is impious either to approach or reproduce them. In fact, her ways are so entirely "*discrepant*" from any arts of men "that there is no arguing from the *one* to the *other.*" This simultaneously condescending and hierarchical respect for the Virgin Queen allows Stubbe to confine her care to high priests like himself, who will carefully record her symptoms without presuming either to explain or to predict their courses. Since Nature is a haughty, wayward, mystifying woman, she must be pacified to the best of our ability, not understood as an equal or "entertained" as a spouse.[12]

Bacon clearly considers this condescending "obedience" not only obfuscatory and presumptuous but ultimately as tyrannical as its "queen," and followers like Sprat fully agree. Refusing to regard his "spouse" as untouchable, Sprat carefully observes all "the footsteps and progress of *Nature*" in order to comprehend the "natural ways by which things are produced." This careful scrutiny avoids mystifications of the Stubbesian type because the new philosopher "indevors rather to know, than to

[11] Kate Aughterson, "'Strange things so probably told,'"160–4, 168–9, 173. Cf. Catherine Gimelli Martin's "The Ahistoricism of the New Historicism."

[12] Henry Stubbe, *The Plus Ultra Reduced to a Non Plus* (London, 1670), 74–5.

admire" nature's paths, fully recognizing that "There is scarce any thing that renders a man so useless, as a perverse sticking to the same things in *all times,* because he has sometimes found them to be in season." Instead, he corrects his preconceptions by reminding himself that nothing is either irrevocably certain *or* uncertain, and knowledge is not a unilateral "given" but a true meeting of minds.[13] Since mankind can only overcome its "inconveniences" by keeping both modern assumptions and ancient authority at a distance, Bacon's *Masculine Birth of Time* also insists that the neglect of this Metis-like duty disastrously re-enthrones the idols that "beset and block" mens' minds with habits so "deeply implanted and, as it were, burned in" to the human imagination that no "clean and polished surface remains" in its mental mirrors. Scouring the mirror therefore requires both secret and drastic measures: "A new method must be found for quiet entry into minds so choked and overgrown," whereby "frenzied men … exacerbated by violent opposition … may be beguiled by art." Once again, this method is *both* femininely "mild" and masculinely strong. In order to afford "no occasion of error," it "must have the inherent power of winning support and a vital principle which will stand up to the ravages of time, so that the tradition of science may mature and spread like some lively vigorous vine." Once the vine bears fruit, an implicitly androgynous science will at last be able "to select … followers … worthy to be adopted into her family" *(PFB*, 62).

Hence as suggested above, Bacon's unstable binary oppositions in the last instance require the "secret" introduction of a third party or intermediary in the Art/Nature dialectic, a third party predictably representing perfected scientific method as a "handmaiden"–a Metis, Atalanta, or other "artful" bride–who prepares the mentally confused bridegroom to see nature in a new light. Directing him to the fertile soil of "things themselves," this "lively and vigorous" feminine "vine" is at once "mild" and "beguiling" enough to tame the "madness" of the masculine mind and to win it away from the siren-song of false method. Bacon's fruitful splitting of the feminine mind into true and false versions at this point in *The Masculine Birth of Time* clearly sets a vital female vine capable of straightening her male "elm" against both the "concubines" of the scholastics and Stubbe's wayward and unpredictable Nature. In the process, it makes the masculine mind by far the most problematic party to the marriage. Grown crooked and deformed by subjecting itself to Stubbe's other idol, Aristotle, the potential bridegroom has been perverted by the sterile and "crafty trifling" taught in his "art or manual of madness … [which has] made us slaves of words" (*PFB*, 63). Aristotle's innumerable descendants have been so eager to pay due homage to his narcissistic male Logos that they have further multiplied his false "precepts and propositions" by "relying on the restless agitation of their own wit." Finally, their petty quibbles like "darksome idols from some subterranean cave … spun as it were spiders' webs which he [and they] would have us accept as causal bonds, though they have no strength nor worth" (*PFB*, 63). These are chains from which only a very different mind can free them.

Yet instead of freeing his pupils from these chains, Plato simply abetted his master-student by exalting "vague inductions" that "took men off their guard and

[13] Sprat, *History of the Royal Society,* 334–5.

weakened their native sinews" by giving "out the falsehood that truth is, as it were, the native inhabitant of the human mind and need not come in from the outside to take up its abode there." As we have seen, this inward-turning and implicitly misogynistic logocentrism also betrayed Nature's "holy name" by neglecting the "careful sifting" that her true handmaiden, science, or art, could have placed at her service. By refusing the "artificial instruments and aids" that arduous "feminine" labors could have supplied, Plato and Aristotle then begat pseudo-scientists like Paracelsus who forsook Nature's "wild" but real labyrinth for the comfortable limits of their preordained systems. Thus like Atalanta, the Paracelsians or false artists not only robbed themselves of the prize properly belonging to real art, but like Hippomenes, they sacrificed a true bride or "help-meet" to gain an unwilling concubine. This aspect of Bacon's critique again centers on his critique of the schoolmen, who by abandoning a fruitful spouse for a sterile courtesan combined the errors of the mythical Narcissus and Pygmalion with those of the myth-making Plato. All foolishly believed that they could "invent" a spouse from the substance of their own imaginations but wound up embracing a mere shadow or statue instead (*SEH*, III: 284; cf. *PFB*, 66). In either case, passive adulation or worship from afar is no real tribute to femininity but merely a male fantasy. Realizing this, Bacon believes he can at once meet man's needs and magnify nature's fertility by demystifying the traditional "Aristotelian conception" wherein man "can help Nature only in fulfilling her own designs," not in improving them like a good husband, counselor, or help-mate.[14]

Early in the Baconian era, Bacon's most astute followers began to capitalize on his startling reformulation of traditional gender and nature mythology. Dialectically opposing a redeemed to a fallen counterpart of the feminine mind, they "engendered" new Atalantas for every new Hippomenes and new Ariadnes for every new Theseus. Their self-consciously reformist ambitions also led the truly great Baconians to follow their master in favoring craft or skill-based myths. In the process, even Medea is transformed by finding a new Jason who needs her to expel the cobwebs of the past, so that like the fully liberated Proserpina in Bacon's *The Wisdom of the Ancients*, the female principle is no longer damned for her Eve- or Pandora-like curiosity of "taste." Thus where Theseus and his male allies failed, Bacon's "Proserpina or Spirit" successfully escapes from the underworld by marrying her own mater, "matter." This feminine power is utterly distanced from the Ancient male hero and instead associated with the mythical golden bough traditionally "grafted" by female sibyls or prophetesses, who practice a true "art, not any medicine or method which is simple or natural," that is, not a chance discovery of goats or ibises (*SEH*, VI: 761). Only these hybrid yet also traditionally feminine arts and crafts can wind us out of the labyrinth by teaching the entire human race how to become the "more severe and laborious sort of inquirers into truth" that Erasmus had identified with Hercules (*SEH*, III: 284).[15]

---

[14] Riejer Hooykaas, *Religion and the Rise of Modern Science* (Grand Rapids, MI: William B. Eermans, 1972), 64.

[15] Sarah Hutton, "Persuasions to Science: Baconian Rhetoric and the *New Atlantis*," in *Francis Bacon's "New Atlantis,"* 55. Hutton cites Brian Vickers's translation of *The Advancement of Learning* but the text is precisely the same.

Much the same analysis appears in Bacon's rhetoric, which accuses the Ancients of deliberately establishing a " kind of contract of error between the deliverer and the receiver" of knowledge. Like marital failure, this contractual error is the reciprocal result of the credulity of servants and the vainglory of masters, but the male desire of "glory" is particularly responsible for making an author reluctant "to lay open his weakness" and his "disciple not to know his strength." Bacon's solution to the problem of making learning more reciprocal again makes it more feminine or hand-maidenly: like "a thread to be spun on," knowledge should be both "delivered and intimated … *in the same method wherein it was invented*," for only then will it depart from "anticipated and prevented knowledge, [where] no man knoweth how he came to the knowledge which he hath obtained." This ignorance will naturally vanish once both master and pupil can "revisit and descend unto the foundations of his knowledge and consent; and so transplant it into another as it grew in his own mind. For it is in knowledges as it is in plants: if you mean to use the plant, it is no matter for the roots; but if you mean to remove it to grow, then it is more assured to rest upon roots than slips. So the delivery of knowledges (as it is now used) is as of fair bodies of trees without the roots; good for the carpenter, but not for the planter; but if you will have sciences grow … look well to the taking up of the roots" (*SEH*, III: 404).

This companionate approach to empirical knowledge is also identified with Bacon's aphoristic method, which by cutting off "the discourse of illustration" and removing all authoritative connections, comments, and examples, "trieth [both] the writer, whether he be superficial or solid," and the reader, who must dig deeply into the roots of the matter to "regrow" the plant-like thread of "knowledge broken." For like a broken bough, slip, or root, this knowledge "invite[s] men to enquire farther; whereas Methods, carrying the shew of a total," convey only an illusory sense of security and mastery (*SEH*, III: 405).

This famous passage at once summarizes the essential rationale of Bacon's program and connects it to all of the metaphors and myths so far surveyed. Like Prosperina and her mother Ceres, the program requires him to be a prophetic sower of vital seeds: "I bear myself soberly and profitably, sowing in the meantime for future ages the seeds of a purer truth" (*SEH*, IV: 104). Also like the mature mustard seed in the biblical parable, when properly cultivated his "seminal ideas" will branch out into a vast tree of learning whose top reaches the highest heavens. Afterward, there will be a long and arduous harvest whose fruits, like the seeds of the mythical Psyche, must be carefully winnowed and sifted to remove the wheat from the chaff. Then at last the fibers of truth can be carefully spun into their final product, the Ariadnian threads winding us out of the labyrinth. Woven into "seminal" books acting "as skeins or bottoms of thread which may be unwinded at large when they are wanted" (*SEH*, IV: 472), knowledge will be both broken and whole. However, at this point both Bacon and his followers tend to stabilize their gender mythology, so that the "seminal ideas" wound into his "books" of nature and scripture remain as masculine as the potent dragon seeds sown by the founder of Thebes:

> the images of men's wits and knowledges remain in books, exempted from the wrong of time and capable of perpetual renovation. Neither are they fitly to be called images, because they generate still, can cast their seeds in the minds of others, provoking and causing infinite actions and opinions in succeeding ages … that whatsoever motions the spirit of man could act and perform without the organs of the body … might remain after death. (*SEH*, III: 318)

In arguing for a similarly dialogical and ultimately androgynous model of truth, one of Bacon's most faithful and astute followers repeats these sentiments almost verbatim. Taking both Bacon and his early champion, Lord Brooke, as witnesses to the fact that "Truth may have more shapes than one" (*CPW*, 2: 563), Milton's famous defense of the freedom of the press and opinion declares:

> Books are not absolutely dead things, but doe contain a potencies of life in them to be as active as that soule was whose progeny they are; nay they do preserve as in a viol the purest efficacies and extraction of that living intellect that bred them. I know they are as lively, and as vigorously productive, as those fabulous Dragons teeth; and being sown up and down, may chance to spring up armed men … [A]s good almost kill a Man as kill a good Book … the pretious life-blood of a master spirit, imbalm'd and treasur'd up on purpose to a life beyond life. (*CPW*, 2: 492–3)

Yet although in either case this masculine "seed" remains magisterially potent, in neither case does it simply sow a feminine "field" of passive matter. Instead, like the matter vitalistically "married" by Prosperina or spirit, it cherishes and preserves "the pretious life-blood of the master spirit" until, Metis-like, it produces a fully armed and androgynous offspring. A closely related passage in *Areopagitica* identifies its cherishing handmaiden/mother as Isis, the faithful sister and wife of Osiris, who by regathering his shattered body gradually restores the perfect image or "mirror" of divine Truth. Although she cannot restore his male member, she conserves the rest until his fullness of being or "presence" returns with the true master Spirit, Christ.[16]

Bacon specifically associates the empirical aspect of this recovery with the work of feminine words, which in the form of true eloquence frees masculine reason by winning "the Imagination from the Affection's part." The false, Hagar-like concubine "Affection" will thus be subdued as words rejoin things and Eloquence creates a new "confederacy between the Reason and the Imagination against the Affections," the most fallible member of the triad (*SEH*, III: 410). Yet with the affections serving as a kind of Hagar to the Sarah of the redeemed Imagination, even the fallen "female" passions can now be enlisted in support of this new contract or "confederacy," as Bacon calls it. Thus once again, the idolatrous queens and concubines of the past succumb to a wife who is both an industrious handmaiden to her spouse and a strong mistress to the lesser members of her household. Bacon needs this helper because he still traditionally conceives of male virtue and wisdom as abstractly contemplative "light," which requires a feminine support system for its active work or "fruit." For as

[16] Milton's prose works (abbreviated *CPW)* are cited from *The Complete Prose Works of John Milton*, eds. Don M. Wolfe et al., 8 vols. (New Haven: Yale Univ. Press, 1953–82).

"seminally" vital as his ideas are, the contemplative mind still retains the quintessential male task of statically mirroring the mind of God. This means that while Paolo Rossi correctly observes that Bacon typically thought of action as masculine and words as feminine, his feminine words are not therefore passive. They must actively nurture the imagination and its rhetorical tools while also preparing to conduct the investigative knowledge-gathering work of a Psyche or an Isis.[17] This "relay-system" fully explains why Bacon prefers to describe empirical work in terms of feminine arts and crafts which require a subtle, receptive, yet also humble mind with enough of the "masculine" discipline to direct the household responsibilities assigned by her head.

In part, this system also reflects Bacon's unusually synthetic solution to the traditional Elizabethan ambivalence about rhetoric. Although fully aware of its dangers, his new scientific program is far too dependent on free-ranging communication to ignore the benefits of the verbal arts. His solution to their potential misuse is thus once again to isolate two very different kinds of wordy women who symbolize the ambivalent capacity of words either to reinforce or to overrule the rational judgment. Of course, detaching the positive from the negative uses of well-honed words is a commonplace from at least Plato onward, but Bacon uniquely enlists words on the side of the passions rather than against them. As a result, he insists that just as "the end of Logic is … to secure reason, and not to entrap it; [and] the end of Morality is to procure the affections to obey reason, and not to invade it; the end of Rhetoric is to fill the imagination to second reason, and not to oppress it." By further allying the passions or affections with the mediating or triangulating imagination which allows them to be ruled by reason, Baconian rhetoric ceases to be Plato's voluptuary art and becomes an ally of "real" knowledge in the wars of truth. Aristotle's *Rhetoric* had of course sanctioned using the same ally for deliberative, forensic, and epideictic purposes, but never for the discovery of scientific truth, which remained in the province of pure logic. Not content simply to follow Ramus's reintegration of rhetoric and logic, Bacon thus "re-marries" the traditional trilogy of passion, imagination, and reason in new ways. As "an agent or nuncius in both provinces, both judicial [or rational] and ministerial [or affective]," the Imagination and the will informing it (for "Imagination ever precedeth Voluntary Motion") become flexible and serviceable, not simply fallen. While recognizing that "this Janus of Imagination hath different faces," and "usurpeth no small authority in itself, besides the duty of the message," Bacon makes it an ambivalent handmaiden pointing either toward Truth and Good or toward falsehood and error, the "province" of the unruly affections. Yet his optimistic faith that faithful "Sarah" will usually prevail over unruly "Hagar" is not simply naïve; as in marriage, so in other negotiations no malign influence can lead us astray unless we first conduct "a failed negotiation within ourselves" (*SEH*, III: 409).

The same optimism dominates not just Milton's *Areopagitica* but his closely contemporary anti-prelatical tracts, which agree with Bacon that because the redeemed Imagination and her winning words can "*move great love and affection*" toward virtue, they can trump the inherent weaknesses of false logic and sophistic

[17] Paolo Rossi, *Francis Bacon: From Magic to Science*, trans. Sacha Rabinovitch (London: Routledge & Kegan Paul, 1968), 82.

rhetoric. Pure logic alone could never do so, since as Bacon insists, its fondness for "sharp disputations and conclusions" cannot sympathetically sway "the will of man" (*SEH*, III: 409–10). Thus the poet willingly follows him in featuring a more practical handmaiden, a reformed soul or true Psyche who helps reason to "cull," winnow and sift the "confused seeds" of truth. With this bride at reason's side, "though all the windes of doctrin were let loose to play upon the earth, so Truth be in the field … Let her and Falshood grapple; whoever knew truth put to the wors, in a free and open encounter" (*CPW*, 2: 514, 561)?[18] This optimism would be inexplicable without the Baconian belief in the human mind's capacity to homeopathically cure itself by the same "infection" that diseased it: passionate, eloquent, but also "sensible words" will enlist her erstwhile enemies on the side of virtue. By tracing out man's "diseases and infirmities … which are no other than the perturbations and distempers of the affections," good poets and rhetoricians not only teach "how affections are kindled and incited" but also how they are "pacified and refrained" (*SEH*, III: 437–8; cf. *CPW*, 1: 817–18). Of course, rhetoric's best "doctors" never entice their patients into falling in love with their nurse, which would mean mistaking the antidote for the cure and the inferior for the true bride; or taking "the current tokens or marks of Popular notions of things" for the thing itself (*SEH*, III: 388).

Ambivalent appropriations of the feminine mind

As a powerful rhetorician, Bacon forcefully communicated both this analysis and its characteristic tropes to a wide variety of followers. John Amos Comenius was one of the first to develop and popularize his linguistic assumptions in his popular *Reformation of Schools,* which cautions against the vanity of "sterile," self-indulgent "circumlocutions, and windings, and turnings of expression, which fetch not out the kernell, but only make a few assayes upon the shell." He thus recommends a typical Baconian "cure of this last and worst disease" of learning: by "sowing and sprinkling abroad the seeds of the true knowledge, and feare of God through all this field," his harvest workers will gather in "various heapes" or fragments of truth awaiting their "due forme and order" as the Solomonic "Universall Temple of Wisdome" is completed. As in *Areopagitica,* the temple remains under construction (*CPW*, 2: 555–7), but Comenius anticipates a much more mystical, apocalyptic culmination: rather than entering into a working marriage, his universal Truth simply plucks off all her veils after being "courted" through prayer and pansophy.[19] As Carolyn Merchant and Brian Easlea have variously shown, eroticized images like these tend to restore the old conquest or "mistress" model of gender relations. Abandoning Bacon's handmaidenly model of the feminine mind for the traditional

[18] Blair Hoxby argues in *Mammon's Music: Literature and Economics in the Age of Milton* (New Haven: Yale Univ. Press, 2002) that like Milton's other personifications, these are based on the analogy of free exchange and gradual equilibrium over time (26–7, 34–5), a Baconian ideal promoted by Milton's Cambridge tutor, Joseph Mede (31–56).

[19] John Amos Comenius, *A Reformation of Schooles* (London: Michael Sparks, 1642), 12, 21–2.

masculine glory of penetrating "naked truth," "vulgar Baconians" seek a courtesan who will freely "open" herself to the sons of science without any of the obligations of holy wedlock.[20]

In his first tribute to Bacon's *Advancement of Learning*, the young Milton wrote a Cambridge oration that also fails to grasp the importance of the marital metaphor and simply anticipates the masculine birth of time and truth:

> when universal learning has once completed its cycle, the spirit of man, no longer confined within this dark prison-house, will reach out far and wide, till it fills the whole world and the space far beyond with the expansion of its divine greatness. Then at last most of the chances and changes of the world will be so quickly perceived that to him who holds this stronghold of wisdom hardly anything can happen in his life which is unforeseen or fortuitous. He will indeed seem to be one … to whom, lastly, Mother Nature herself has surrendered, as if indeed some god had abdicated the throne of the world and entrusted its rights, laws, and administration to him as governor. (*CPW*, 1: 296)[21]

Yet in a little more than a decade, the poet's keen interest in exploring the Judeo-Christian origins of marriage and divorce led to a striking revision of this paradigm. Since he agrees with Bacon that the essence of marriage is equal and delightful "conversation," his divorce tracts declare that any marital contract is invalid if the feminine partner passively surrenders to an unwanted or unequal relationship. In other words, whenever reciprocal conversation proves impossible, the marriage contract can be consensually canceled on the proto-modern grounds of mutual incompatibility.[22] Significantly, Milton's discovery of this radical new paradigm made him think of himself as a new Theseus who has found the thread of natural liberty which "windes" us out of the "labyrinth of servitude" to canon law (*CPW*, 2: 514, 2: 240). In *Paradise Lost,* he not only makes unhindered marital conversation the very essence of paradise, but gives Eve the handmaidenly role of discursively supplementing Adam's more "intuitive" or contemplative knowledge through her active husbandry of their garden and her natural bent toward empirical

---

[20] See Merchant, *The Death of Nature* and Brian Easlea, *Fathering the Unthinkable: Masculinity, Scientists and the Nuclear Arms Race* (London: Pluto Press, 1983), and for a summary of the conventional idiom of "naked truth" in the new science, see Barbara Shapiro, *A Culture of Fact* (Ithaca: Cornell Univ. Press, 2000), 58–62. Although Comenius has sometimes been considered more utilitarian, he is also more interested in mystical Neoplatonic harmonies than Bacon; on this point, see Dagmar Capkova, "Comenius and his ideals: escape from the labyrinth," in *Samuel Hartlib and Universal Reformation*, eds. Mark Greengrass, Michael Leslie, and Timothy Raylor (Cambridge: Cambridge Univ. Press, 1994), 75–91.

[21] On Milton's debts to Bacon and on Bacon's contributions to other defenders of the English revolution, see Christopher Hill, *The Intellectual Origins of the English Revolution* (Oxford: Clarendon Press, 1965). The most thorough study of Milton's relationship to the Baconians of the Hartlib Circle is still Charles Webster's *The Great Instauration: Science, Medicine, and Reform 1626–1660* (New York: Holmes & Meier, 1976).

[22] On monism see Stephen Fallon, "The metaphysics of Milton's divorce tracts," in *Politics, Poetics, and Hermeneutics in Milton's Prose*, eds. David Loewenstein and James Grantham Turner (Cambridge: Cambridge Univ. Press, 1990), 69–83.

inquiry.[23] As several liberal feminists note, Milton's "erotics" of truth thus shuns the conquest model in both marriage and social contract theory, so that according to Francis Barker, *Areopagitica* follows a paradigm ironically similar to that of his political nemesis Thomas Hobbes.

Yet despite its many strengths, Barker's postmodern analysis of the "wars of Truth" waged between these two Baconians goes awry precisely by overlooking the original source of the marriage trope employed by both. Misreading Milton's rewriting of Atalanta's race as a conquest trope, he claims that when Milton insists that as soon as Truth "gets a free and willing hand," she will open "herself faster than the pace of method [or Hippomenes] and discourse can overtake her" (*CPW*, 2: 561), thus depicting a struggle between a rapist (empiricism) and its feminine victim (Truth). Barker nevertheless concedes that Truth's curious "amalgam of classical amazon, renaissance republicanism and puritan valour" lies at the heart of a "complex and dynamic economy of truth" which engages not just one but "various sets of eroticized relations among its uttering subject or subjects, the reading subject, the overt and virtual addressees."[24] In order to compete with Milton, Hobbes's *Leviathan* therefore needs to outline a competing "erotics." It does this by stressing the dangers both of Milton's unfettered and triumphantly "free" competition between method, discourse and Truth, and the endless desire fueling it. Since competing desires can only lead to the endless "war of all on all," desire itself must be instead made into a handmaiden serving the purposes of social stability. Regulated by Hobbes's Sovereign, fallen or "feminine" desire is far less free than her Miltonic counterpart yet ironically just as centrally important. Because desire can never be eradicated, she must be tamed by recognizing the rational superiority of social contract theory both for her own purposes and those of society as a whole. Without the vital impetus of desire, no man or human institution "can … any more live … than he, whose Senses and Imagination are at a stand"; but by regulating it, competitive individuals can safely pursue happiness or "Felicity." The "handmaiden" of hope leads the continual search and progress of this desire "from one object to another; the attaining of the former, being still but the way to the later." Yet by providing this framework for non-violent competition, the social contract successfully harnesses "the object of mans desire," which "is not to

[23] See Milton's *Paradise Lost*, rpt. in Merritt Y. Hughes, *John Milton: Complete Poems and Major Prose* (New York: Odyssey Press, 1957), Book 4: 288–311, 657–80; Book 5: 500–12; Book 7: 59–130; and Book 8: 1–178). All future references to *Paradise Lost* will appear in the text parenthetically as *PL*. For liberal feminist commentary on Milton's depiction of gender relations, see Diane McColley, *Milton's Eve* (Urbana, IL.: Univ. of Illinois Press, 1983) and Catherine Gimelli Martin, *The Ruins of Allegory: "Paradise Lost" and the Metamorphosis of Epic Convention* (Durham, NC.: Duke Univ. Press, 1998), chapter 6. On the Baconian aspects of Milton's epic, see Karen Edwards, *Milton and the Natural World* (Cambridge: Cambridge Univ. Press, 2000) and Martin, "'What If the Sun Be Centre to the World?': Milton's Epistemology, Cosmology, and Paradise of Fools Reconsidered," *Modern Philology* 99 (November, 2001): 231–65.

[24] Francis Barker, "The Wars of Truth," in *The Culture of Violence: Tragedy and History* (Chicago: Univ. of Chicago Press, 1993), cited 130–33. Given his Foucauldian outlook, Barker of course regards this new "regime of discourse" (132) in largely anti-empirical and "unfree" disciplinary terms, but his textual insights provide a kind of negative verification of the approach taken here.

enjoy once onely, and for one instant of time; but to assure for ever, the way of his future desire."[25] This dynamic "marriage" between social rationality or patriarchy and feminine pleasure or matriarchy thus turns a lawless "Gaia" into a constructive handmaiden of both consumer culture and the sovereign state.[26]

Nevertheless, even the masculinist Hobbes does not actually advocate a traditional male conquest model of social order. Quite the contrary, he proposes that since all human beings are vulnerable to fraud, or force, or both, all men are more or less created equal. No man "naturally" has sufficient "Power to keep … all in awe" to his will, so social stability can only be achieved through the mutual binding "science" of the social contract. This contract in turn allows all men and women to pursue happiness in peace free from continual fear of losing the possessions and power they insatiably desire, which they now achieve by material rather than martial means. In that sense, all the Sovereign's subjects must "femininely" relinquish the liberty they retained in the state of nature and become "Metic" negotiators. Hobbes even implies that successful sovereigns originally convinced their military competitors to recognize the feminine within by skillfully using words and metaphors to convince them of their need for security and the "softer" fruits of peace. Thus while Hobbes considers "feminine" metaphors and arts of rhetoric even more dangerous than most other Baconians do, he still assigns verbal eloquence the handmaidenly task of political persuasion. This step indicates his careful reading of his former master, Bacon, for while both wish to establish "generall, and infallible rules" of procedure, both recognize that without "the arts grounded upon words" there can be no masculine science of proper definition, not even in the logical, axiomatic, and "artificial" language that Bacon, Hobbes, and the Royal Society hoped to create.[27]

While Hobbes himself was notoriously skeptical of the Royal Society's Baconian defenses of empirical science, he thus clearly accepts Sprat's famous argument that all extravagant "swellings of style" should be rejected in favor of a "primitive purity, and shortness, when men deliver'd so many *things,* almost in an equal number of *words* … bringing all things as near the Mathematical plainness, as they can." Sprat compares this ideal style to the practical "language of Artizans, Countrymen, and Merchants," which is more "masculine and durable, because it consists of *Images* that are generally observ'd, and such visible things which are familiar to mens minds."[28] Yet as Barker points out, Hobbes's self-interested clarity conceals "what counts as 'sense' or 'intelligibility'" within its highly discursive and (like Sprat's) actually highly metaphorical structure. Thus in reality, "The fabulous contract itself is" strangely "subsequent to and consequent upon the language in which it can be conceived and framed: the discourse which is outside power [practical prudence] constitutes an absolute power. It is not simply that mathematical or geometrical

[25] Thomas Hobbes, *Leviathan,* ed. C. B. Macpherson (Hammondsworth: Penguin Books, 1968), 160.

[26] For an intellectual history of this Hobbesian contribution to modernity, see Albert O. Hirschman, *The Passions and Interests: Political Arguments for Capitalism before Its Triumph* (Princeton: Princeton Univ. Press, 1977).

[27] Hobbes, *Leviathan,* chapter 13, 183–85.

[28] Sprat, *History of the Royal Society*, 113, 415.

science (or Hobbes's own reasoning, which he would have be of a similar order) … but another mode of discourse from that either of the natural passions or the laws of any duly-constituted commonwealth" finally constructs Hobbesian theory. It is that while this theory "contrasts markedly with the positive figurations of Truth in Milton," they both create systems in which Barker finds them both create schemas wherein the only solution to unruly desires is to submit them to the coercive regime of contractual competition.[29] Yet Barker's Foucauldian perspective here occludes the fact that both contracts are still as open-ended as Miltonic marriage. Although Hobbes makes the "feminine" party submit to the masculine "head" or sovereign, she thereby gains greater powers of negotiation within the social contract arrangement. Barker is therefore right to observe that Hobbes's sovereign "marriage" of ruler and ruled remains a heuristic fiction, but wrong to forget that their relationship remains completely dependent upon "her" handmaidenly prudence. Without it, the war of all on all returns along with real rather than merely metaphorical violence. Milton ironically reflects his commitment to essentially the same paradigm when he portrays Eve, not Adam, ending the violent separation that follows their fall by helping him to see that only a mutual recommitment to waging eternal warfare upon their real enemies without and within (*PL,*10: 966–1053) will restore their embryonic society.[30]

Ironically, the "wars of truth" theme first broached in *Areopagitica* and later summed up in the final books of *Paradise Lost* becomes a quasi-official Royal Society maxim in the same year that Milton published his great epic. Both Sprat's *History* and Cowley's "Ode to the Royal Society" show that these wars should no longer be fought on the open plains of Troy or Sparta but only in the tangled wilderness of the new Promised Land.[31] In Cowley's "Ode," converted brides like Rehab join the triumphant "Hebrew" march out of this labyrinth as she imitates Ariadne in helping to "wind" her lost friends to higher ground. Like the many related role models discussed above, this handmaidenly work is ultimately Baconian in provenance. Neither nude nor coyly concealed from the male gaze, the sensuous feminine mind can now be used to symbolize a prudential civic pragmatism capable of adapting the wilderness of male "wit" to the practical uses of society. Since it is still only symbolic, this service is naturally not equivalent to women's liberation, yet by disrupting the logocentric masculinism of earlier tradition, it clearly has real-world effects. Although much is genuinely wrong with modern science and technology, the scientific historian Riejer Hooykaas shows that the Greek devaluation of human art in favor of an inviolable and isolated "Mother Nature" was much more detrimental to real women:

---

29 Barker, *The Culture of Violence,* 136–8.

30 As Joseph Summers long ago showed, Eve's handmaidenly humility anticipates the birth both of Christ and of Christian charity, which as in Bacon, gradually overthrows both the male chauvinism and the "effeminacy" of the Ancients. See his *The Muse's Method: An Introduction to "Paradise Lost"* (Cambridge: Harvard Univ. Press, 1962).

31 Sprat, *History of the Royal Society,* 32–3; cf. 41.

> In the first century B.C. a Greek poet sang the praises of the newly invented waterwheel, which was to deliver the women from the hard labour of grinding corn and which seemed to announce a return to the supposedly leisurely life of the Primitive Age. Nevertheless, this mechanical mill remained a rarity during the next twelve centuries. Perhaps because of the abundance of "animated tools", both human and animal, the need for non-animated, self-moving instruments was hardly felt. It was certainly no lack of technological skill which hindered the transition from theory to practice, but rather the contempt for manual labour. Science was something that belonged to the head alone and not to the hands.

Hooykaas traces this snobbery to "rationalism" not in the empirical, Baconian sense but in the Platonic/Aristotelian sense. By deifying nature, underestimating art and devaluing manual work of all kinds–male but especially female–Greek rationalism devalued experimental verification of its deductions and insisted that artificial activities can give no real insight into natural events. This form of "naturalism" also led Greek philosophers to consider any investigation of the working methods of mechanics far beneath their dignity, and their science far above the needs of mechanical workers or women who performed their menial services. For this work *by definition* placed them in inferior classes completely outside of the "civilized" world of the Greek polis.[32] By ending that truly artificial divide, Bacon did more than insure the possibility of scientific progress: he helped to assure the future progress of women. His similarly rhetorical reconciliation of feminine with masculine powers pointed the way toward more real cooperation between the sexes and less rigid divisions of labor in marriage.[33]

These conclusions should lead us to reconsider the Faustian Bacon conjured up by his early twentieth-century critics. Late Romantics like C. S. Lewis, L. C. Knights, and Harold Fisch, and Frankfort School Marxists like Max Horkheimer and Theodor Adorno resemble his feminist critics in forgetting that the deposed Lord Chancellor never sought profit for his own sake but only for the good of humankind. As Hooykaas points out, they thus obscure the humane Christian apologist revealed by the concluding prayer of Bacon's *Historia Naturalis*, and the fact that their "contemplative, almost medieval, vision" of "Mother Nature" would once again turn her into "a stagnant pool." For in fact, the modern enthusiasm for a spontaneous, magical, or mystical communion with nature merely rehearses the Greek view of nature and the Stoic principle of complete conformity to its dictates, while Bacon's own course was far less ideologically extreme. His *via media* conserved the higher ideals of the medieval position by encouraging the ethical application of science and the reunification of "faith and works" both as valuable "contributions to the commodities of life" and real "tokens of truth." It also considered "a purely utilitarian science … [as] of little use. If scientists concentrated on application, without first establishing a solid basis of pure science, they were going up a blind alley," so that like Atalanta, they

32 Hooykaas, *Religion and the Rise of Modern Science,* 81–3.

33 For a seventeenth-century feminist's proposal of how to facilitate wifely "husbanding," see Bathsua Makin's *Essay to Revive the Ancient Education of Gentlewomen, in Religion, Manners, Arts & Tongues* (1673).

would "break off their course in order to get the golden apple and so miss victory."[34] Instead in Bacon's utopian vision, Atalanta's "feminine" crafts are no longer lowly or servile. Valued by artisans and aristocrats alike, they assist real women to become invaluable help-mates who actually "husband" their households rather than passively ornament or preserve them as places of male rest. Finally, they create dialogical relationships in which neither man's nor woman's work is ever really done.[35]

[34] Hooykaas, *Religion and the Rise of Modern Science,* 72–4.

[35] Critics have long supposed that Bacon influenced Shakespeare's *The Tempest;* if so, the work ethic that Prospero requires of Ferdinand as a down-payment on the egalitarian relationship he demands for Miranda forms a striking contrast to the male/female relationships in earlier comedies like *A Midsummer Night's Dream* and *The Taming of the Shrew.*

Chapter 5

"The Very Idea!" Francis Bacon and E. O. Wilson on the Rehabilitation of *Eidos*

John C. Briggs

> There are found some minds given to an extreme admiration of antiquity, others to an extreme love and appetite for novelty; but few so duly tempered that they can hold the mean, neither carping at what was well laid down by the ancients, nor despising what is well introduced by the moderns. (*SEH*, IV: 59–60)

> The choice between transcendentalism and empiricism will be the coming century's version of the struggle for men's souls… Where it settles will depend on which view of the world is correct, or at least which is more widely *perceived* to be correct… If empiricism is disproved, and transcendentalism is convincingly upheld, the discovery would be quite simply the most consequential in human history… The matter is still far from resolved.
>
> (E. O. Wilson)[1]

The Ideas of the Forms

The scientific kinship of Francis Bacon and E. O. Wilson extends beyond their interest in founding new sciences. Despite appearances, they share a persistent and deep respect for the old learning, which they consider to be much more than a historical curiosity, a basis for entertainment, or a mere set of tools for intellectual advancement. For both men, the persistent claims of ancient learning–especially what Wilson refers to as "transcendentalism" and what Bacon finds in ancient wisdom–confront the modern sciences with the project of reconciling, testing, and perhaps discarding preconceptions in *both* kinds of learning. Since each man hypothesizes and seeks the unity of knowledge, each accepts the challenge of assessing competing or divergent forms of learning.

Once we begin to compare Bacon's esteem for the old learning with Wilson's neo-Darwinian respect for the persistence of certain *ideas*–of certain transcendental claims he ostensibly rejects but then partially rehabilitates–both men's unifying projects become more urgent and complex.

---

1 E.O. Wilson, *Consilience: The Unity of Knowledge* (New York: Alfred A. Knopf, 1998), 240, 258.

An interesting focus for such a comparison is the *way* in which Bacon and Wilson treat ideas. They each have an interesting *idea* of ideas (a substantial conception, at least, of what ideas are) embedded in their understandings of the new scientific learning. Neither man focuses his discussion on the ideas in themselves. Indeed, they typically veer from explicit treatment of the topic. Bacon prefers to use the term "notion", and Wilson refrains from discussing ideas as ideas except in a few scattered passages.

Nevertheless, Bacon associates ideas with formal causes, which are at work in Platonic, Aristotelian, and eclectic Neoplatonic traditions. He is famous for pointing out the shortcomings of those traditions, but he appropriates their metaphysical aspirations to serve his conception of the Laws of the Forms, which in function imitate and (in Bacon's view) surpass the old doctrines of the ideas and the traditional formal causes. The Laws of the Forms offer the power to transform everything into everything else (*SEH*, III: 357).[2] E. O. Wilson, for his part, is deeply skeptical of non-empirical "transcendentals," though he is by the same token curious about the evolutionary endurance of certain ideas and their effect on the survival of cultures. A crucial instance in his book *Consilience* has to do with the ideas of the moral philosophers of the Scottish Enlightenment, who, as it happens, were admirers of Bacon and contributors to a form of social cohesion (forms of common law and moral reasoning thought to be allied with common sense and evident in lasting social institutions and practices) that for a neo-Darwinian is a tantalizing socio-biological fact not adequately explained by Darwinian empiricism.

Although both men strive to separate the physical and metaphysical realms, they each entertain the possibility of testing the two realms' interaction. Their books address audiences they assume have broad interests and a concern in the outcome. They raise fundamental questions about the nature of scientific evidence, idea, passions, judgment, and choice, and the influence of these things on the course and substance of scientific inquiry. If we do not read them with respect for what *they* respect, and if we do not exercise an awareness of the difficulty they face in pursuing their hybrid projects amidst doctrinaire defenders of narrowly empirical or transcendental claims, we misread them.

Bacon frequently expresses his opposition to the old natural philosophies with an eloquent, polemical energy. His well-known doctrine of The Four Idols ingeniously associates the false ideas of the old non-progressive sciences with the aberrations of heretical religion. The false coinages of language, the foibles of tribal prejudices, the misleading systems of philosophy, and the delusions of idiosyncratic (idiotic) habits of thought–all are associated in his thinking with the worship of the golden calf. They turn investigators' eyes and dispositions toward self-indulgent pleasures, away from the arduous toil commanded by true scientific inquiry.

Bacon's deeper doctrine of anti-idolatry is not a thoroughly secularizing iconoclasm. It entails a deep interest in religion and metaphysics. The religious or at least

---

[2] See John Channing Briggs, *Francis Bacon and the Rhetoric of Nature* (Cambridge: Harvard Univ. Press, 1989), Chapter one.

quasi-religious dimensions of his project have been discussed elsewhere.[3] Here we will focus on his notion that there are certain ideas that are not idols but true forms, transformative principles that are at once ideas, formulae and powers that are substantially manifested in material things. To wonder in this way about Bacon's Laws of the Forms is to depart from the opinion of Bacon's great Victorian editor, James Spedding, who makes his skepticism clear in the introduction to his edition of Bacon's works: "I am, for reasons that will be hereafter mentioned, much disposed to believe that the doctrine of Forms is in some sort an extraneous part of Bacon's system. His peculiar method may be stated independently of this doctrine" (*SEH*, I: 28). Spedding cites the *Valerius Terminus* (*Of the Interpretation of Nature*) as his primary authority for Bacon's own dismissal of the Forms. But in that treatise we find one of Bacon's most emphatic statements of their importance in the new science's rehabilitation of "received philosophies." The new learning's motion of thought in discovering scientific laws of transformation is not only "not much other matter" than the old philosophers' attempts to establish axioms for understanding such changes; it is "more nearly also in that which [the old philosophers] term the form or formal cause." Thus Bacon says he shares the ancients' high praise for anyone who might reach an understanding of such a cause, and he proposes to reach that understanding with powerful methods the ancients did not have:

> … Aristotle's school confesseth that there is no true knowledge but by causes, no true cause but the form, no true form known except one, which they are pleased to allow; and therefore thus far their evidence standeth with us, that both hitherto there hath been nothing but a shadow of knowledge, and that we propound now that which is agreed to be the worthiest to be sought, and hardest to be found.[4] (*SEH*, III: 239)

If, following Spedding's lead, we focus prematurely on Bacon's famous analysis of The Four Idols as the defining doctrine of the New Learning, we might easily miss Bacon's distinction between idols and ideas that have to do with the Laws of the Forms. Paolo Rossi's landmark study of the role of magic in Baconian science develops three salient observations regarding the paradoxes in Bacon's encounter with ancient and Neoplatonic principles and practices, which are the seedbed of Bacon's notions of the forms. First, Rossi documents the impressive degree of Bacon's involvement in the various occult and Neoplatonic traditions–an intellectual context that had been distorted by exaggeration or ignored by previous interpreters of the new learning. Second, he distinguishes between that debt and Bacon's development of new notions of induction, experiment, and collaboration, what Rossi and most of Bacon's modern interpreters consider the most substantial legacy to the new learning. Thirdly, he observes that Bacon was frequently not clear in what he said about the difference between the old learning and the new. Bacon's contribution to modern

[3] For my own contributions to this tangled issue, see *Francis Bacon and the Rhetoric of Nature*, *passim*, and "Bacon and Religion" in *The Cambridge Companion to Bacon*, ed. Markku Peltonen (Cambridge: Cambridge Univ. Press, 1996), 172–99.

[4] A similar passage can be found in Book II of *The Advancement* (*SEH*, III: 355).

science remains, according to Rossi's analysis, problematic in important ways: "[O]ne of the major difficulties confronting the Baconian scholar is the presence of both mechanist and dynamic vitalist conceptions of reality in Bacon's physics."[5]

We could say the same thing about Bacon's doctrine of the forms. In Rossi's view, Bacon largely succeeded in subordinating Neoplatonic elements to the rest of his thought so as to overcome "the dominion of magic," the worship of false idols that corrupted natural philosophy. But Rossi is too good an observer of Bacon not to notice that toward the end of his life he was "finally entangled once again in its snares." Although "the manner and rules of the race had been changed for ever," Rossi concludes, we cannot be so assured that the legendary founder of the new sciences saw that triumph in rigorously modern terms.[6]

Bacon's lapse, according to Rossi, is embodied in the thousand sections of the *Sylva Sylvarum*, the last work of his career, and for many years the most widely circulated. In that forest of forests, we read accounts of experiments and observations that are often transcribed from old sources. Amidst accounts of experiments to undermine belief in the old learning's love of secret virtues and properties, Bacon discusses and favors explanations that partake of the old language of vitalism:

> 457. It is reported, that trees will grow greater, and bear better fruit, if you put salt, or lees of wine, or blood to the root. The cause may be the increasing the lust or spirit of the root; these things being more forcible than ordinary composts. (*SEH*, II: 488)

Here the "lust or spirit of the root" is a life force drawn or pressed into action by fertilizers traditionally thought to have embodied an aspiring essence or spirit of life. The observation is in one sense matter-of-fact, and yet it does not detach itself from the old learning's anthropomorphic language of desire and disposition: here one desire answers another, and the trees grow.

A second passage shows Bacon's dependence upon Neoplatonic habits of mind in his thinking about how to determine the engendering effects of snow that has lain long on the fields. This entry relies upon the hierarchical and quasi-occult notion that snow is not only different from, but superior to, ordinary moisture: it comes from the air and is analogous to sea-foam. As snow is to ordinary moisture, foam is to ordinary seawater. Born of the clouds, snow is finer–more watery than water–and so fructifies the ground more effectively than other forms of moisture:

> 663. Snows lying long cause a fruitful year; for first they keep in the strength of the earth: secondly, they water the earth better than rain; for in snow, the earth doth (as it were) suck the water as out of a teat; thirdly, the moisture of snow is the finest moisture; for it is the froth of the cloudy waters. (*SEH*, II: 545)

---

[5] Paolo Rossi, *Francis Bacon: From Magic to Science*, trans. Sacha Rabinovitz (Chicago: Univ. of Chicago Press, 1968), 13.

[6] Rossi, 35.

The passage suggests an experimental regimen to confirm its hypothesis, which is mixed with notions that are at home in the lore of popular and esoteric Neoplatonism. The earth has strength that can be kept in reserve, and snow is the "finest" of moistures, like the foam on the sea. It is purer partly because its whiteness is a mixture of air and water. The earthly and the transcendental somehow exist in one being (Bacon described this operation in the *Valerius Terminus* as a transition from transparency to whiteness when the watery form is "freed from water, and advanced to a clear body, and still tied to air" [*SEH*, III: 236]). Snow works better as a fertilizer than other forms of moisture because like foam, it has somehow taken on an exquisite, working fineness in the nimbused air.

Bacon's habit of mingling of transcendental and empirical causes culminates in the thousandth and last of his entries, which is a set of speculations about the power of words and thoughts. The entry ends with a conjecture on the influence of something analogous to a "universal limbus"–a ubiquitous medium that transmits and works upon the aspirations of men's minds:

> 1000. The delight which men have in popularity, fame, honour, submission and subjection of other men's minds, wills, or affections, (although these things may be desired for other ends), seemeth to be a thing in itself, without contemplation of consequence, grateful and agreeable to the nature of man. This thing (surely) is not without signification, as if all spirits and souls of men came forth out of one divine limbus; else why should men be so affected with that which others think or say? (*SEH*, II: 672)

For Bacon, such delight cannot be adequately explained in terms of mere reward or usefulness, the fluid mechanics of unimaginative empiricism. Whatever the substantial benefits the new learning offers for alleviating the human condition, Bacon considers the phenomenon of limbic influence to be a thing in itself, not contingent upon lesser causes. The influence is not neutral or merely mechanical. Men are capable of seeking rewards with a delight that has no primary regard for recompense. That delight is for Bacon a fact, and because he seeks to explain it with empirical credibility despite its ephemeral nature, he must paradoxically resort to the idea of the limbus to fix that fact in the mind. The analogy is provisional, and yet in its reliance upon a quasi-occult mechanism of influence involving active and passive proclivities it is reminiscent of Neoplatonic habits of thought Bacon never fully abandons, and indeed attempts to rehabilitate.[7] For Bacon the idea of the limbus has empirical purchase: it helps explain an observable and important aspect of human

[7] For example, see Ficino's description of the "third essence." The "seat of the rational soul … is that which Timaeus of Locri and Plato in the De Mundo [the *Timaeus*] described as compounded from undivided and divided nature. This is what implants itself in things mortal without itself becoming mortal. For just as it implants itself as a whole and is not split asunder, so it withdraws as a whole and is not dispersed. And because it controls bodies while it also clings to things divine, it is the mistress [*domina*, or female master] of bodies, not their companion [*comes*]. This is the greatest miracle in nature. For the remaining things below God are each individually something singular in themselves, but this essence is all things together. It possesses within itself images of things divine on which it depends, and these images are the

experience that apparently he cannot otherwise understand. It is not presented here as an idea to be jettisoned as soon as a more suitable empirical explanation has been devised. It is the thing itself, which the new learning might eventually elucidate by discovering the more fundamental laws of its operation.

It is important to dwell upon the explicit doctrine of the new learning: Bacon's contention that "Natural Magic" will be rehabilitated. The goal of the new learning, he writes in the *Advancement*, is to "revive and reintegrate the misapplied and abused name of Natural Magic; which in the true sense is but Natural Wisdom, or Natural Prudence" (*SEH*, III: 351). The old discipline is not to be discarded, but developed free of those who have abused it, and embedded into the new sciences. Natural Wisdom and Natural Prudence are the real Natural Magic. Here Bacon's syntax is richly, ambiguously suggestive: wisdom and prudence are to absorb magic and yet be transformed by it in its true form. The new learning's wisdom and prudence are supposed to act much like formative ideas upon the old natural magic: they will draw it toward a better version of itself. In that very process, however, they must expose themselves, at least for a time, to the older tradition's goetic arts of coercion and delusion. Are those arts transformed, or adapted to new ends?

As we look into this more esoteric level of Bacon's understanding, the Laws of the Forms act in some ways like Ficino's *genus*, which is a Neoplatonic archetype, a formative idea. Brian Copenhaver describes its operation in terms of the familiar Neoplatonic notions of influence, imitation, and ascent toward a characterizing idea or *genera*: "Every genus contains one highest member and one only, which causes all other members of the genus to belong to it–to possess the features that characterize the genus. God causes the first member (*primum*) of the genus, which in turn causes the rest of the genus." The "universal genera," such as the beautiful and the true, act in a similar way.[8] Plotinus's original understanding of the relation of form and matter gives to form a similar determining role: it has power to invest matter with an image of itself. In the words of another commentator on Plotinus's theory of forms, the most orthodox Neoplatonists thought of matter as "in no case anything but a sort of reflecting surface upon which the Form is able by means of the word to project a picture of itself, a sensible picture adapted to our modes of sensible apprehension."[9]

In his specialized writings on form and matter, as well as in the more accessible *Advancement*, Bacon sometimes seems to renounce Neoplatonic understandings. He vehemently resists the traditional notion that the Forms have the power to bring matter into existence (or actuality) from a merely potential or undifferentiated state. He insists that form and matter be understood together: "[I]n nature nothing really exists beside individual bodies, performing pure individual acts according to a fixed

reasons and paradigms of the lower entities which in some sense it produces … It can with justice, accordingly, be called nature's center, the mean of everything in the universe, the succession or chain of the world, the countenance of all things, and the knot and bond of the world," Marsilio Ficino, *Platonic Theology*, trans. Michael J. B. Allen and John Warden, 2 vols. (Cambridge: Harvard Univ. Press, 2001), I: 242–3.

[8] Brian Copenhaver and Charles B. Schmitt, *Renaissance Philosophy* (Oxford: Oxford Univ. Press, 1992), 151.

[9] Charles Bigg, *Neoplatonism* (New York: E. and J. B. Young, 1895), 205.

law … And it is this law … that I mean when I speak of *Forms*" (*SEH*, IV: 120). This is a practical objection to the old learning more than a theoretical one. Bacon's more basic objection is to the habit of separating form and matter in the *process* of investigating nature. He does not argue that forms do not exist, and indeed, as we have seen, grants them a virtually metaphysical status. He stigmatizes the bad habit of separating the forms from matter, which leads to the idolizing of ideas and the neglect of knowledge and deeds that would benefit mankind. The forms are indeed "eternal and immutable," Bacon says, "in the eyes of reason, at least, and in their essential law," and can be investigated in Metaphysics, a discipline with remarkably respectable status in the new learning even though it includes a practical component called "Magic." Bacon welcomes it into the new learning, in its purified state (*SEH*, IV: 126).

In direct investigations of nature, the combination of form and matter must be more rigorous than before. The inquirer must treat form and matter together so as to steer clear of Plato's error of "trying to apprehend the Forms as absolutely abstracted from matter" (*SEH*, IV: 360). In other words, the inquirer must not, as Plato does, "make over the world to thoughts" and mix natural philosophy with theology, at the expense of practical, charitable action (*SEH*, V: 468).

The subtlety of Bacon's distinctions needs close attention. He is not claiming that the Forms are reducible to matter. The Forms have always been combined with matter: the "first matter" of the Creation was united with the "first form" (*SEH*, V: 468). At the same time, they are not merely configurative principles. The atomists were wrong to assume that the world was made up of similar seeds that were arranged in various ways to make the different forms of matter (*SEH*, IV: 320). The new learning recognizes that the true Forms will "generate and superinduce" a "nature" upon a body (*SEH*, IV: 121), in imitation of God's treatment of created matter without having to defer to inherent limits in the stuff being given form. They will also proceed "from what in nature is constant and eternal and universal"–those things being, for Bacon, the Laws of the Forms themselves, which are such "natures" and genera that were created when God made matter. As both effects and forms, these are the phenomena of color, weight, fluidity, and other qualities that come to characterize the thing upon which they are superinduced.

This understanding of the transformative power of Forms threatens to reintroduce the Platonic excess that Bacon has specifically criticized: the notion that the Ideas or Forms so control and dominate passive matter that the Forms are literally all that matters. Although he looks forward to a time when the sons of science might wield the Laws of the Forms with an almost divine power to transform matter, he carefully stipulates that in the process of discovering those forms careful attention must be given to the eccentricities of material bodies (*SEH*, IV: 124). He emphasizes that he agrees with that ancient understanding of matter that assumes it has a "principle of motion" (*SEH*, V: 467). He eloquently, repeatedly, reminds his reader that a knowledge of causes must conform to the laws of nature that the new learning promises to discover.

Bacon's Neoplatonic terminology of genus and form, which emphasizes the power of transformative coating, investing, and superinduction, meets the threat of slipping back into idolatry with a deeper doctrine of form that he discloses in *De Augmentis*, his scientific elaboration of the *Advancement*:

> [Ordinary] Physic carries men in narrow and restrained ways, imitating the ordinary flexuous courses of Nature; but the ways of the wise are everywhere broad: to wisdom (which was anciently defined to be the knowledge of things divine and human) there is ever abundance and variety of means … [W]hosover knows any Form, knows also the utmost possibility of superinducing that nature upon every variety of matter, and so is less restrained and tied in operation". (*SEH*, IV: 362)

It is precisely this innovation in the new sciences that Bacon is pointing toward when he says that a mastery of superinduction "opens broad roads to human power, such as (in the present state of things) human thought can scarcely comprehend or anticipate" (*SEH*, IV: 122). The importance of this idea explains the persistence of his admiration for Plato's thought, despite his excoriation of the philosopher's intellectual practice. (Plato's "doctrine concerning Ideas" was prescient, he says, in its claim that "Forms were the true object of knowledge" [*SEH*, IV: 360, IV: 320, III: 57]). It helps us understand why he insists that nature's "fundamental and universal laws … constitute Forms" (*SEH*, IV: 123), and why he claims that "the invention of Forms is of all parts of knowledge worthiest to be sought, if it be possible to be found" (*SEH*, IV: 360).

This is to say that Bacon is doing much more than endorsing the traditional Neoplatonic position that contemplation, discovery and wisdom are superior to the active life of commerce, strength, application and "the first fruit of inventions" (*SEH*, VI: 744, VI: 720, III: 299, IV: 115),[10] or that he is insisting that "[t]he images of wit and learning are more valuable than the monuments of power or of the hands" (*SEH*, III: 318). The new Laws of the Forms are at once the negation, transformation, and universalization of Neoplatonic understandings of the ideas. They empower the sons of science to remake the natural forms by obeying their deepest laws of action, treating everything in created nature as convertible into everything else. Such laws simplify, and strangely enlarge, Neoplatonic understandings for the sake of a radical transformation of the world.

In Ficino's blend of theological and natural philosophy, the things below God, like the philosopher himself, are best understood in terms of perfective desire. Writing a century after Ficino wrote his translations and commentaries on Plato, Bacon would undermine and yet absorb such Neoplatonic thinking. Ficino's *Theology* had, by the time of Bacon's writing of *The Advancement*, proliferated throughout Europe and England in thirty editions.[11] It assumed and attempted to demonstrate that all things look upward, informed by an essence that descends upon and animates the aspiring body, somehow incorporating itself into that body without losing its immateriality.[12] Natural bodies in this process become, in a sense, transcendental: material embodiments of something that draws them out of and toward themselves in a perfective cycle. Something of the strange richness and complexity of this process is captured in the poetry of Bacon's contemporary Edmund Spenser, whose *Mutabilitie Cantos* describe Nature passing judgment upon Mutability's claim that all change is mere disorder. Change, Spenser's God of Nature concludes, is actually eternal. It is not

10 Rossi, 63–4.

11 Copenhaver, 145.

12 Marsilio Ficino, *Platonic Theology*, I: 242–3.

only "eterne in mutability;" it is the mask of a perfective spiral of enlargement, dispersion, and restoration, a returning of things to their original yet higher selves. Aspiring choice toils with, but is not defeated by, a determining fate:

> I well consider all that you have sayd,
> And find that all things stedfastnes doe hate
> And changed be: yet being rightly wayd
> They are not changed from their first estate
> But by their change their being doe dilate:
> And turning to themselues at length againe
> Doe worke their owne perfection so by fate (VII.vii.58)[13]

Bacon does not depart from Ficino and Spenser in assuming that the Laws of the Forms that govern these motions are indications of human contact with the divine. But he pushes the notion that though these forms are not direct evidence of God's purposes, they make available the appropriation, not just the imitation, of the principles God uses to give the various forms to created matter in Genesis. Thus although Bacon joins with Plato and the Neoplatonic tradition in praising the knower of the forms as a god (*SEH*, III: 239), he has in mind a version of the all-powerful biblical Maker (and the possibility of imitating that maker). It is in this sense that the new sciences would be able to transform the world.

The divine principles of purpose and action behind such transformations, because they are so godlike in this biblical sense, are strange to Neoplatonism. Properly understood as part of Bacon's teaching for the wise, they are profoundly hidden, even cryptic, though not inaccessible to the new sciences. To put it another way, they are simultaneously unknowable to the uninitiated mass of men and open to the ones who see what power over the forms entails.

Bacon carefully explains that God's policy, like a king's, is to act as a perfect monarch who moves men and things without disclosing his purposes (*SEH*, IV: 364–5). The new sons of science can serve and imitate that policy–as indeed they do in Bacon's *New Atlantis*. Again, the pursuit of such a combination of perfect power and cryptic purpose is largely alien to Neoplatonic ambitions, which are no stranger to esoteric doctrines and aspire to various types of mastery and make use of hidden purposes and potentialities but do not share the new learning's doctrinal embrace, only partially qualified by Bacon's cautions, of a power to transform all things. Alchemists of the old school sometimes fantasized about turning the sea to gold; but the Neoplatonic magus was not bent on a doctrine of universal transformation, of the power to turn anything into anything else. Moderation, often honored in the breach, was at least an ethical reference point. Shakespeare's Prospero struggling against his desire to take a complete revenge upon his foes exhibits a restraint more characteristic of the magus than Marlowe's Faustus despairing over the boredom that attends his indulging of every wish. Castiglione's luminously Neoplatonic *Book of the Courtier*

[13] Edmund Spenser, *The Faerie Queene*, ed. A. C. Hamilton (London and New York: Longman, 1977), 734.

models Spenser's principle of perfective desire for the human sphere: formed by the *idea* of the perfect courtier, the human subject is tempered by the practical discipline of courtly moderation for the sake of embodying an unaffected grace and temperate soul. Innovation for its own sake is profoundly contrary and destructive to this courtly preoccupation. Facility, though highly developed in the courtier's discipline, is a power in the service of graceful character rather than the reverse. The human is the model of the physical, rather than the other way around. Bacon's restraint, on the contrary, is strategic and primarily directed toward objects–all the better to exercise the true power of the forms at a time when the veiled king wishes to exercise his power, display his will, but not reveal his mind.

The new learning, modeling itself after biblical paradigms, was more likely to disregard moderation as a matter of principle. Rather than moderation, Bacon focuses on the simpler and more methodologically problematic virtue of charity, which in his project must depend upon a new and arduous form of religious care. This charity is not the same as kindness or conventional Pauline generosity. It is an art of self-mortification and concealment that must be strong enough to smash idolatry and counter the temptation to misuse the God-encrypted powers that become available to a master of the forms.[14]

In these Baconian formulations, courtly moderation becomes encrypted behavior. The Laws of the Forms are universal yet must not be revealed to the uninitiated. The phenomena of nature manifest those laws to those who know them: the sons of science observe in nature not merely the evidence of deeper patterns but the thing itself, the encrypted power of the divine maker. Glimpses of such revelations are not available to all. Those who have not undergone the discipline of self-denial and charity are not qualified to have access to the Laws of the Forms. To honor these truths and protect them from abuse, the courtier of the new sciences must seem to be moderate even though he is learning the means of transforming the world.

The great irony of Bacon's legacy is that these aspects of his innovation were hardly the reason for his later popularity. In the age of liberation from monarchy he became a republican hero. His most prominent admirers, including the Scottish Enlightenment philosophers and some of the American Founders (most notably Franklin and Jefferson), saw in his works a liberatory doctrine for the aspiring common man. They did not misread Bacon. Their enthusiasm was undoubtedly stirred in part by the grandeur and excitement of his eloquent introduction to *The Advancement*, with its evocation of the prospect of liberal learning, the alleviation of human suffering, and experimental, cooperative inquiry freed of interference from theological or political authorities. The idea that learning was of ancient respectability, and that it could advance with dedicated effort without depending upon birth or calling or submitting to political or theological authority became–in the course of events that led to the rise of mixed and republican regimes–an innovation of such magnitude that it is difficult to underestimate its animating power on those who took it in. What Bacon had written in shadowy layers of meaning for a courtly elite

[14] See Briggs, *Francis Bacon and the Rhetoric of Nature*, 1–40.

became a foundation upon which self-inquiring republican citizens believed they might better themselves and the prospects of republican government.

In this sense, therefore, the new Baconians were not Baconians. Although they were disposed to use certain Baconian methods and they adopted much of his general outlook regarding the prospects of science, they did not appropriate his pivotal, esoteric philosophy regarding the Laws of the Forms. They did, however, use and transform his general doctrine of ideas. Jefferson may have done so when he penned his famous declaration of the meaning of self-government: "We hold these truths to be self-evident." What modern skeptics might consider an anti-Baconian proposition about unsubstantiated ideas is two things at once: a) the affirmation of a pre-existent principle, with the biblical or at least deistic sanction of "Nature's God", and b) a venturing of an hypothesis informed by Lockean and other surveys of the natural grounds of reason and government. The spirit of the sentence is Baconian in its presentation of an axiom to be tested. It is what is the case, presented to be honored and tried by the ordeal of revolution.

This confluence of principle, hypothesis, and observed fact of course permeates the Scottish moral philosophers' discussions of the moral sympathies, which influenced Jefferson's early education and perhaps had purchase on his phrasing of the Declaration. (We will pursue another example of this pattern in the next section, where E. O. Wilson extends Bacon's legacy by discerning and emphasizing principles that are not stereotypically Baconian.) Honoring Bacon's legacy, Reid, Hutcheson, and Smith hypothesize and substantiate the existence of those sympathies in human judgment and social interchange by means of broad-ranging observations of long-established patterns in moral thinking. Perhaps taking a cue from the way Bacon assembles his observations of courtly behavior in the *Essays*, they treat their observations not only as inductive building blocks but as potential instances of the thing itself, embodied signs of the working power of moral sympathy. The moral sympathies in that tradition are combinations of the transcendental and the material, of moral ideas and inherited dispositions. They are emphatically present in daily life, and derived from divine prerogative. The Scottish philosophers do not, as Bacon does, cloak these ideas by dividing their focus into open and esoteric doctrines for naïve and sophisticated readers. Without Bacon's courtly guile and strategic inhibitions in the service of a world-transforming science, they present their observations with a Baconian confidence in the significance of their evidence as the discovery of the thing itself, and as a guide to moral conduct.

E. O. Wilson's *Consilience* and the Baconian legacy of ideas

We are now ready to pursue more thoroughly the influence of Bacon's legacy on the treatment of ideas in *Consilience: The Unity of Knowledge,* by the Harvard sociobiologist E. O. Wilson. An admirer of Bacon, Wilson frames his project in a comprehensive spirit. Consilience would harmonize and seek to unify the study of conduct, political life, and religion as well as the pursuits that animate the various fields of modern natural philosophy. A new, more comprehensive understanding of evolution,

Wilson hopes, will guide this synthesis. Like Bacon, Wilson honors and defends many of the contributions of the old learning. He takes up this process where Bacon left off: the post-Baconian period, up to the dawn of the neurosciences and the most modern evolutionary theory. And like Bacon he attempts to identify and stigmatize the idolatrous flaws in those forms of learning. He projects his critique into the post-war debate over the relative value of science and the humanities.

Wilson returns to that controversy, what C. P. Snow called the debate between "the two cultures," with a Baconian respect for its significance that sets him apart from those who would reduce it to academic politics or a false dilemma. As a student of Bacon, Wilson knows that the question of the relation of the sciences to the arts and humanities, with all their attendant subjects of human interest, is an old one, and at least as important today as it was when Bacon wrote *The Advancement of Learning*.

Wilson's project of consilience derives the modern enterprise of science from Bacon's innovation, especially Bacon's rejection of intellectual idolatries and his embrace of empiricism, methodical experimentation, and scientific collaboration. It looks ahead to a harmonious unification of science and the humanities that would not only create a body of enlightened knowledge and alleviate the human condition, as Bacon intended, but would democratize access to the new scientific knowledge and secure the evolutionary survival and prosperity of the human species. Toward that end, Wilson launches his criticism of transcendentals–the legacy of ideas he associates with the misguided sciences born in the Enlightenment as well as theology, philosophy, and the humanities–not only to establish the scientific superiority of empiricism, but to intervene in the evolutionary process to improve the chances for human survival.

It isn't that all transcendentals are in all ways bad. Wilson asserts repeatedly that he has great respect for the heritages of religion, philosophy, the arts, and their contributions to the persistence of human culture. The problem, he argue, arises from two opposite abuses. Either the transcendental and empirical truths have been mixed up by benighted practitioners, or they have been unwisely, dogmatically separated. The *Philosophes* relapsed into "deadly abstractions." They separated reason from passion until the sciences became "bloodless"–so indifferent to ethics and religion they became increasingly conducive to totalitarianism and the denial of human freedom: "If scientific inquiry diminishes the conception of divinity while prescribing immutable natural laws, then humanity can lose what freedom it already possesses." The twentieth-century's logical positivists, on the other hand, could not agree upon the difference between empirical fact and generalization.[15] Both groups pursued objective scientific knowledge but ended up mistaking or confusing the empirical and transcendental realms. They either ignored or failed to discover the real grounds for their interaction, which Wilson attempts to disclose in a new "consilience" of the different kinds of knowledge.

The true relation between transcendentals and the empiricism of the new evolutionary and neurological sciences must be established by a new level of intelligence that Wilson stipulates must be "wise." *Wisdom* is a crucial term in Bacon's work,

[15] Wilson, 15, 34, 63.

denoting an insight into the hidden laws of nature that goes to the heart of their laws of operation. The new intelligence would be capable of understanding the neurological and evolutionary foundations of intelligence. It would also, in Wilson's conception of scientific progress, gain an ability to predict the consequences of certain decisions it would need to make–consequences bearing upon the prospects of the species' survival and prosperity.[16] Thus the diction of Wilson's book shows his preference for *mind* over *brain* (a pattern evident in the space he devotes to each in his index). *Mind* is the more capacious term for the seat of a wisdom that must achieve a new level of self-knowledge (an understanding of the evolutionary origins of its own minute structures and operations) as well as a new, almost prophetic perspicuity–an ability to penetrate the vagaries of the evolutionary prospect.

There is an urgent need to develop such wisdom, Wilson argues, because human beings are developing the capability to make evolutionary choices that gain control of evolution:

> Homo sapiens, the first truly free species, is about to decommission natural selection, the force that made us ... Evolution, including genetic progress in human nature and human capacity, will be from now on increasingly the domain of science and technology tempered by ethics and political choice.[17]

It would seem that ethics and political choice would have to entail ideas, or at least appeals to some sort of transcendentals, to distinguish their contribution from those of science and technology. But what could these ideas be, in Wilson's unity of knowledge, if not reflections and applications of evolutionary thinking? What would ethics and understandings of politics have to offer, if anything, in their own right?

Wise intelligence, along the lines that Wilson is proposing, must somehow take responsibility for the evolutionary biological process that has not only created it but has *made it* responsible. It cannot do so, it would seem, without access to something outside of its own biology. Otherwise, its taking of responsibility is an ephemera–a mere transcendental in the worst, unscientific sense of that word. Somehow evolution has developed the first "truly free species" independent of evolution's power. How then does one conceive of freedom, or responsibility, in a radically evolutionary world? How is it possible for freedom and responsibility to be encoded in the biological and cultural heritage of the human being? Wilson is keenly interested in such questions:

> Natural selection, in short, *does not anticipate* future needs. But this *principle*, while explaining so much so well, *presents* a difficulty. If the principle is universally true, how did natural selection *prepare the mind* for civilization before civilization existed? That is the great mystery of evolution: how to account for calculus and Mozart. [emphasis mine][18]

---

[16] Wilson, 240 and *passim*.

[17] Wilson, 276–7.

[18] Wilson, 48.

This is a particularly vexed point for Wilson, one that gives us a glimpse of the core of his thinking and his difference from Bacon in his treatment of ideas. On the one hand, he is not willing to grant that considerations of ethics and politics can ever trump or even stand outside of biology when evolutionary biology is correctly understood. On the other, he relies upon wise considerations of ethics and politics to guide the wise intellect that now must begin to take responsibility for the course of evolution. The dilemma simultaneously devalues ideas (and most certainly any notion that transcendentals are operative in natural processes in any meaningful, i.e. empirical, sense) *and* raises them to new heights as some of the highest products of eons of evolutionary activity. In this sense they are embodied powers working upon the destiny of the genome, influencing the choices of the wise intelligence and from there the radically material evolutionary process of variation and selection.

There is a logic (or, as Wilson might say, a cognitive pattern) in these notions that points toward a particular understanding of human nature. Wilson surrounds his argument for consilience with an idea of *what it means to be human* amidst evolutionary forces. For him, a human being has the power to choose, and indeed can choose to enhance that power through a consilient science of mind. A human being can lose the exercise of that power, and perhaps the power itself, when the wrong choices lead to tyranny over the mind as well as the body, and it seems also evolutionary debasement. Thus wise choices are those that do not destroy or abuse that power to choose by making human beings into moral monsters, cyborgs, or gods.[19]

Like Bacon, Wilson is concerned to preserve human being and human possibility, which have to do with a sense of human scale. On the one hand, both men promote what looks like the opposite: a humility that reduces, hardens, and disperses the investigator's nature so that it can resist human tendencies to find truth in the mere images of things. In doing so, they also magnify the powers of observation so that what is now incomprehensibly great and minuscule can come into view by means of a new kind of sight. On the other hand, however, both men argue that evolution itself evolves, and new sciences come to be known through those who can activate the human propensity to learn new things and connect them to the old. Remarkably, both men resist abandoning forms of belief through the images and forms of things that give proportion and meaning to the human consciousness as it copes with change.

Wilson wants the new science of mind to journey out beyond the limits of ordinary human nature, but then to return home to "that narrow sliver" of "organismic time where human consciousness, being organismic itself, is forced to exist." Having completed a "science-led grand tour of space-time," the creators of a new science of mind are to "arrive home in a world for which the evolution of the brain prepared us." They find there, as well as in their researches, what Wilson calls the "peculiarly human."[20] It is that which makes the human organism, shaped by an evolutionary capacity to evolve, able to resist and perhaps mold the determining and accidental forces that would otherwise dissolve it into a machine, a beast, or a god.

---

[19] Wilson, 248.

[20] Wilson, 237.

Though he assumes that evolution controls these developments, Wilson repeatedly returns to the notion that evolution cannot be adequately understood in this new era of brain research and the decoding of the evolved operations of human thought, without such an idea of what it means to be human. We have returned by a roundabout route to something close to Bacon's axiom that "man" is the one being whose form is not within the power of men to make out of something else (*SEH*, III: 350). Wilson too is somehow drawn back to the power of this idea, though not its explicit theological meaning. As the one being of whom it can be said that he shadows forth his maker *and* an idea of his being, Man–the human being–is for Bacon beyond human manipulation, at least with regard to what makes him most human. Bacon and Wilson seem both to agree that what makes him most human is his godlike yet flawed freedom, especially his freedom to aspire to wisdom, not just to power. It is a freedom to make mistakes not only of evolutionary tactics but also with regard to judgments about the nature of Man.

Behind Wilson's deepest scientific questions is, I am tempted to argue, a tacit acknowledgment of the power and significance of the idea of what it means to be human, both in what it refers to (human nature) and what it is (a descriptive and yet formative cause of the thing it names). Once we recognize the persistence of that idea in Wilson's thought (if we dare do a little experimental brain-scanning in a conciliatory spirit), it offers a way to begin to conceive of some of the deeper mysteries of the evolutionary process in an unconventional light.

Of course, consilience is itself is a meta-empirical concept in keeping with Wilson's recognition that Bacon placed metaphysics above all, and that the Lord High Chancellor defined science ("broadly," Wilson says with qualified but carefully recorded acknowledgment and appreciation) as mystery-gathering poetry, and poetry as science. If the laws of physics govern everything, as Wilson asserts,[21] his evolutionary thinking is disturbed by an aspiration for Baconian capaciousness. This strange preservation and activation of a few Neoplatonic genes in this modern evolutionist is something that scholars of Bacon can point out, and wonder at.

One way to appreciate the particular importance of ideas for Wilson's project is to survey his discussion of the "moral sentiments," which the philosophers of the Scottish Enlightenment assumed were harmonious with Bacon's project and tantamount to the foundations of ethical, political, and religious thought and action. Wilson refers to moral sentiments in several crucial passages, though by placing his reference in quotation marks he ostensibly dismisses their standing as independent ideas–like those of discredited natural law and not-so-evident self-evident truths. Still, he proposes that they are "hereditary predispositions," empirically discoverable components and operations of persistent organisms and cultures:

> So perhaps we can do better [than natural law], by taking empiricism more seriously. Ethics, in the empiricist view, is conduct favored consistently enough throughout a society to be expressed as a code of principles. It is driven by hereditary predispositions in mental development–the "moral sentiments" of the Enlightenment philosophers–causing broad convergence across cultures, while reaching precise form in each culture according to

[21] Wilson, 23, 270, 266.

historical circumstance. The codes whether judged by outsiders as good or evil, play an important role in determining which cultures flourish, and which decline.[22]

In the work of Thomas Reid, Hutcheson, and Adam Smith, one finds a belief in an inborn and teachable sympathy, or an inherited moral sensibility. That belief, which draws from particular ideas of the origins and ends of human nature, enables the philosophers to investigate the persistence of moral understanding, its mechanisms of operation, and the ways in which it perpetuates itself through self-reflection and teaching. In the process of their investigations, the philosophers themselves of course participate in, and contribute to, the reflection and teaching that perpetuate their subject of inquiry. Whether or not they solve the problem of determining the source and principles of moral reasoning, they offer Wilson a way to envision Darwinian moral knowledge as a fusion of inherited dispositions and ideas that by means of wise choices literally *in-form* civil society with persistent characteristics. They keep open the possibility that what *is* and what *ought* to be are the same in the case of ideas and attitudes whose persistence is conducive to a morally prosperous and hence evolutionarily productive civil life.

The Scottish Enlightenment philosophers assumed that the moral sentiments were more than the results of purely inductive experience: they were inborn, hence God-given. For that reason, some called them "self-evident," in the sense that they existed for all to recognize without need of additional proof. Self-evidence was of course not the same as universal acceptance; moral culture was needed to draw out the moral sense and dispose the power of choice. But these philosophers tended to understand Bacon's system of induction in the context of its power to discover the nature and operations of the moral sense as something *within* human nature as well as beyond it. In the evidence of the moral sense's active existence, they found its connection to ideas of just and appropriate attitudes and behavior, which they drew out and examined as means of disposing the moral sense to form itself more consistently toward those ends. Thus the sentiment or emotion blended, from the beginning, with the idea of what it was and what it was for.

Emotion, rather than being essentially opposed to reason, was understood as a moral phenomenon, often in need of guidance and reprimand, but seeking in these things confirmation of its own nature–as moral sentiment in and of itself. As Adam Smith put it in the opening sentence of *The Theory of Moral Sentiments*, the claim can be made that there is a deep connection between natural sympathy and moral reasoning: "How selfish soever man may be supposed, there are evidently some principles in his nature, which interest him in the fortune of others, and render their happiness necessary to him, though he derives nothing from it except the pleasure of seeing it."[23] In Wilson's terms, it is a form of altruism that does not necessarily derive evolutionary benefits for itself yet is conducive to the survival of the genus.

[22] Wilson, 240.

[23] Adam Smith, *The Theory of Moral Sentiments* (Oxford: Clarendon Press, 1976), 9. See also Douglas Sloan, *The Scottish Enlightenment and the American College Idea* (New York: Columbia Teacher's College Press, 1971), 90–1 and *passim*; John Witherspoon, *Lectures on Moral Philosophy* (Newark: Univ. of Delaware Press, 1982), 78–9 and *passim;* Thomas Reid, *Essays on the Intellectual Powers of Man* (University Park: Pennsylvania Univ. Press, 2002), 547–5.

For Wilson, the moral sentiments are "moral instincts." They are of great interest to someone who is renowned for arguing that there is altruism among ants. The moral sentiments in human beings are "derived from epigenetic rules, hereditary biases in mental development." Emotion and judgment depend upon and help to enact these predispositions by means of choices that affirm, adjust, or break from previous practices, with the results–if they are successful over time–becoming encoded, through natural selection, in the instinctual apparatus itself.[24] It is as though there were a biological, empirically verifiable means of demonstrating that moral sentiments were part of the biological being, that they arose in the interaction of biology and culture and could be embodied in that being when they were articulated in successful evolutionary action. Wilson considers them so valuable to the genome and to human culture that he hypothesizes an instinct to preserve *them*:

> I predict that future generations will be genetically conservative. Other than the repair of disabling defects, they will resist hereditary change. They will do so in order to save the emotions and epigenetic rules of mental development, because these elements compose the physical soul of the species. The reasoning is as follows. Alter the emotions and epigenetic rules enough, and people might in some sense be "better," but they would no longer be human. Neutralize the elements of human nature in favor of pure rationality, and the result would be badly constructed, protein-based computers. Why should a species give up the defining core of its existence, built by millions of years of biological trial and error?[25]

Yet it might. Wilson concedes that the human mind "did not evolve to believe in biology." The choice to alter the genome would be by definition extra-biological, if only because mastery entails the power to change what is mastered.[26] For Wilson, only an intelligent understanding of human nature and the consequences of such an action should govern such a decision. But its dependence upon extra-biological reference points throws it back into the realm of ideas, and idolatry.

If wise understanding were to be based upon the most advanced observations and concepts of biology, it would still have to be comprehensible and operative, by Wilson's democratic standards, to an audience that is *biologically predisposed* not to believe or act upon the principles and findings of advanced evolutionary biology. Hence the mixed character of Wilson's book, which devotes most of its chapters to topics that are not ostensibly biological so as to serve this biological insight about the limitations of biological knowledge. Since the human mind "*evolved to believe* in the gods," at least until recent times (emphasis added), he must appeal to that mind precisely because its evolved belief in magic and religion has biological authority.[27] "Accepting [magic's] emotive power, we know something important about human nature," not merely that human beings have long been distracted by magic but that that distraction has had an empirical, evolutionary purchase on their persistence that

---

24 Wilson, 251, 254–5.

25 Wilson, 277.

26 Wilson, 262, 120.

27 Wilson, 262.

thus in some cases affected the genetic basis of human dispositions and of human culture.[28]

In these parts of Wilson's argument, as we have seen, there are flights of inference that both lower and enhance the status of religion, magic, and attendant ideas of human nature. How do these approaches to the evolutionary force of ideas engender conviction among those who do not accept that evolution should be the ruling paradigm? The most suggestive, almost hyperbolic instance of this double action is Wilson's passage on the free-willing self and the nature of the soul:

> The self, despite the illusion of its independence created in the scenarios [that compose the conscious mind], cannot exist apart from the body, and the body cannot survive for long without the self. So close is this union that it is almost impossible to envision souls in heaven and hell without at least the fantastical equivalent of corporeal existence. Even Christ, we have been instructed, and Mary soon afterward, ascended to heaven in bodies–supernal in quality but bodies nonetheless. If the naturalistic view of mind is correct, as all the empirical evidence suggests, and if there is also such a thing as the soul, theology has a new Mystery to solve. The soul is immaterial, this mystery goes, it exists apart from the mind, yet it cannot be separated from the body.[29]

Although the argument is empirical–ostensibly and mainly reductive–Wilson freely expands his frame of reference to include considerations he considers transcendental. He calls the free self an "illusion" and discusses the soul in terms of dogma. Yet he determines that the illusion, when it is embodied, is indispensable to the person's biological survival. And he reasons analogically that Christian teachings about the soul, insofar as there is such a thing as the soul, help establish a biological principle without contradicting theology: what we know of the soul, if it exists, is that it is embodied, be it in a "supernal" or organic mode.

Are such arguments about the nature of evolution convincing? Bacon is well known for making ingenious use of religion and Neoplatonic conceptions of the ideas to project his thinking in circumstances that threatened to brand it as a dangerous innovation that was false to religion and the old learning. Upon closer examination, his project seems to be steeped in religious and Neoplatonic ideas, so much so that it is often difficult to disentangle them from what we now refer to as Baconian science. In a turn that is strange to modern expectations, he maintains (whether from convention or genuine belief is not our concern here) the conviction that human beings are creatures of God, not their own makers or the product of fate. And he continues to entertain, to the end of his life, Neoplatonic and magical notions of nature that interest the relatively benighted general reader in the enterprise of the new learning. These survivals of the old learning not only make his works more plausible to benighted audiences; they reflect much of what they are about, and go some distance to justify his suggested innovations of accepted relations between religion and learning, magic and science.

---

28 Wilson, 237. Compare this transvaluation of magic to Wilson's stigmatizing use of the term in an earlier passage. Real change takes place, he writes, "not by magic" (213).

29 Wilson, 119.

Wilson's work makes a vaguely similar intellectual gesture, but the result is more obviously problematic. Bacon took pains not to tamper directly with the idea of divine prerogative, and indeed incorporated it into his idea of nature as a divine code. Neither did he reduce all magic or all forms of ideas to illusions. Wilson more pointedly, and sometimes adamantly, separates evolution from divine direction and frankly labels non-empirical transcendentals as illusions sometimes necessary for survival. His thorough investment in the modern scientific enterprise along with his belief in the democratic destiny of scientific discovery and the exchange of information make it difficult for him to imitate Bacon's courtly ways of layering and sometimes concealing deeper meanings so as to persuade various audiences of the reasonableness and compelling prospects of his cause.

Facing the challenge of making the new biology the foundation for decisions affecting the genome and the planetary environment, Wilson sends his own brand of mixed signal. He asks his audience of non-biologists to accept a form of consilience in which evolutionary biology takes precedence over religion, the tradition of substantive ideas, and the humanities in exchange for respect for their contributions thus far in the evolution of human beings. On the other hand, he changes the meaning of evolution by elevating aspects of culture such as religion, ideas, the arts, the humanities, and even certain forms of magic, to a functional status in some ways coequal and superior to that of the genes and the process of selection.

Material explanations will "never" in Wilson's view decisively remove the idea of the self and free will, which it seems are essential to the biological persistence of persons and cultures because "the individual mind cannot be fully known and predicted, the self can go on passionately believing in its own free will. And that is a fortunate circumstance. Confidence in free will is biologically adaptive. Without it, the mind, imprisoned by fatalism, would slow and deteriorate."[30] There is a crucial place, in fact an indispensable role, then, for transcendentals in scientific persuasion and evolution–as long as they are not transcendentals in their own right. Old, "self-evident" ideas, sometimes misunderstood in terms of non-evolutionary natural law, are responsible not only for "appalling failures" such as slavery and genocide, but also "noble successes"; Wilson needs them at least as much as he distrusts them. Greater intelligence and wisdom would have led, he argues, to more consistently positive evolutionary consequences if those transcendentals had somehow been understood–understood convincingly by those who have been taught to doubt the existence of transcendentals–in a different (Wilsonian) light.[31]

Success in this endeavor, as we have seen, depends for Wilson upon ideas of human nature and the human prospect that biology must validate but cannot adequately supply. To be scientifically respectable and persuasive, he must therefore rely upon transcendentals that he believes are embodied in the genes, the evolutionary history of human cultures, and by extension the contributions to culture that speculations about evolutionary biology make in the present. But this conviction is a

[30] Wilson, 120. For the argument that biological and scientific analysis will never fully master the phenomena it studies, see 67–8.

[31] Wilson, 239, 240.

faith, or at least a philosophical and moral argument that is dependent, for its persuasive expression, on Wilson's eloquent neo-Darwinism. Thus re-enter religion, the ideas, the arts, and the humanities in the courtly dress of neo-Darwinism. Wilson is in this sense a new Bacon, a courtier scientist attempting to unify persuasively what is empirical and transcendental in an evolutionary wisdom that knows it must be completed with a wisdom not normally considered scientific. Once unearthed and rehabilitated, will the *eidos* of that completing wisdom guide the new sciences in unexpected directions? Will the sciences become all the more consilient in this process by mastering the rest, or will they instead find themselves consilient in being *transformed by the transcendentals*, becoming strangely dependent upon forms of wisdom outside their expertise? The question remains.

Chapter 6

Francis Bacon and the Unity of Knowledge: Reason and Revelation

Jerry Weinberger

In his perceptive account of the modern Enlightenment and its subsequent and most recent tribulations, E. O. Wilson says that the grand architect of the project was Francis Bacon. According to Wilson, Bacon heralded the ideal of comprehensive, unified, and transformative science. In the Baconian scheme, science was defined more broadly and differently from today's ordinary conception to include a foreshadowing of the social sciences and parts of the humanities. To him experiment meant not just controlled manipulations in the manner of modern science. It was all the ways humanity brings change into the world through information, agriculture, and industry. He thought the great branches of learning were open-ended and constantly evolving, but he nonetheless focused eloquently on the underlying unity of knowledge.

Above all, Bacon was keenly aware of human psychology as crucial to science, as regards the impediments to seeing the world as it really is and as regards the ways and means of genuine discovery. It would not be exaggeration to say that Bacon was the first critical philosopher. For Bacon, it was essential first to understand the defects and pretenses of the human mind in order to liberate its real powers. Left to its own natural and historical devices, the human mind is powerfully inclined to grasp the world as our hopes would have it be, rather than as it actually is. The most pernicious effect of the "Idols of the Mind," in Bacon's view, was the persistent inclination to think of the world as a collection of formal causes held together by some final cause. This long-lived but sterile delusion in fact takes the world as a kind of god and presumes that it is especially here for us, as if we human beings somehow deserve to have a perfectly ordered home and road-map to life handed to us on nature's plate. But the spell of the Idols can be broken, and one first step was the coming of the "true faith" whose teaching was that the world is but the created object of divine art, will, and power. The final step–Bacon's Great Instauration–would free the mind from the idols, generate a true science of nature and give to human art, will, and power the means to conquer nature for the relief of "man's estate."

Bacon was, in Wilson's view, a rationalist all the way down. Once the mind is disciplined first by critique and then by method, and once as a result the latent courses of nature are revealed by experimental science, the human *estate* can be relieved. Bacon seems to mean that when the pressing needs of the body are relieved, so too are the tribulations of the soul–and so too are dispelled the vain illusions that roil politics and society. From this lofty rationalist hope, says Wilson, the Enlightenment project eventually came upon hard times. The Enlightenment first "stumbled" when it failed to resolve the conflict between reason and revelation, and this slip was followed by the related failure to "colonize ethics." The unity of knowledge which Bacon proclaimed was in fact cleft by the unbridgeable chasm of religion and morality, and natural science–with its reductionism and immutable laws–was thus indictable as a threat to human freedom and dignity. The resulting two cultures–the humanistic and the scientific–parted ways and, with the advent of postmodernism, are now actively at war.[1]

Wilson's nut-shell history is for the most part accurate, although it remains to be seen if the struggle between the West and radical Islam will take some wind out of the postmodern sails: it's not so easy to be flippant about such universalistic modern ideas as human rights when fanatical obscurantists are blowing up those who, as the postmodernists in their comfort really do, take them for granted. Wilson pokes some paradox-revealing fun at Derrida's famous *Il n'y a pas de hors-texte* (if it's true then we can't know what it means), but it seems to me that he misses an important point that Derrida and modern natural science have in common: neither has a single good argument to make against the claim that there is indeed something beyond the text (for postmodernism) or nature (for science) that really matters–the One True God who reveals his will, law, and providence to humankind.[2] This want of an argument is more embarrassing to Derrida than to science, since science can prove many things and Derrida can prove none. But then again, perhaps not: this want of an argument reminds us that faith and postmodernism are natural allies, since where the voice of reason is still, the call of inspiration is freed to claim its own special authority.[3] And faith–the call of the One True God–is the most serious impediment to the Enlightenment vision of victorious and unified knowledge.

Wilson argues that the Enlightenment's first stumble–its failure to defeat orthodoxy on behalf of deism (or atheism, if one wants to be tough about the matter)–sprang not from a rational flaw, but an emotional one. According to Wilson the Enlightenment was unable to address the psychological roots of orthodox faith. Enlightenment deism could not win the day because a pure reason is unappealing because it is bloodless, and "ceremonies stripped of sacred mystery lose their emotional force, because celebrants need to defer to a higher power in order to consummate their instinct for tribal loyalty."[4]

---

[1] E. O. Wilson, *Consilience* (New York: Knopf, 1998), 23–8, 32–4, 38–40. I have filled out this account with reference to Bacon's criticism of classical, and thus pagan, teleological approaches to nature. See *Advancement of Learning*, *SEH*, III: 350–5; *De Augmentis*, *SEH*, IV: 363–5; *Novum Organum*, *SEH*, IV: 64–6; and my "Metaphysics and Religion: Francis Bacon's Critique of the Ancients," in *Educating the Prince: Essays in Honor of Harvey Mansfield*, eds. Blitz and Kristol (Lantham: Rowman & Littlefield, 2000).

[2] Wilson, 40–4.

[3] As is the claim of mere will, which explains the embarrassing-to-postmodernists political engagements of Heidegger and Paul de Man.

[4] Wilson, 32–4.

Now, this very mode of argument, it seems to me, suggests that the real problem of the Enlightenment was *precisely* rational and *not* emotional. Wilson, the Enlightenment rationalist, can assert that the problem was psychological because he more fundamentally assumes a natural seed of religion (to use a phrase of Hobbes). All this proves, however, is that if one begins by assuming that inspiration is a natural phenomenon one can give it a natural explanation. This is of course to beg the argument in question: that the moral law is the miraculously revealed will of God and that the evidence of this miracle–the experience of the divine spirit moving within–is open only to the faithful and cannot be refuted by circular and dogmatic arguments that appeal only to the natural possibilities of matter in motion. Were Calvin alive today (and were he capable of a sense of humor), and if he read Wilson's essay, he might well have a justified smile on his lips.[5] The important conclusion is that the Enlightenment proceeded on the mere *faith* that inspiration is a natural phenomenon. And with this the Enlightenment gave up any serious claim for the self-sufficiency of reason and the unity of all knowledge.

According to Wilson, the first Enlightenment stumble occurred a full century after the "grand architect's" proposal for the great instauration of human knowledge. I want to suggest that Wilson's chronology is not correct as regards Bacon. Which is not to say that Bacon "stumbled" on the matter of orthodox faith. On the contrary. A case can rather be made, I wish to suggest, that Bacon was, perhaps more than any other great figure of the modern project, aware of the basic problem of faith, and of the possibility of an Enlightenment stumble, from the outset. I want to suggest that he considered this problem with greater seriousness than did any other thinker of the Enlightenment. The important point is not that Bacon's synoptic mind made him less than a "thoroughly modern man," as Wilson says.[6] Rather, the point is that Bacon's apparently Renaissance and humanistic view of enlightenment made for unusual prescience and realism on his part.

In this paper I argue that Bacon broke from the tradition of Christian philosophy that maintained the compatibility of faith and reason. In particular I will argue, in the company of no less a figure than Shaftesbury, that Bacon was a non-believer, even if he pretended to be otherwise.[7] Moreover, Bacon was a non-believer on rational grounds. The fundamental question for Bacon, as for his Enlightenment progeny, was the status of the miracles by which God supposedly communicates his will to humankind. If these miracles can be demonstrated to be impossible, then faith and the experience of divine inspiration can be shown to be merely natural (psychological) phenomena, just as Wilson supposes. Then the Bible can be revealed for what it really is: a tangle of fables, illusions, and outright frauds.

---

[5] See Calvin, *Institutes* (1.16.3); (2.2.11); (3.18.1); (3.21.1.).

[6] Wilson, 27.

[7] In the *Characteristics*, Shaftesbury argues that when Bacon said that inspiration arises from "an imperfection in the creation" he revealed himself to be an atheist. That one comment from Bacon was sufficient to reveal his true view, even though it floated in a sea of otherwise orthodox declarations of belief in the "true faith." No such believer can ever declare that the creation was imperfect; *Characteristics*, ed. Lawrence E. Klein (Cambridge: Cambridge Univ. Press, 1999), 368. Shaftesbury is here commenting on *De Augmentis* 2.13, *SEH*, IV: 325.

For the most part, Bacon's Enlightenment progeny tried to demonstrate the impossibility of miracles by the materialistic metaphysics and evidence of natural science. Bacon differed from his own followers, the ones who "stumbled," because he did not think the believers could be defeated on the grounds of natural science. To this extent he agreed with the tradition of Christian philosophy. But Bacon thought the believers could be defeated on other, but no less rational, grounds: on a critique of the background conditions for the belief in miracles–our common moral opinions and beliefs. Unlike the later Enlightenment rationalists, Bacon could engage the believers on their own turf, so to speak, and in this fact consists the real power of his rationalist critique of religion.

The place to begin is not *The Advancement of Learning*, but rather the much later and posthumous *New Atlantis*, or as Wilson puts it Bacon's famous "utopian fable about a science-based society." The *New Atlantis* is indeed about a science-based society. The scientific academy depicted in the story is said to be the "lanthorn" of Bensalem and is dedicated to the study of the material things above the earth, below the earth, and of course on the earth. Bacon's description is consciously meant to remind us of the charges brought against Socrates and described in the *Apology*: that Socrates was a natural philosopher who studied the things aloft and under the ground and, for that reason, did not believe in the gods.[8] We should not be surprised, therefore, that the *New Atlantis* is as much or even more about revelation than it is about natural philosophy.

The entire story is about the revelation of the hitherto secretive and unknown Bensalem to the rest of the world. In addition to founding the scientific establishment, the lawgiver Solamona had also promulgated laws that prohibited the entrance of strangers and that allowed small and specially appointed parties of Bensalemites to spy on the rest of the world while remaining unknown to those among whom they traveled. These laws, in force for almost two thousand years, are abrogated at the end of the fable when the Father of Salomon's House gives the narrator "leave to publish it [the relation of the scientific establishment] for the good of other nations; for we here are in God's bosom, a land unknown." The *New Atlantis* is, one could well say, Bacon's version of the new Gospel, the Good News that ushers in the modern age.[9]

But the *New Atlantis* is about revelation in another and far more orthodox sense as well. In the account of Bensalem's history Bacon quite astonishingly rearranges the history of science and the history of revelation. King Solamona founds the institute for discovering "the true nature of all things" as a consequence of learning from the Hebrews that God created the earth and all things in it in six days. Natural science thus well precedes the advent of Christianity but depends on the teaching of the Old Testament. And just twenty years after the resurrection and ascension of Jesus, the Bensalemites receive by a miracle not just all the books of the Old and New Testaments then extant, but also some New Testament books (at least Acts, Paul's Epistles and the Revelation to John) that were "not at that time written."[10] In other

[8] *New Atlantis and Great Instauration*, ed. Jerry Weinberger (Arlington: Harlan Davidson, 1989), 71–2; Plato, *Apology* 18 b–c; 19b–d. See Aristophanes *The Clouds* 218–25, 365 ff.

[9] *New Atlantis*, 46, 50–1, 56–7, 83.

[10] *New Atlantis*, 47–4.

words, the revelation of the Bible to the Bensalemites, presented as an episode within the revelation of Bensalem to the world, *includes an event of original revelation of the New Testament word of God as we have it.* In addition to this amazing fact, the Bensalemites are able to read all of the material revealed by way of a second miracle of tongues.

At first glance this mangling of the historical record serves the literary function of emphasizing the near-perfect harmony among natural science, the relief of man's estate and the truths of Biblical revelation. This comports, it would seem, with Bacon's claim in *The Advancement of Learning* that the coming of the true faith, with its argument that the world and all things in it are but created things and not themselves divine, was essential to free the mind and true natural science from the natural teleology of ancient philosophy, which was in fact under the spell of ancient paganism.[11]

It is often argued that Bacon thought religion and natural science could get along just fine. Thus as Perez Zagorin writes, following Paul Kocher, "Bacon's attitude was typical of the English thinkers of his time who took an interest in science. All of them held that religion and science should be kept separate and also that the two were complementary." As a consequence, the "distinction [Bacon] assumed between faith and reason was a very old one, common to many Christian philosophers. Certainly it need not have implied any doubt of the supernatural truths of the Christian religion as known through revelation."[12] Science and religion are based on different kinds of evidence, and the true faith and natural science are not in contradiction and can rule each in its own sphere. Careful inspection of Bacon's texts, however, reveals a much more complicated–one might say obscure–presentation of the issue than is compatible with this neat and common scholarly view.

It is clear on its face that the Bensalemites think science and miracles are compatible. However, they do not at first seem to agree with the view that science and religion inhabit different evidentiary realms. On the contrary. Not only do the Bensalemites receive Christianity by way of an original miracle of revelation, but they expect the scientists of the House of Salomon to *verify* the miracle as genuine.

A wise man of Salomon's House, who as it "fell out" was in one of the boats that approached the pillar of light announcing the newly revealed texts, says in prayer that the "Lord God of heaven and earth 'hast vouchsafed' to those of our order, to know thy works of creation and the secrets of them; and to discern (as far forth as appertaineth to the generations of men) between divine miracles, works of nature, works of art, and impostures and illusions of all sorts." He proclaims the miracle to be genuine and then says that "as we learn in our books that thou never workest miracles but to a divine and excellent end, (for the laws of nature are thine own laws and thou exceedest them not but upon great cause) we most humbly beseech thee to prosper this great sign and to give us the interpretation and use of it in mercy; which thou dost in some part secretly promise by sending it unto us."[13]

---

[11] See n. 1 above and *The Advancement of Learning*, *SEH*, III: 264–8. See also Perez Zagorin, *Francis Bacon* (Princeton: Princeton Univ. Press, 1998), 44–51.

[12] Zagorin, 49.

[13] *New Atlantis,* 48.

Now the only way for science demonstrably to verify a miracle would be for science to have grasped the entirety of nature and to have revealed each and every one of its possible motions. Absent this absolute grasp of nature, it could always be said that we just don't know for sure if an apparently miraculous event is miraculous–we could rather only say that it might be, but that it might also just be a natural event that natural science cannot yet explain. By speaking of the limits of the generations of men the wise man's prayer leaves it unclear as to whether such an absolute science is possible. But whatever the case may be, the Bensalemites would care about the matter only if they were somehow aware that things can be and are thought of as miracles when in fact they are not: sometimes men take things to be miracles when they are in fact natural phenomena, or works of art, or impostures, or illusions.

In the first case simple ignorance of causes is at work. Thus as science progresses, fewer of these honest mistakes will be made. But the case is really different for mistakes based on art, imposture, or illusion. Why so? Because, as we learn in the description of Salomon's House, as science progresses so too does the power of the scientists to create all manner of art and illusion, including the pillar of light that heralded the revelation of scripture to Bensalem. Indeed, according to the narration the scientists of Salomon's House also "have so many things truly natural that induce admiration" that they could "in a world of particulars deceive the senses, if [they] would disguise those things and labour to make them seem more miraculous."[14]

As Bensalemite science progresses, the scientists have ever more power to deceive the senses and thereby to fake miracles, despite the fact that they claim to forbid the use of this power. We cannot reflect on this fact without suspecting that the revelatory miracle and even the gift of tongues, both so central to the *New Atlantis*, are artful fabrications of the Bensalemite scientists. Thus it's not so clear that in Bensalem science, religion and miracles go merrily hand in hand. One might rather say that in Bensalem only the wise–and not the charlatans and false prophets–will have command over the frauds that miracles always are.

Bacon adds to this disturbing suspicion with another important twist. Although the *New Atlantis* is a revelation of the formerly secret Bensalem, and although Bacon tells us quite openly about the scientific capacity to make natural things look like miracles, and tells us with equal candor that fakery in this regard is punished by the scientific establishment with ignominy and fines, some secrecy still remains at the heart of the scientific establishment even after its existence and structure have been revealed.

According to the Father of Salomon's House, the members of the scientific estate "have consultations, which of the inventions and experiences which we have discovered shall be published, and which shall not: and take all an oath of secrecy, for the concealing of those which we think fit to keep secret: though some of those we do reveal sometimes to the state, and some not."[15] The scientists judge by themselves what not to reveal, and of those some are kept even from the highest political authorities.

At the very least, we can say that for Bacon a science-oriented society and telling all and nothing but the truth are not the same things. Of course we have to wonder

[14] *New Atlantis*, 77–8, 80.

[15] *New Atlantis*, 80–82.

what these secrets might be. Perhaps just inventions that are so powerful and dangerous that they can be known about and possessed safely only by the most wise, which does not include even the peak of the political establishment. This must certainly be a plausible candidate. But equally plausible might be some experience having to do with religion–perhaps the "invention" at work at the time of the revelation of the Bible to the Bensalemites. This question of secrecy turns out to bear on the core of Bacon's argument about religion. As it turns out, the secrecy that really matters is not that practiced by the Bensalemite scientists. It might be a good thing for them to keep secrets about faking miracles, if indeed that is what they have done. But that secrecy would not settle the matter of the possibility of miracles. The case is otherwise, I think, for the secrecy practiced by Bacon himself.

In *The Advancement of Learning*, Bacon also has an important place for secrecy in his Great Instauration. In the discussion of the "method of tradition," Bacon describes and accepts the ancients' practice of the "acroamatic" or "enigmatical" method. The ancients, says Bacon, observed the distinction between the exoteric and acroamatic method "principally in the publication of books." Bacon, however, "transfer[s]" the distinction to "the method of delivery." The ancients used the acroamatic method "with judgment and discretion" but in later times it was disgraced by those seeking only to put forth counterfeit merchandise. The intention of the method, however, "seems to be by obscurity of delivery to exclude the vulgar (that is the profane vulgar) from the secrets of knowledge, and to admit those only who have either received the interpretation of the enigmas through the hands of teachers, or have wits of such sharpness and discernment as can pierce the veil" (*SEH*, III: 404–5; IV: 450–1).

Obviously shocked by the thought that Bacon might tell a fib or two, Spedding commented in his Note B to the Preface of the *Novum Organum* that, while in the passage from the *De Augmentis* Bacon alludes to the ancient transmission of knowledge "as a secret from hand to hand," there is "no evidence that Bacon approved of it" (*SEH*, I: 107–13; the quote is from page 112). This conclusion rather ignores the evidence at hand, however, since Bacon clearly goes out of his way to distinguish between the ancients who used this method with "judgment and discretion" and those in later times who have disgraced it. What else can this observation mean but that it is a method that *can* be used with judgment, discretion and intellectual respectability?

In speaking of the enigmatical method as the transmission of knowledge as a secret from hand to hand, Spedding distorts Bacon's account of the method. What Bacon "transfers" to the method of transmission is not the use of unpublished and secret texts passed among some initiated elect, and not even a secret teaching passed among such an elect. It is, rather, the disclosing of a teaching below the obscure surface of an openly published text and available for *anyone* to see who has enough wit to pierce the veil. Moreover, Spedding fails to mention here that according *The Advancement of Learning* government is "part of knowledge secret and retired, in both respects in which things are deemed secret; for some things are secret because they are hard to know, and some because they are not fit to utter" (*SEH*, III: 473–5). And contrary to Spedding's note in the *Advancement*, referring to *De Augmentis* 8: 1–3, Bacon there hardly "omits altogether" the remarks on the secret nature of government (*SEH*, III: 473). On the contrary, in the *De Augmentis* Bacon actually amplifies the subject of secrecy and the art of government (*SEH*, V: 31–2).

He begins Book Eight of the *De Augmentis* by referring to the story of the philosopher's silence before the ambassador of a foreign prince who wanted to sample the wisdom of Greece. Bacon says that since he has come to the art of government and is writing to a king who is a master of the subject, he will be like that philosopher (and Cicero and Pindar) and hold his tongue. Then, after dividing civil knowledge into its three parts (conversation, negotiation, and empire, or government) and discussing the first two (including Machiavelli's "evil arts"), Bacon says again that as regards civil government he has imposed silence on himself (*SEH*, V: 78–9). This time however, Bacon says nothing of the need to be silent before the king, but rather and with obvious irony says that he himself "might not be entirely unqualified to handle such topics with some skill and profit." Even so, he says, "if my leisure time shall hereafter produce anything concerning political knowledge, the work will perchance be abortive or posthumous."

In the meantime, says Bacon, lest the eminent subject of government be left vacant from his account of the sciences, he has "determined to mark as deficient only two parts of civil knowledge, which do not belong to the secrets (*arcana*) of empire, but have a wider and more common nature; and according to my custom I will set down examples thereof." Bacon could not be clearer: the discussion that follows (an almost exact translation of "Of the True Greatness of Kingdoms and Estates") pertains to those aspects of government that are *not secret*–which means that there are parts of the science of government that are indeed secret. Bacon's silence is not just deference to the king; it must be, as he says in *The Advancement of Learning*, that some parts of the science of government are secret and retired because they are too hard to know or not fit to utter.

The important question is thus not whether Bacon knew and approved of esotericism. He clearly did. The important questions are rather whether and why Bacon thought esotericism necessary, and if so whether he practiced it himself. It has not been easy for scholars to grasp these questions firmly by the horns. Thus Perez Zagorin (following Spedding) admits with obvious reluctance that Bacon did not "repudiate" esotericism and that despite not being "in favor of secrecy" Bacon "did want to direct his work to the fit and receptive minds to avoid the misconstruction and abuse of his ideas and continued to feel a lingering allegiance to the principles of esotericism as a way of safeguarding his philosophy from incomprehension or misuse by the hostile or unworthy."[16]

But when push comes to shove, Zagorin rejects with contempt contemporary studies of Bacon that reflect the "unsubstantiated thesis that Bacon was secretly irreligious and that he used an enigmatic method of communication to prepare the way for the introduction of a progressive new political and social order." This view, scoffs Zagorin, reflects the "wearisome method of reading premodern texts of political philosophy based on the often arbitrary assumption that they contain an esoteric

---

16 Zagorin, 43–4; cf. 184. There is alas no evidence to support the apologetic claim that Bacon kept his "allegiance" to esotericism *despite* not being in favor of secrecy.

meaning intended for select, initiated readers, which their authors concealed under the apparent or surface meaning."[17] It's no "arbitrary assumption" to think that a book might be written esoterically, especially since this assumption was clearly shared by Bacon himself. And there is no "initiation" required. All one needs is patience and a skeptical and open mind. The proof then depends wholly on the reading and the handling of the evidence. But one thing for sure, as Bacon himself clearly states: the reading and the evidence will be precisely "wearisome," at least for those wits not patient enough to pierce the veil. That's the whole point.

So if Bacon says that the ancients wrote esoterically and did so with judgment and for good reason, and if he says himself in *Valerius Terminus* that esotericism is "not to be laid aside, both for the avoiding of abuse in the excluded, and the strengthening of affection in the admitted" (*SEH*, III: 248), and if government is a part of knowledge that is secret and retired because it is too difficult and not fit to utter, then it's purely dogmatic and stubborn to scoff at the idea that Bacon wrote esoterically and that he did so at least in part about matters of political and social orders. In the case of government and esoteric writing, what looks like silence will in fact be a teaching to be found behind a veil.

Moreover, Bacon did not think government the only topic for which silence is called. He states just as clearly that it applies to "sacred or inspired divinity" as well. In *De Augmentis* 9: 1, Bacon says that if he proceeds to treat of this divinity he will "step out of the bark of human reason, and enter the ship of the church; which is only able by the Divine compass to rightly direct its course. Neither will the stars of philosophy, which hitherto have so nobly shone upon us, any longer supply their light. So that on this subject also [*quoque*, not *idem*] it will as well be as well to keep silence" (*SEH*, V: 111; see *SEH*, I: 829). Bacon's expression here suggests that he continues the earlier reference to the reasons for practicing silence in the matter of government. But apart from this stylistic consideration, it is odd in itself that he would so coyly recommend silence on a topic about which reason has nothing to say.

We might say that he means simply: reason has nothing to say about divinity, so I have nothing to say and hence my silence on the matter. Or he could mean that talking about a topic on which reason has nothing to say could lead to confusion and error. Why then not say this directly and straight away? And why speak narrowly of "the Church" and not more broadly of religion? If the Church needs the "Divine compass," does it follow that reason does as well? So Bacon could just as easily mean that he keeps silent for the reasons already discussed: that what reason might have to say on the matter is too difficult and not fit to utter, and not fit to utter especially to those who won't like what reason might legitimately have to say about God's revealed word.

This speculation, provoked by the reference to silence and Bacon's coy reference to reason's having nothing to say about the Church, quite reasonably *at least comes to mind*. At the very least, Bacon raises the issue of reason's relation to faith, and he raises it in such a way as to make us wonder, as we do from considering the *New Atlantis*, if reason and faith are as distinct as Bacon often says they are. At the very

[17] Zagorin, 264 n.56; see also 49, 242 n.75

least, if the reasons for secrecy with respect to religion are the same as for government, then it could make perfect sense for Bacon to lie about the separateness and easy coexistence of religion and science.[18]

According to Bacon's account of the enigmatical method, one can keep silent and at the same time have one's say. By using deliberate obscurity as a veil, a writer can hide a truth from those who will not like it and reveal that same truth to those with sufficient wit to pierce the veil. There is, if we look with care, considerable obscurity in Bacon's teaching about reason and revelation. First, Bacon makes a remark that reveals him to be an atheist. The remark is obscure because it floats in a sea of commonplace pieties, but once we see it there can be no doubt about what we are to conclude. I was drawn to the remark by none other than the Third Earl of Shaftesbury, who, in the *Characteristics*, says the following of Bacon on the matter of revelation in *De Augmentis* 2:13: "It was a good fortune in my Lord Bacon's case that he should have escaped being called an atheist or skeptic when, speaking in a solemn manner of the religious passion, the ground of superstition or enthusiasm (which he also terms a panic), he derives it from an imperfection in the creation, make, or natural constitution of man."[19]

Shaftesbury has a good point as regards Bacon's atheism: Were Bacon an orthodox believer he could not have made the argument about the religious passion having a natural cause in the ill-constitution of the human being. But lest we think that Bacon thought such a natural explanation settles the matter of revelation (as his protégé Hobbes seems to have thought), Bacon indicates powerfully to the contrary in the *New Atlantis*, to which we have to return.

The central event of the story in the *New Atlantis* is the original event of revelation that brings Christianity to the scientific Island of Bensalem. The crucial element of the story, however, is Bacon's claim, put in the mouth of one of his characters, that the Bensalemites believe that only a scientist, who understands the courses of material nature, can authenticate the miracle and so mark it off from a natural occurrence, an illusion, or a deliberate fraud. But as we've already seen, Bacon also tells us in the same text that as science progresses, and so gets better able to discern miracles, it also secures increasing powers to manufacture illusions and make natural events seem miraculous. So at the very least, the non-scientists have to believe what they are told, not assent to what they know. And so for them the evidence for the miracle is the conscience within, not the knowledge of matter in motion.

But then of course the same has to be so for the scientists as well–if we think for just a minute about Bacon's bizarre claim about scientific verification. That verification could work only at the absolute end of science–we could know scientifically that, when Jesus walked on the water, a real miracle occurred, only if we could rule out any natural explanation for the fact. And to do that, we would have to understand all the courses of nature. That's a tall order, and until it is met both the scientists and the non-scientists can only rely on the evidence of conscience (the spirit) moving within.

---

[18] Zagorin, 49.

[19] *Characteristics*, 366–8. Shaftesbury says this after concluding that More could be so accused as well, were it not that he was a "revered father of the church."

But even if that order were met, there would still be room for the evidence of conscience. It could not be ruled out that the regular courses of nature are themselves the product of miraculous power. So that miracle could be revealed only by the evidence of conscience moving within. And if the scientists discovered a fraud, they could not rule out the possibility that God miraculously caused the fraud to serve some providential end. Again, that miracle too could be revealed only by the evidence of the conscience moving within. And likewise for some probabilistic medical cure. Could not God cause the odds to favor this person rather than that person–don't perfectly sensible people pray for the scientific doctor to be the tool of God's miraculous intervention? The vulgar Bensalemites think science can verify miracles. But a moment's reflection shows them to be deluded on this point: first, because they may be prey to scientific fraud; and second, because it's not the voice of nature, but rather the voice of conscience, that ultimately discerns between such frauds and the true word of God.

So here is where we stand: on the one hand, Bacon makes the atheistic argument from the natural cause of inspiration. But on the other hand, he takes great pains in the *New Atlantis* to make us see that the evidence of science and the evidence of faith are indeed not the same. The scoffer says the believers are nuts by assuming that their visions have only material causes. That bit of question begging won't convince the thoughtful believer (say, Calvin) who can say that the scoffer's lack of faith simply blinds him to the evidence that really matters: the miraculous moving of the spirit within, the miraculous call of conscience that when experienced is as real as anything made, say, of meat, water, blood, and nervous tissue.

The problem with the scoffers, so it seems, is that they cannot see, much less leap, the evidentiary divide between reason and faith. And Bacon, the scoffer, tells us about this fact. As it turns out, however, Bacon sees a way to leap the evidentiary divide and, in the name of reason if not natural science, sees a way to engage the believers on their own turf: the evidence of conscience moving within. To see how, we have to consider his more-than-a-little obscure discussions of natural theology and theology as such.

Bacon often argues that mixing religion and philosophy was the great error of the ancients. It is certainly an error to think that nature is somehow itself a god, as the ancient pagans thought and ancient philosophers effectively assumed with their teleological approach to nature. To prevent this error, it is important, says Bacon, to acknowledge with the true faith that since nature and God are entirely different things (nature is created, God is not), we can learn little, and should expect to learn little, about God from the study of nature, and vice versa.

In *The Advancement of Learning*, Bacon tells us that natural theology is the "knowledge concerning God which may be obtained by the contemplation of his creatures." But all this knowledge can achieve is to "convince atheism." It can convince us that a God exists, but it tells us nothing about divine will, "because no light of nature extendeth to declare the will and true worship of God. For as all works do shew forth the power and skill of the workman, and not his image; so it is of the works of God; which do shew the omnipotency and wisdom of the maker, but not his image" (*SEH*, III: 349–51).

Bacon scarcely finishes his comment, however, before then qualifying it so far as to establish a patent contradiction: "Wherefore by the contemplation of nature to induce and enforce the acknowledgement of God, and to demonstrate his power, providence, and goodness, is an excellent argument and hath been excellently handled by diverse." Whoa! Providence and goodness are no minor aspects of God's will and they surely bear on why God made the world and human beings. In *De Augmentis* 3:2, Bacon makes this contradiction even stronger: "Wherefore that God exists, that he governs the world, that he is supremely powerful, that he is wise and prescient, that he is good, that he is a rewarder, that he is an avenger, that he is an object of adoration–all this may be demonstrated by his works alone" (*SEH*, IV: 340–2).

It makes a big difference to know that God is good and that he rewards and punishes, as opposed to being a malevolent spider whose arbitrary sting we should do what we can to avoid. "On the other side," continues Bacon, "out of the contemplation of nature and elements of human knowledge to induce any conclusion of reason or even any strong persuasion concerning the mysteries of faith, yea, or to inspect and sift them too curiously and search out the manner of the mystery, is in my opinion not safe." So reasoning on nature does, but does not, disclose absolutely crucial aspects of divine will. What, we thus wonder, is the remaining domain of mystery?

To see our way clearly at this point we have to consider more carefully the account of theology in *De Augmentis* (*SEH*, V: 111–19). Bacon starts by making a powerful case for the difference between faith and reason: "the more discordant therefore and incredible the Divine mystery is, the more honor is shown to God in believing it, the nobler is the victory of faith; if we will truly consider it, it is more worthy to believe than to know as we now know. For in knowledge man's mind suffers from sense which is the reflection of things material, but in faith the spirit suffers from spirit which is a worthier agent; Wherefore we conclude that sacred theology ought to be derived from the word and oracles of God, and not from the light of reason, or the dictates of reason" (*SEH*, V: 112–13).

If the scoffer (such as Wilson and Hobbes) merely presumes a natural seed of all religion, then the evidence of faith is simply assumed away; for the faithful that evidence, which does not even count to the dogmatic scoffer, is everything. Bacon then says that this distinction between faith and reason holds not just for God, creation, and redemption, but also for the "more perfect understanding of the moral law," which makes us love our enemies, do good to those who hate us, and so on, so that we will be like God who gives rain to the just and the unjust. Of these words about the perfection of the moral law, says Bacon, it could well be responded as did Virgil "'that they do not sound human'; since it is a voice beyond the light of nature." To explain what he means here, Bacon recurs (again) to "the heathen poets" who describe laws and moral doctrines "as if they were contradictory and malignant to the liberty of nature." The wise men of Greece–according to Bacon's allusion, Socrates, Pythagoras, and Diogenes–were described by Dendamus as being "excellent men" but with the fault of having "too great reverence and veneration for a kind of phantom, which they called laws and manners."[20] "Wherefore it must be confessed,"

[20] See Plutarch, *Alexander*, 65.

concludes Bacon, "that a great part of the moral law is higher than the light of nature can aspire to" (*SEH*, V: 113).

Bacon's argument here is very compressed, we might even say obscure, and depends on his allusions to others. But if we think about it carefully, it becomes clear that he refers here to the ancients' distinction between nature and convention (law), and to the dispute about whether there is or is not some non-materialist natural foundation or measure of what the materialistic natural philosophers called "merely" conventional and which public opinion took to be the commands of the gods.

Our natures strive for self-preservation and pleasure while the moral law demands self-denial and hardship. That moral law is either mere convention (says the materialists); or, with Socrates, is based on a proper metaphysical understanding of nature (*eidos*); or, says public opinion, is based on the will of the gods. Bacon then weighs in on just this dispute–*roughly* on the side of the wise men such as Socrates and against the materialistic natural philosophers: "Nevertheless what is said, that man has by the light and law of nature some notions of virtue and vice, justice and injustice, good and evil, is most true."

To explain this fact, Bacon distinguishes between two different meanings of the "light of nature." The first sense is that light as it "springs from sense, induction, reason, and argument according to the laws of heaven and earth." The second sense is that light as it "flashes upon the spirit of man by an inward instinct, according to the law of conscience; which is a spark and relic of his primitive and original purity. And in this latter sense chiefly does the soul partake of some light to behold and discern the perfection of the moral law; a light however not altogether clear, but such as suffices rather to reprove the vice in some measure, than to give full information of the duty. So then religion, whether considered with regard to morals or mysteries, depends on revelation from God."

With this complex set of moves, Bacon makes the following argument: so far as the knowledge of the motions of earthly and celestial matter is concerned, reason as the light of nature discloses nature in opposition to the commandments of the moral law. But we can and do have unassisted knowledge–which we can also call the light and law of nature–of morality as the law of conscience. This "light of nature" cannot be from the Socratic (Platonic) metaphysical *eidos* as the form of material nature, however. Of Plato's forms, Bacon has said explicitly that they were an infection of natural philosophy with theology (*SEH*, III: 293, 355–6). Where Plato looked in vain to the forms of physical nature, Bacon looks to the call of conscience.

Even though that conscience emanates ultimately (but also supposedly) from God, we can grasp it with the light of our natural reason. We can thus grasp virtue and vice, justice and injustice, good and evil, by the unassisted efforts of reason. This latter is the flash of the light of nature "upon the spirit of man by an inward instinct." We can, with unassisted natural reason, grasp and make sense of our moral intuitions. We need divine revelation–God's word and oracles–because our rational moral knowledge is, though true, not fully informative of our particular duties. This unassisted and natural knowledge of morality, not the contemplation of material nature, is the only natural and unassisted knowledge that reveals God to be good, a rewarder and an avenger. The Greek wise men were correct to think that we can have natural knowledge of the moral law; but that knowledge comes from thinking about the law itself, and not, as the wise men erroneously assumed, from thinking about the eidetic forms of material nature.

But why does Bacon again make us work so hard to make sense out of his obscure presentation of religion and the rational light of nature? I think for the following reason: if the problem of religion concerns its potential conflict with reason, and if it is therefore prudent to practice the acroamatic (enigmatical) method of transmission about the matter, then his obscure exposition of this crucial issue has the tendency to make us impatient with the obscurity and conclude (with Bacon's other pronouncements and with scholarly opinion and even with the *New Atlantis*) that reason and science stand on one side of an evidentiary divide and faith and inspiration stand on the other. If we move patiently however, we can see that as Bacon proceeds the divide becomes fuzzy to say the least.

Once we finally figure out what Bacon means by the light of nature–that we do have rational knowledge of matter in motion *and* rational knowledge of morality as the substance of conscience–we discover that we do have important natural, i.e. *rational knowledge*, of the will of God: that God is good and is a rewarder and an avenger. We likewise have rational knowledge of what things are good and bad and which things deserve reward and punishment. So what then remains as a mystery that must simply be believed? It is the perfection of the moral law as revealed by the Gospels: that we should love our enemies, do good to those who hate us, and approve of the fact that God sends his rain to the just and unjust alike.

What is Bacon saying here? Certainly not that the God of the Gospels refuses to judge. We know that God does. We learn this fact from the same chapter of Matthew from which Bacon quotes the perfection of the moral law: those whose righteousness does not exceed that of the scribes and Pharisees will never enter the kingdom of heaven, and those who do not remove a bodily member that causes them to sin will have their whole bodies cast into hell.[21] We can understand these latter words and oracles of God because we understand them with the unassisted light of reason. What we do not know is why God is to judge and not us, and why God lets the unjust in this world drink along with the just or even, as more often happens, drink while the just die of thirst. But God does judge as the righteous might do on this earth. God's ways and means on this earth are a mystery. But God's will *in the end* is not: justice.

With this conclusion we come to the crucial step in Bacon's argument about faith and reason as it transpires in *De Augmentis* 9:1. In the sequel to the conclusion that "sacred theology" should be derived from the divine word and not from the light of nature or reason, Bacon immediately relaxes the strict distinction between reason and faith. He argues that the Christian faith is, as regards the use of reason in spiritual things, midway between the heathens, who had no constant belief or confession, and the religion of "Mahomet, that interdicts argument altogether." Thus the Christian religion "both admits and rejects the use of reason and disputation, but according to just limitations" (*SEH*, V: 113–14).

[21] Matt. 5: 20, 29–30.

In the very next paragraph, however, the distinction between reason and faith is more than merely relaxed. It turns out that revelation is impossible unless it is intelligible to the light of reason. Bacon identifies two uses of reason in religion: one regarding the explanation of the mystery, and the other regarding the inferences to be derived from it. Of the former, Bacon says: "we see that God vouchsafes to descend to the weakness of our apprehension, by so expressing his mysteries that they may be most sensible to us; and by grafting his revelations upon the notions and conceptions of our reason; and by applying his inspirations to open our understanding, as the form of the key to the ward of the lock. But here we ought by no means to be wanting to ourselves; for as God uses the help of reason to illuminate to us, so we should likewise turn it every way, that we may be more capable of receiving and understanding his mysteries; provided only that the mind be enlarged, according to its capacity, to the grandeur of the mysteries, and not the mysteries contracted to the narrowness of the mind" (*SEH*, V: 114).

With this last comment, Bacon makes a very simple but telling point: if God speaks to men, he doesn't just babble or speak in some divine language that is beyond human understanding. And what he says, as well, has to matter to men for them to care: we would not I think hearken to a God who did nothing but sing "Three Blind Mice" (although we might fear him on account of his obvious insanity). It's all well and good for Bacon to say that we have to enlarge reason to the grandeur of the mysteries rather than contract the mysteries to the narrowness of the mind. But the fact is that the full scope of reason is the full scope of reason; however much we strive to grasp the mysteries, we still must do so within the horizon of rational intelligibility–they still have to matter and be relevant to our cares.

So what do we know of revelation by the light of reason and nature, according to Bacon? However much the ways of God on this earth may be mysterious–however much he lets his rain fall on the unjust and asks us to forgive and love our enemies–in the end his care for and miraculous communication with us, by way of the conscience, concerns one perfectly intelligible matter: justice, rewards for virtue, and punishments for vice.

The movement of the spirit within is a mystery. But what that spirit says is not. There can be no mystery in the ultimate end of this divine government of the world: the righteous who may suffer on this earth are rewarded in heaven; the evil who prosper on this earth and remain closed to salvation are punished after death. We may be able to conceive of a God who is just a cosmic Mafia boss; but we really don't believe, Bacon forces us to conclude, that divine rule is *in the end* anything but just (thus in the *New Atlantis,* the wise man of Salomon's House says, when he verifies the miracle, that it operates to a "divine and excellent end.).[22] With this conclusion, Bacon has made his way to the believer's own evidentiary turf: the voice of conscience and its promise of justice. About that promise reason is perfectly well able to consider whether it does or does not make sense.

In *The Advancement of Learning*, Bacon tells us that divine government, just like human government, is secret: "We see the Government of God over the world is

---

[22] *New Atlantis,* 48

secret, insomuch as it seemeth to participate of much irregularity and confusion" (*SEH*, III: 474). Now, if the knowledge of human government is secret and retired either because it is too hard or not fit to utter, which of these reasons applies in the case of God's secret government of the world? We would at first certainly think, from considering Bacon's expression, that it must be secret because it is too hard for us, although not so for God, to whose mind what is dark and shady to us is a crystal. We don't know the ways and means of God's Providence in this world: we don't know why the good suffer and the evil prosper on this earth, even if we can be sure that the inscrutable divine means work ultimately toward righteous ends. Moreover, we can't think it secret because it is in fact not fit to utter. Bacon has already forced us to consider that we cannot really believe that God is a malevolent Mafia boss or spider.

That said, could there be something else about divine government that might be unfit to utter, if not by God, then by Bacon? If so, then it would be necessary to transmit it by the method that avoids the "abuse in the omitted" inflamed by what they would not like to hear. Does Bacon ever indicate such a teaching? I think he does. And as it turns out, Bacon suggests that the voice of conscience, which on its own terms must claim to make sense, really doesn't.

In both *The Advancement of Learning* and *De Augmentis*, in the discussion of the "architecture of fortune," Bacon contrasts Machiavelli's evil arts of advancement to the good arts of advancement. In the course of the long and detailed argument, Bacon cuts right to the heart of the issue between moral virtue commonly understood and Machiavellianism, which Bacon sums up as the principle that one should "seek not to attain virtue itself, but the appearance only thereof; because the credit of virtue is a help, but the use of it is cumber." Replies Bacon: men should not practice the wicked arts but should remember that worldly success is but vanity and vexation of the spirit and "that being without well-being is a curse, and the greater being the greater curse, and that all virtue is most rewarded and all wickedness most punished in itself" (*De Augmentis* 8:2, *SEH*, V: 76–7; *The Advancement of Learning, SEH*, III: 472). He then explains this by saying "as the poet excellently says":

> Quae vobis, quae digna, viri, pro laudibus istis
> Praemia posse rear solvi? Pulcherrima primum
> Dii moresque dabunt vestri.
> ("What's worthy, you men, what prize for such laudable deeds?
> How can we pay you? Gods must give you the finest rewards first–
> and your own character.").[23]

This quote from Virgil is meant to illustrate the distinction between virtue practiced for the external rewards it brings and virtue practiced for its own sake: that is, virtue practiced as its own reward or for its goodness "in itself." As trivial as the distinction may seem, it is crucial for the refutation of Machiavelli that Bacon, supposedly, intends. If virtue is not good for its own sake (or does not consist in indifference to all

---

[23] *Aeneid* 9: 252. Slightly altered from Virgil, *Aeneid*, trans. Edward McCrorie (Ann Arbor: Univ. of Michigan Press, 1955), 196.

rewards and in the sacrifice of one's interests to others), then it is the means to other goods and two disturbing things follow: first, if virtue is simply for the sake of reward, the nobility of virtue is clearly diminished. Second, if virtue is only a means for something else, then it may or may not and *often is not* the best means to that something else–in which case it is, as Machiavelli says, better to seem to be virtuous than actually to be so. Hence Bacon's insistence on the distinction between virtue as a means and virtue practiced for its own sake. The problem is, however, that *the Virgilian lines make no such distinction.*[24]

In the poem, the aged Aletes praises the Trojan warriors by saying that the first and best rewards will be paid to them by the gods *and* by their own worth, and the rest will be paid by Aeneas and Ascanius. There are thus three kinds of rewards: divine, virtue as its own reward, and rewards provided by men. According to Virgil in the quote, the divine rewards and virtue's being its own reward are the same–both are "first and best." Now if virtue is good simply for its own sake, then divine rewards, whatever they are, cannot be the same and must be something else, some kind of external and extra-terrestrial add-on.

But if we assert that virtue for its own sake and divine rewards are really the same, then again some disturbing consequences follow. At the most simple level of observation, one might conclude that the Virgilian quote points to agreement with Machiavelli: virtue *is* but a means, and the only means, to external rewards. According to this conclusion, a God who rewards and punishes "virtue" and "vice" would be a divine Machiavellian who rewards those who think that virtue is the only means to other goods and punishes those who, through no fault of their own, think it is not the only such means and happen not to know that God disagrees. (No fault, since would any sane person want to disagree with a punishing Machiavellian God?).

Or, to observe more subtly, one might conclude that the quote points to a confusion in our understanding of moral virtue. On the one hand we think that its nobility consists in sacrifice. Virtue at its peak consists in devotion to and sacrifice for some worthy end, and we think that virtue is debased if it is really for some gain. But the noble sacrifice is itself worthy of praise and reward; and apart from any reward (which may or may not come) it is better, it is a gain, to be worthy by virtue of the sacrifice. To the moralist, that worthiness is the highest good. But if so, then the worthy deed is, really, no sacrifice at all. In this light, the term "rewarded in itself" is an oxymoron, like "round square." What we love about virtue, to which we are exhorted by the divine law, just isn't what God and our conscience lead us to think it is.

At the very least, we have to admit that Bacon's remarkable and twice-committed misinterpretation of the quote from Virgil–it subverts the very distinction it is meant to exemplify–leads us to wonder about the coherence of our common and most basic moral opinions and beliefs. These opinions and beliefs comprise the very background condition for our being able to hearken to a God whose miracles disclose divine governance of the world. This background condition at first makes perfect sense to

---

[24] See an earlier and slightly different version of this argument in my essay "On the Miracles in Bacon's *New Atlantis*" in *Francis Bacon's* New Atlantis: *New Interpretive Essays*, ed. Bronwen Price, (Manchester: Manchester Univ. Press, 2002).

us. But it is, it turns out, subject to criticism by unaided reason alone. One could well guess that such criticism bears directly on our ability to take supposed miracles seriously. God is just and commands us to practice virtue for its own sake. But the very notion of such virtue makes no sense. And if we act for the sake of rewards in this life and in heaven, then God might as well be called Niccolo.

In the case of the evidence of the spirit moving within, such criticism might make that motion seem more like the whisperings of false or deluded hopes. One might suspect in this case that the non-scientists in Bensalem (and even after the perfection of science) might not want to hear this criticism–they might not want to discover that the reason they hope will ground their religion might rather subvert the faith on which they really depend. And we can't be sure that even the scientists would be so keen to hear it, even with the advent of perfect science.

In the very best and most unlikely case, if God burned a bush and then spoke miraculously to a scientist at the end of science, that scientist, having thought through our moral opinions and beliefs, might well reply: "What? What God said doesn't make any sense and I can't possibly really comply with his command as he said it. If I act for the sake of virtue for its own sake, then I haven't acted for the sake of virtue for its own sake. He was talking nonsense and might as well have been singing 'Three Blind Mice.' I know he burned the bush. But maybe that he spoke to me is a figment of my then existing moral confusions. And if he *really did* speak to me and if all that's at stake is the reward, then I've got good reasons to fear him and to obey, but not a single one to revere him." That's not a pretty sight. And until we actually see and verify the burning bush, it's not likely that any thoughtful person could really believe in such a god. For Bacon, reasoning about divine rule is secret, not because it's too hard, but because it's not fit to be uttered.

All this brings us back to Wilson's question about the unity of knowledge. To the extent that the basic fissure in the intellectual globe is the one between religion and reason, we could say that Bacon concluded both for and against the unity of knowledge. According to Bacon, the divine voice cannot but speak to reason. That's why the Bensalemites hope in vain that natural science will verify their religion. And with perfected natural science, natural science could in part verify the religion: it could prove that the bush being on fire is outside the bounds of natural causes. But natural science has nothing to say about what a divine voice might say. It does not follow, however, that in the end reason and faith occupy wholly different realms of experience. For the divine voice is addressed to our moral opinions and beliefs, and these can be subjected to the critical light of reason.

Bacon thought the fissure between faith and reason could be bridged, but that the construction would not be in the interests of faith as anyone really experiences it. Therefore, he did not think this bridge could or ever would become common knowledge, given the powerful moral longings of the human heart and mind. Thus as regards miracles, the "experience" to be hidden was not some scientific fraud, but the non-scientific and non-metaphysical confrontation of faith and reason that takes place on common ground: our fundamental moral opinions and beliefs. So while the scientists "discern" between true and spurious miracles, and do so on the basis of their knowledge of material nature (and here to discern clearly means to "verify"), they also "discern" the true miracle in assuming the essential goodness (and above all, justice) of its purpose. This "discernment" is, however, subject to a radical and rational critique that has nothing to do with material nature.

The Enlightenment figures who thought they could bring about a wholly secular, rationalized, and scientific world did not really know on what grounds it could be built. In every case they challenged faith with metaphysics and materialistic presuppositions, which Bacon knew could never refute those who claim miraculous inspiration by the one true God. But Bacon, who knew how to engage the believers on their own moral turf, did not think himself likely to convince them and did think it unwise to try. Not likely, because a refutation of faith on these grounds must disappoint the righteous as well as the faithful. Unwise, because a refutation of faith on these grounds, however powerful, would, given the longings of the human heart, arouse equally powerful indignation. Our moral opinions and intuitions *may* be so much nonsense. But we love them with all our hearts and all our souls. Bacon, the father of the scientific project for the conquest of nature, was not really a founder or partisan of the Enlightenment.

Chapter 7

The Hidden Life of Matter: Techniques for Prolonging Life in the Writings of Francis Bacon[1]

Guido Giglioni

Introduction: Bacon's notion of *demortua natura*

The attempt to prolong the length of human life is a key component in Bacon's grandiose program of reforming learning, transforming nature, and improving man's condition. Extending man's lifespan was central to Bacon's interests not only because of the project's religious significance, its social and pragmatic implications, and because of the fact that inquiry requires longevity (life is short, while art long, as the well-known Hippocratic aphorism put it),[2] but also for very specific natural-philosophical reasons: Bacon viewed the attainment of a successful method for delaying the natural process of ageing as a striking validation of his new metaphysics of matter, a metaphysics based on the idea of a living substratum endowed with natural appetites. Indeed, Bacon's proposal diverges from traditional techniques of

[1] I am particularly grateful to David Marshall for revising the English of my text, and to Julie Robin Solomon and an anonymous reader for their helpful comments.

[2] Quoted in Bacon, *Historia vitae et mortis*, *SEH*, II: 105; V: 217. In the same work, Bacon describes his theoretical investigations on the prolongation of life as a contribution to the "common good" so that physicians can become "the instruments and dispensers of God's power and mercy in prolonging and renewing the life of man." And he continues: "For although we Christians ever aspire and pant after the land of promise, yet meanwhile it will be a mark of God's favor if in our pilgrimage through the wilderness of this world, these our shoes and garments (I mean our frail bodies) are as little worn out as possible." (*SEH*, II: 103; V: 215). In *Valerius Terminus* (*SEH*, III: 222), Bacon defines the "discovery of all operations and possibilities of operations from immortality (if it were possible) to the meanest mechanical practice" as the true end of knowledge. Such discovery coincides with the "restitution and reinvesting (in great part) of man to the sovereignty and power (for whensoever he shall be able to call the creatures by their true names he shall again command them) which he had in his first state of creation." In *New Atlantis*, "the prolongation of life, the restitution of youth in some degree, the retardation of age" are presented as the great enterprises of the natural philosophers in Salomon's House (*SEH*, III: 167). On the theme of the prolongation of life in Bacon's *New Atlantis*, see: Charles Webster, *The Great Instauration: Science, Medicine and Reform 1626–1660* (Bern: Peter Lang, 2002); Richard Serjeantson, "Natural Knowledge in the *New Atlantis*" in *Francis Bacon's* New Atlantis: *New Interdisciplinary Essays*, ed. Bronwen Price (Manchester and New York: Manchester Univ. Press), 2002, 82–105.

prolonging life precisely because of the emphasis he places on the material properties of life rather than on supposedly vital qualities irreducible to matter (innate heat, primigenial moisture, vegetative power, and nutritive soul).[3]

We know that the third of the six parts of the *Instauratio magna* was to have contained meticulously crafted natural histories serving as materials for properly executed inductions. The *Historia vitae et mortis*, published in 1623, was to be one such history. There Bacon expressed his dissatisfaction with the state of both medical and chemical knowledge regarding the prolongation of life: "For both the common phrases of physicians concerning radical moisture and natural heat are deceptive and the extravagant praises of chemical medicines raise men's hopes only to disappoint them."[4] As is often the case with Bacon's criticisms, this censure of traditional learning is severe and biased. It must be admitted, however, that nowhere are his claims to originality and innovation more justified than in questions related to the prolongation of life, for his lambasting of traditional physiology drew upon a new theory of matter and vital functions. Instead of focusing on ineffable vital properties, Bacon announced a way of lengthening the course of human life founded on a new metaphysics of living matter. Matter is alive, argued Bacon, because regulated, structured, and maintained by primal appetites. Appetites account not only for all manifest actions and reactions in matter, but also for its ineradicable latent tendency to self-preservation and indefinite persistence.

What is more, Bacon's major criticism of the conventional medical wisdom amounted to no less than the rejection of the idea of a natural ineluctability of death shared by traditional physicians. It may seem that Bacon is here as elsewhere writing with the bombast of a Lord Chancellor, but it ought to be conceded that his position on the prolongation of life is of a piece with the rest of his natural philosophy. The life of man is but part of the life of matter and, consequently, we may alter a man's life-span if we know and control the inner workings of matter and their effects. "Natural philosophy," Bacon explains while interpreting the myth of Orpheus, "proposes to itself, as its noblest work of all, nothing less than the restitution and renovation of things corruptible, and (what is indeed the same thing in a lower degree) the conservation of bodies in the state in which they are, and the retardation of dissolution and putrefaction" (*SEH*, VI: 647–8, 721).

3 On Bacon's metaphysics of matter and primal appetites, see Guido Giglioni, "The Darkness of Matter and the Light of Nature: Notions of Matter in Bacon and Comenius and Their Theological Implications," *Acta Comeniana* 17 (2003): 1–23. On Bacon's notion of matter, see Maxwell Primack, "Outline of a Reinterpretation of Francis Bacon's Philosophy," *Journal of the History of Philosophy* 5 (1967): 123–32; Graham Rees, "Matter Theory: A Unifying Factor in Bacon's Natural Philosophy?," *Ambix* 24 (1977): 110–25; *idem*, "Francis Bacon's Biological Ideas: A New Manuscript Source," in *Occult and Scientific Mentalities*, ed. Brian Vickers (Cambridge: Cambridge Univ. Press, 1984), 297–314; *idem,* (assisted by Christopher Upton), *Francis Bacon's Natural Philosophy. A New Source* (Chalfont St. Giles: The British Society for the History of Science, 1984).

4 Bacon, *Historia vitae et mortis*, *SEH*, II: 105; V: 217. See also *De vijs mortis*, *OFB*, VI: 273; *De sapientia veterum*, *SEH*, VI: 682; 760; *Filum labyrinthi*, *SEH*, III: 496–7; *The Advancement of Learning*, *OFB*, IV: 96–104. For the criticism of astrological and magical medicine, see *Historia vitae et mortis*, *SEH*, II: 153, 157–8.

The "conservation of bodies in the state in which they are" is the conceptual core of Bacon's theorizations on longevity and prolongation of life. His insistence on the primacy of the inanimate properties of matter over its animate outgrowths and protrusions means not that the animate and inanimate states are two different metaphysical conditions of being resulting from the presence or absence of the venerable principle of life (i.e., the soul), but that animate is merely an accidental and transient effect of a stable and permanent nature, what Bacon calls *demortua natura*. All matter is ultimately inanimate, that is, a product of original simple motions or primal drives, and only a part of it evolves into more complex structures such as sensory organs, elaborate combinations of motions, and complicated interlacements of spirits of various nature. If inanimate matter is the domain of the self-identical persistence of being, animate matter encapsulates all the factors that unsettle this paragon of indefinite duration. Life, in Bacon's cosmos, starts when a source of discomfort interrupts the self-identity of being.[5] It takes on the aspect of a predatory entity. Unlike inanimate spirits, vital spirits prey upon the substance of the body in which they happen to reside. We are thus confronted with the paradox that what is usually meant by life is in fact for Bacon a destructive reaction, the acceleration of a process of decay, and an aimless endeavor to compensate for a sense of hunger, whereas death and *demortua natura* are the spontaneous and natural tendency to recover the original state of rest and fulfilment. Life is uneasiness and irritability, death is appeasement, which means that the traditional notions of life and death undergo a conceptual reversal: life is death and death is life.

There is still another reason why Bacon's project differs from traditional ways of addressing the question of long life. Contrary to the old way of looking at natural ageing as ultimately irreconcilable with techniques for artificially prolonging life, Bacon did not rule out the possibility that new procedures can be devised to change the course of nature. The whole system of ancient prophylaxis was based on the idea of fostering the natural course of life by gently and judiciously altering only those factors and operations allowed by nature. Regimen depended on the control of the so-called six non-naturals, that is, air, food and drink, sleep and waking, motion and rest, excretions and retentions, dreams, and the passions of the soul.[6] The phrase "non-naturals" denoted precisely that domain intermediate between unmodifiable nature and modifying unnatural agents (diseases), a domain that was deemed to be modifiable because open to man's action and interventions. The sphere of the non-naturals

---

[5] Bacon, *Historia vitae et mortis*, *SEH*, II: 108: "duplex est duratio corporum, altera in identitate simplici, altera per reparationem."

[6] On the non-naturals, see Galen, *Thrasybulus, sive utrum medicinae sit an gymnasticae hygieine*, in *Opera omnia*, ed. Carolus Gottlob Kuehn (Leipzig: Cnobloch, 1821–33), V: 837; *idem*, *De opt. corp. const.*, in *Opera omnia*, IV: 742–3; *idem*, *De san. tuend.*, in *Opera omnia*, VI: 77–80. For the history of the six non-naturals, see L. J. Rather, "The 'Six Things Non-Natural': A Note on the Origins and Fate of a Doctrine and Phrase," *Clio Medica* 3 (1968): 337–47; Saul Jarcho, "Galen's Six Non-Naturals: A Bibliographical Note and Translation," *Bulletin of the History of Medicine* 44 (1970): 372–7; Jerome Bylebyl, "Galen on the Non-Natural Causes of Variation in the Pulse," *Bulletin of the History of Medicine* 45 (1971): 482–5; Peter Niebyl, "The Non-Naturals." *Bulletin of the History of Medicine* 45 (1971): 486–92.

had always been a delicate matter for the nature-centered cosmology and medicine of the ancients. The implicit question was: how far can nature be altered? More specifically, with respect to longevity, was it really sensible to change the course of nature in order to prolong life if nature was thought to be the wisest agent in the cosmos? These doubts were even more legitimate if death represented a natural tendency, inscribed in the very process of generation.

The principal aim of this essay is to examine Bacon's ideas concerning the prolongation of life and to demonstrate how it was that in his treatment of the subject the emphasis shifted from moisture and heat to spirits and natural appetites, from animate to inanimate bodies, from the art of ageing well to the technique of artificially restoring and preserving bodies. I will start by presenting a synthetic account of the classical views on natural heat and primigenial moisture that were still current among seventeenth-century practitioners in order to set a conceptual counterpart to Bacon's alternative. The central theme of the essay will be a discussion of Bacon's idea of the preservation of life and its relationship to his notion of matter as a vital substratum driven by inexorable appetitive forces. It will become apparent that Bacon shifted his focus from the fleeting quality of animal life to the active continuance of matter, from the ephemerality of individual lives to the durability of the primal appetites of nature.

Overlapping aims: prolongation of life, conservation of health, vital flourishing, and natural longevity

The complementary doctrines of innate heat and primigenial moisture long represented the foundations of western anatomo-physiology. Not only were innate heat and primigenial moisture two basic principles underlying therapeutics and prophylaxis, they were also inextricably intertwined with overarching views of life and death. In *De iuventute et senectute*, Aristotle set the tone of the discussion on living and dying by comparing the flame of life to the flame of fire: "we can see two ways in which fire is destroyed–by dying out and by extinction. By dying out we mean a decay arising from itself, and by extinction a decay caused by opposites: one is death by old age and the other by violence. But in fact both forms of destruction arise from the same cause; for in both cases fire dies from the failure of what feeds it, that is when it cannot get food for the heat."[7] The maintenance of life depends on a balanced fueling of the innate heat. In order to avoid either the extinction or the exhaustion of the vital flame, the system needs constant refrigeration and appropriate supply of food. The impossibility of real self-renewal results from the unrestrainable consumption of the primigenial moisture and its irreplaceability, from the disintegrating action of external factors (especially air), from the impossibility of a thorough assimilation of food, and above all, from the irreversibility of the very process of generation. Dying amounts to an inescapable tendency to degradation, desiccation, and calcification. In *De sanitate*

---

[7] Arist., *De iuvent.*, V: 469b, trans. W. S. Hett (Cambridge and London: Harvard Univ. Press, 1957), 423. See also *De respirat.*, VIII: 474b; XVII: 478b.

tuenda, Galen sums up the whole argument by saying that our body has two sources of deterioration, "one intrinsic and spontaneous, the other extrinsic and accidental," namely, the decay of vital moisture and the injuries of the environment. The drying and withering of life is the "innate destiny of destruction of every mortal creature." There are lesions and diseases in our bodies that are inevitable and intrinsic "having as their root the sources of generation." From this point of view, ageing is comparable to a disease, the very germ of destruction intrinsic to the process of generation. Galen hypothesizes that, "if it were practicable to restore exactly what has been evacuated," the renewal of lost life would be possible in principle, for "if there were no failure, either in replacing the wastes or in the excrements remaining within, the animal would be healthy and would flourish for a long time." But in fact there is always a residue of unassimilable otherness resulting from the deficiencies of nutrition: the ingested food can never be thoroughly converted into the substance of the living being. Galen's conclusion is in line with the tenets of the ancient cosmology based on the incessant coming-to-being and passing-away of natural beings: "it is not possible to make incorruptible (*aphtharton*) that which is engendered."[8] Avicenna sanctioned this view and provided logical order to the theses that the moisture of the body is progressively consumed by the vital flame and that this moisture cannot be restored because nutrition is only a palliative for the ineluctable decay and an inadequate substitute for the original life.[9] This model was received by seventeenth-century philosophers and physicians without significant changes. It encapsulated the essential ambiguity that heat was both the source of life and the germ of destruction. As Bacon put it: "the complication arising from the fact that consumption and repair are both the works of heat, is the greatest obstacle to longevity" (*SEH*, V: 333).

An important question related to prophylaxis and the duration of life was concerned with the relationship between the length and quality of life. Both physicians and philosophers had long been wondering whether the quantitative and qualitative aspects of man's existence could be reconciled, that is, whether a long life with no prospect of being satisfactory and fulfilling could still be defined as a good life, namely, the expression of vital enhancement, ethical and emotional flourishing, and intellectual vigor. Classical and Renaissance views on prophylaxis, based as they were on the dialectic between vital heat and primigenial moisture, sustained a fundamental paradox: life is fuelled by the consumption of food and moisture, but this process is itself the beginning of biological decay in the body. The unavoidable outcome of this dialectic is death, and the irony is that the very process of living accelerates the end of life. Ever since Galen, physicians had tried to resolve this paradox by reconciling the pursuit of health with the attainment of longevity, by finding ways of favoring the expansion of vital faculties without nurturing premature aging.

In order to understand this theoretical background it is important to bear in mind a series of distinctions made by medical writers since antiquity, the most general of which was the distinction between three main states: the good state of the body and

8 Galen, *De san. tuend.*, in *Opera omnia*, VI: 3–7, 63; Robert Montraville Green, trans. (Springfield, Ill.: Charles C. Thomas, 1951), 7–10, 38.

9 Avicenna, *Liber canonis* (Venice: Paganinus de Paganinis, 1507), ff. 52v–53r.

its maintenance, the pathological alteration of this state and its restoration, and the prolongation of the good state. The three states corresponded to three sections of the medical art: hygiene, therapeutics, and techniques for prolonging life. Galen elaborated on the subject by adding further distinctions. Within the whole concept of "restorative art," including both the technique of healing and that of preservation, he introduced three subdivisions in the category of preservation: the art of maintaining the peak of perfect health, the art of reinforcing the health of people who have just recovered from illness, and, between the two, the art of making the state of "healthiness" stable.[10] Obviously, however clearly these distinctions were drawn, the respective notions tended to shade into each other. Galen acknowledged a certain latitude in the condition of the body called health. Health, he wrote in *De sanitate tuenda*, varies throughout the life of an individual and from individual to individual.[11] On the other hand, in order to distinguish his medical project from rival healing programs and to reinforce, at the same time, the philosophical and systematic dimensions of his medical project, he minimized the distinction between the health, strength, and beauty of the body, presenting such distinction as a rhetorical gambit put forward by physical trainers, cooks, and beauticians, while laying stress upon the preeminence of *euexia*, that is, the general condition of physical and moral well-being. The physician was thus invested with ethical duties, over and above his therapeutic and prophylactic functions. The way Galen settled the question presupposes an arena of competing and alternative solutions carried on by philosophers, sophists, and rhetoricians. His ideal of *euexia*, though, could not hide the fact that moral flourishing did not necessarily imply health, that beauty could hide a disease, and that a long life could be disjoined both from health and happiness.

Bacon proves to be well-acquainted with this medico-philosophical debate. "The knowledge that concerneth mans bodie," he writes in *The Advancement of Learning*, "is diuided as the good of mans bodie is diuided, vnto which it referreth. The good of mans body, is of foure kindes; *Health*, *Beautie*, *Strength*, and *Pleasure*." Following in Galen's footsteps, Bacon distinguishes several arts of dealing with care of the body: "*Medicine, or Art of Cure*: *Art of Decoration*; which is called *Cosmetike*: *Art of Actiuitie*, which is called *Athletike*: and *Art Voluptuarie*, which *Tacitus* truely calleth *Eruditus Luxus*." Given the intricate overlapping of different competences, he acknowledges that the "Subiect of mans bodie" is both "most susceptible of remedie" and "most susceptible of errour" (*OFB*, IV: 96). In the *Historia vitae et mortis*, while noticing that "the same things which conduce to health do not always conduce to longevity" (*SEH*, II: 160; V: 267), he recognizes that moral flourishing can foster vital expansion. One way of extending the length of life in Bacon's theory, as we will see in the next section, is to curb the predatory nature of the vital spirits, and moral stability,

[10] Galen, *Thrasyb.*, in *Opera Omnia*, V: 863–4. On the relationship between medicine and prophylaxis in the early-modern period, see Richard Palmer, "Health, Hygiene and Longevity in Medieval and Renaissance Europe" in *History of Hygiene*, eds. Y. Kawakita, S. Sakai, and Y. Otsuka (Tokyo: Ishiyaku EuroAmerica, 1991), 75–98; Heikki Mikkeli, *Hygiene in the Early Modern Tradition* (Helsinki: Academia Scientiarum Fennica, 1999).

[11] Galen, *De san. tuend.*, in *Opera omnia*, VI: 25; Green, trans., 16–17.

through the control of passions and vital impulses, seems to affect the reactivity of the spirits and to promote longevity. Of all affections, writes Bacon, hope is "the most useful, and contributes most to prolong life, if it be not too often disappointed, but feed the imagination with the prospect of good" (*SEH*, II: 172; V: 279). Owing to their correct approach to passions, rhetoricians, who "busied themselves rather about light of speech than the darkness of things," live longer than Aristotelian and scholastic philosophers because their "subtle, acute, and eager inquisition" "fatigues and preys upon the spirits," and therefore shortens life (*SEH*, II: 172; V: 280).

In the historical digression on the long-lived men of antiquity, Bacon attaches "a character or biographical notice, true and very short, but such as in [his] judgment has some bearing upon longevity." The patriarchs lived more than one hundred years (Abraham, "a chaste man, and of a quiet life," Jacob, "patient, gentle, and cunning," Sarah, "a woman of a fair countenance, and of a noble spirit, an excellent wife and mother, and no less distinguished for her frankness than for her duty to her husband," Moses, "a man of courage," and Job, "a model of patience"). The fabulous kings of Arcadia were very long-lived because of their "mode of life pure and uncorrupted." Solon, the Athenian lawgiver, "one of the seven wise men," and Numa, the Roman king, "peaceful, contemplative, and devoted to religion," both lived for more than eighty years. Such hermits as St. Paul, living in a cave and passing time in meditation, died very old. A temperate diet and a moderate disposition like the Pythagorean lifestyle increased a man's life-span as is witnessed by the long lives of Pythagoras and Apollonius of Tyana. By and large, simplicity of life and frugal habits seem to prove the connection between temperance and length of life. "There are periods indeed in every country," says Bacon, when men are longer or shorter lived. Longer generally, when they are less civilised, live on simpler diet, and are more given to bodily exercise; shorter, when they are more civilised and given more to ease and luxury" (*SEH*, II: 132–48; V: 243–56).

On the other hand, Bacon cannot fail to notice that a fair examination of the historical record indicates the unreliability of the classic equation: moral temperance, healthy bodily temperaments, and long life. Among the biblical patriarchs, Levi, "a man of revengeful nature," lived until his one hundred and thirty-seventh year. The Greek poet Anacreon, "a man amorous, voluptuous, and a wine-bibber," lived beyond eighty. The Persian king Artaxerxes, "a man of a dull intellect, averse to important business, loving glory much, but ease more," lived ninety-four years. Terentia, Cicero's wife, "a woman oppressed by many sorrows" and "often troubled with the gout," lived for one hundred and three years. Ambitious and high-spirited popes like Paul III, Paul IV, and Gregory XIII all lived more than eighty years. Guillaume Postel, "a man of disordered brain and unsound mind" lived nearly one hundred and twenty years. Even the famous London "Royal Hospital" for the insane, St. Mary of Bethlem, shows that there are madmen who reach a great age (*SEH*, II: 132–48; V: 243–56). Not to mention the fact that some diseases, once they have been properly treated, may prolong the life of the patient appreciably, so much so that, following the proverb that "to recover health is to renew youth," it would not be inappropriate "to induce some artificial diseases, as is done by strict and emaciating diets" (*SEH*, II: 157; V: 265. See also *SEH*, II: 119). More specifically, such mental illnesses as melancholy prolong life because grief and sadness, "if devoid of fear," "contract the spirits, and are a kind of condensation" (*SEH*, II: 164, 171; V: 271, 279). The historical inductive survey confirms Bacon's hypothesis that a successful technique for prolonging life has to go beyond the simple expansion or contraction of vitality toward a real understanding of the very roots of life in matter. Bacon is convinced that he found such a root in matter's indelible desire constantly to reestablish the original state.

Bacon's notion of inanimate life (natural appetite) and its implications for the theory of the conservation and restoration of bodies

In the ancient characterization of the living being, it is beyond argument that no form of human art can counteract the natural tendency to death. In *De sanitate tuenda*, Galen had declared that the idea of an indefinite prolongation of life was the result of a specious reasoning: "if it were possible always to keep the constitution of the body moist, the doctrine of the sophist, which I discussed at the outset, announcing that he would make anyone who obeyed him immortal, would be true. But since, as was shown, it is impossible for the body to escape the natural road to dryness, for this reason it is inevitable for us to grow old and perish."[12] In *Thrasybulus, sive utrum medicinae sit an gymnasticae hygieine*, Galen had demonstrated that the art of prophylaxis "has no power to create a whole human body; but it is able to repair its parts, like the art of mending garments." Given man's inability to defer the inevitable process of decay, the only thing he can do is to assist nature in its tendency to restore life. The primary agent remains nature, and the physician, in a pure Hippocratic garb, is the servant of nature because the restoration of life is simply another form of recreating life. "[I]f we had the capacity to create an animal in the first place," Galen argues, "the principles involved in that initial creation would be no different from those involved in its restoration. In reality, though, we are not primarily responsible for either of these tasks: it is Nature who first constructs the animal and subsequently heals it if it is sick. And so its method in that original construction of flesh is no different from its subsequent method; nor is its method for the absorption of nutrition in the foetus different from its subsequent method; this function, along with nutrition in general, the separation and excretion of substances–all the functions, in short–are performed in the same way now as in the initial creation."[13]

Bacon, however, rejects precisely the assumption that restoration as a recreation of bodily life is beyond any human effort. Accordingly, he does not rule out the possibility that man can live indefinitely: "Whatever can be repaired gradually without destroying the original whole is, like the vestal fire, potentially eternal." Previous philosophers and physicians, argues Bacon, have looked at death as "something that could not properly be repaired." Their opinions, he continues, are "very frivolous and

[12] Galen, *De san. tuend.*, in *Opera omnia*, VI: 399; Green, trans., 244.

[13] Galen, *Thrasyb.*, in *Opera omnia*, V: 853; P. N. Singer, trans. (Oxford: Oxford Univ. Press, 1997), 75–6.

ignorant."[14] Bacon's principal innovation is to look at the notions of vivification, regeneration, and prolongation of life through the lens of inanimate bodies. Inanimate beings illumine the complex operations of animate beings: "whatever the native spirit (which exists in all tangible bodies whether with or without life) and the ambient or external air do to bodies inanimate, the same they try to do to bodies animate, though the presence of the vital spirit in part disturbs and restrains these operations, and in part intensifies and increases them exceedingly. For it is very evident that many inanimate bodies can last a very long time without repair, but animate bodies without aliment and repair at once collapse and die out like fire. The inquiry therefore should be twofold; regarding first the body of man as a thing inanimate and unrepaired by nourishment; and secondly as a thing animate and nourished" (*SEH*, II: 207; V: 219. See also V: 214). What inanimate bodies display is their resistance and continuance by virtue of primordial appetites. By looking at the higher levels of animal organization from the point of view of the original drives inherent in inanimate bodies, Bacon reconfigures the whole discussion of life and death in matter in a dramatic fashion. Instead of concentrating on the ineffable and unrepairable vital nature of the primigenial moisture, Bacon shifts the emphasis to the unlimited duration of inanimate things. In his opinion, the investigation of the nature of life has been led astray by the idea that vital properties are more visible in elaborate structures such as animals endowed with complex organs and organisms capable of developing from seeds. Instead, life reveals itself in rudiments and outlines of vital processes more than in their final state. As a methodological cue, Bacon suggests to look for life where one would least expect to find it. In the *Historia et inquisitio de animato et inanimato*, he delineates a plan of research centered on such elusive beings as mistletoe, moss, scattered menstrua, and animalcules. He considers putrefaction to be a particularly revealing phenomenon placed as it is between life and death, health and disease. If matter as such, regardless of whether animate or inanimate, manifests life, it is the very distinction between artificial and natural that disappears. The study of the primordial and undifferentiated outlines of life, Bacon argues, indicates that living organisms of this kind "can be generated at pleasure and artificially"(*OFB*, XIII: 228–35). Just as he appreciates the heuristic role of monsters and sports of nature in order to give a full explanation of the regularity of nature, in the same way he focuses on the properties of *demortua natura* and *natura durabilis* to make sense of life and change.

In order to understand the crux of Bacon's argument, it is thus important to make it clear that the traditional divisions in animate and inanimate beings undergo a palpable conceptual redefinition in his natural philosophy. For Bacon, an inanimate thing is not something devoid of life. When he says that, in order to understand the nature of vital processes in a living organism, we need to look first at inanimate bodies, he is not suggesting that life is the result of a mechanical process. On the contrary, he is

[14] Bacon, *Historia vitae et mortis*, *SEH*, II: 106; V: 218: "they looked for death in something that could not be properly repaired, imagining that there was some primitive and radical moisture which was not really repaired, but which even from childhood received a kind of spurious addition and not true repair; and that with time this grew worse and worse, till at last it ended in none at all."

assuming that the life phenomenon instantiated by each complex organism originates from the life of the whole material universe. Matter as a whole is pliant, inherently capable of motions, and pervaded by appetites. Biology, and not mechanics, serves here as a point of reference. The distinction between animate and inanimate does not presuppose denial of the living nature of matter. It simply denotes that matter may be divided into two main ontological parts that are differentiated according to the quality of the spirit that inhabits a specific bodily configuration and according to the nature of the perception that pervades that portion of matter (whether *perceptio* or *sensus*). In Bacon's natural philosophy, each body contains a spirit, which is like a mark of life impressed upon matter, or better, an elementary trace of the natural appetite of matter. The quality of the spirits distinguishes the various degrees of life. Inanimate bodies are characterized by the presence of "intermixed" (*intermisti*) spirits scattered throughout the mass of tangible bodies. Vegetable bodies result from the action of "branching" (*ramosi*) spirits. Animals, finally, are set in motion and made sentient by the operation of spirits gathered in cells (*cellulati*) and connected through a network of channels (*OFB*, VI: 319; *SEH*, II: 215).[15] The distinction between animate and inanimate, therefore, does not rule out the presence of an original life and desire in matter. What distinguishes animate from inanimate bodies is a difference between activity and sluggishness in appetite, between thinness and grossness in matter, and between the ways in which spirits organize themselves while shaping the body (whether more or less concentrated or more or less scattered).

In Bacon's view, it is of no consequence whether the investigation is conducted starting from the presence of pneumatic matter in tangible bodies or from the intimation of perception in the operations of nature. In both cases, the result of the inquiry points to the same conclusion: there is no tangible body without spirit and there is no activity without perception. This demonstrates the existence of forms of sentient matter and natural perception antecedent to the development of complex structures and to the exercise of conscious perceptions and volitions.[16] The difference between animate and inanimate beings is not the result of an ontological gap determined by the presence or absence of an unspecified and indefinable principle of life called soul. The difference is to be understood simply in terms of levels of complexity in the organization of matter. Which means that animate and inanimate are accidental states of matter, whereas its desires and tendencies to self-preservation are the true sources of life in the universe. Bacon makes it clear that such desires and tendencies are ubiquitous in matter, but sometimes they are not exhibited and remain latent until the necessary conditions for their development are supplied: "It is certain that this desire

[15] On Bacon's notion of spirit, see Daniel P. Walker, "Francis Bacon and *Spiritus*" in *Science, Medicine and Society in the Renaissance: Essays to Honor Walter Pagel*, 2 vols., ed. Allen G. Debus (London: Heinemann, 1972), II: 121–30; Graham Rees, "Francis Bacon and *Spiritus Vitalis*," in *Spiritus. IV Colloquio Internazionale*, eds. Marta Fattori and Massimo L. Bianchi (Rome: Edizioni dell'Ateneo, 1984), 265–81.

[16] Bacon, *Abecedarium novum naturae*, *OFB*, XIII: 186 (on the universal presence of spirits); *De augmentis scientiarum*, *SEH*, I: 610–11; *Sylva sylvarum*, *SEH*, II: 602 (on the universal presence of natural perception).

for assimilation, which in a gross body is bound up, as was said, and rendered ineffectual, is somewhat liberated and excited by heat or spirit in its neighbour, and in the end actuated by it. And this is the only reason why inanimate bodies do not assimilate, and animate bodies do (*SEH*, II: 197; V: 304–5). More specifically, lifeless (*demortui*) spirits are characterized by two desires: "one of multiplying itself, the other of going forth and congregating with its connaturals." Bacon explains that "every spirit seated amongst the grosser parts dwells unhappily; and being in such a solitude, where it finds nothing like itself, it the more strives to make and create something similar; and to increase its quantity, it works hard to multiply itself, and prey upon the volatile parts of the grosser bodies." The action of life and the ensuing tendency to ageing and decay are ultimately a direct consequence of the natural appetite of matter. Spirits desiccate and attenuate the substance of the tangible bodies in their effort to escape from the place in which they are detained. As a result, matter contracts and wrinkles. It is no accident that the fold (*plica*) is the mark of life in Bacon's metaphysics of matter. What is more, the whole process can be seen as a result of the inherent responsiveness of matter and of the environment it creates: "the going forth and escape of the spirit into the air is a double action, arising partly from the appetite of the spirit, and partly from the appetite of the air; for the common air is a needy thing, and seizes everything with avidity, as spirits, odours, rays, sounds, and the like."[17]

Bacon is rephrasing the traditional picture of preying heat and desiccating air in terms of natural impulses embedded in matter. This conceptual redefinition has practical implications of paramount importance because, as Bacon points out in his *De vijs mortis*, "the very actions and virtues of things (as they are called) are nothing other than the secret impulses (*clandestinj impetus et motus*) of those very same invisible bodies, and unless we acquaint ourselves rightly and distinctly with these impulses no one can hope for doctrines leading to works." It is imperative to look after the "sources and means" of the natural actions, that is to say, "the appetites and affects of nature, from which they flow," and, on that basis, "gather information and evidence as to how much and by what means those actions may be stimulated, contained, incited or abated; and from all of these things will eventuate an open, quite clear and really well-made road to practical knowledge about conserving bodies" (*OFB*, VI: 321, 281). Bacon's metaphysical investigations into the appetitive nature of matter provides a spin-off for technology: bio-medical techniques such as blood transfusions and infusions, plans for artificial digestion, procedures to induce "artificial" diseases and inventions for conservatories of snow (*SEH*, II: 157, 180–1, 190, 220–1, 222, 223, 229, 242).

We can say that in Bacon's program man's longevity is only an aspect of a larger natural-philosophical project concerned with the conservation of natural bodies. In *De sapientia veterum*, Bacon reads the fable of Deucalion and Pyrrha as an allegory of "restoration." Deucalion and Pyrrha managed to restore the human race after the universal deluge by throwing the stones of the earth–"the mother of all things"–behind their backs (instead of throwing the bones of their mother, as they initially thought they were

[17] Bacon, *Historia vitae et mortis*, *SEH*, II: 217; V: 325; *De vijs mortis*, *OFB*, VI: 287; *Sylva sylvarum*, *SEH*, II: 600.

supposed to do). The "secret of nature" that is concealed by the myth is that "the renewal and restoration of things" cannot be effected by using their same remains, but by going back "to more common principles" (*SEH*, VI: 661–2, 737). It becomes now clear what the nature of these common principles capable of renewing life to which Bacon is referring obliquely and figuratively is. After what has been argued so far, we can interpret these "common principles" as the primal and constitutive appetites of matter, appetites which, as Bacon writes in *Descriptio globi intellectualis*, are the universal "physical reasons" that apply to every domain of the natural world, both celestial and sublunar (*OFB*, VI: 113). Every natural form is for Bacon the result of a composition of motions. If we come to know the motions, we can by composing and decomposing motions alter the forms of nature; since motions are ultimately appetites, by knowing and acting on natural appetites we can alter reality. The theoretical deconstruction of bodies into their elementary appetites (the so-called "simple natures") allows the practitioner to reconstruct the world of nature into the world of man.

A redefinition of death and life

Feasible plans for prolonging life represented one of the many points of contention between ancients and moderns, between traditional practitioners and neoterics in the early-modern period. Up to a certain point of his career, René Descartes cultivated the dream of indefinitely extending the limits of man's life. Likewise, the main appeal of the alchemists' *lapis philosophorum* lay in the promise of procuring rejuvenation and perpetual health. Bacon shared the same operative optimism while distancing himself from facile solutions. Such operative faith in artificial ways of prolonging life and such emphasis on the possibility of acting concretely and effectively on the course of human existence are the components that seem to be missing in previous attempts. Following in Galen's footsteps, Avicenna had declared that the art of maintaining the health of the body must not be confused with the art of averting death or of securing longevity. The ancients had concentrated less on the art of preserving life, more on the teaching of moderation and of a regimen based on the control of food, sex, emotions, and habits. The idea of natural longevity dominated that of artificial prolongation. Indeed, more than prolonging life, it would be more correct to say that the ancients devoted themselves to delaying death. Confronted with the choice between a long but wary or intense but short life, the old philosophers and physicians preferred the pragmatic solution of living to a decorous old age, cheered by a methodic use of practical wisdom and distance from the disgrace of physical and mental decay. In line with this tradition, for instance, Cardano still insisted on the uncertainty of the "science of longevity" and questioned the reality of the fabled centenarians described by Homer, Hesiod, and Moses,[18] the same centenarians used by Bacon as an inductive basis for his theory of long life.

Instead of confining himself to the meticulous practice of balancing heat and cold, moisture and dryness in the attempt to reach an equilibrium among the many factors

[18] Cardano, *De sanitate tuenda*, in *Opera omnia*, ed. C. Spon (Lyon: Huguetan & Rivaud) 1663, VI: 248a.

affecting human life, and rather than acting on the nutritive faculty through a whole array of empirical procedures such as diet, balneotherapy, and exercise, Bacon considered his plans for prolonging life to be a direct consequence of a philosophical and cosmological agenda centered around the redefinition of the notion of animate and inanimate and, through these, of life and death. In his view, only from a sound grasp of the nature of living matter could a plausible technique for prolonging life emerge. He dismissed as a form of theoretical and practical idleness the ancients' belief in the existence of insurmountable limits in extending the length of human life. Their remissive attitude, he wrote in *De sapientia veterum*, "seems to show that methods and medicines for the retardation of age and the prolongation of life were by the ancients not despaired of, but reckoned rather among those things which men once had and by sloth and negligence let slip, than among those which were wholly denied or never offered" (*SEH*, VI: 672, 749). Later in the same work he pointed out that "the ancients regarded the conservation, and to a certain extent the restoration, of natural bodies as a thing not desperate, but rather as abstruse and out of the way" (*SEH*, VI: 682, 761).

If the ancients had preferred to confine themselves to a workable amelioration of the life span naturally allotted to each individual, the moderns nourished the audacious dream of freeing themselves from the constraints imposed on their lives by nature. Whereas the ancients had maintained the progressive irreplaceableness of radical moisture and the inevitable defectiveness of nutrition, Bacon did not exclude the possibility of transforming the cyclical yet inevitably retrogressive process of life into a perpetually self-renewing "vestal fire." Anticipating ideas that would lead some of the members of the Royal Society to try experiments of blood transfusions, Bacon hypothesized practicable scenarios of bodily regeneration. "If it were possible for young spirits to be put into an old body," he suggested, "it is probable that this great wheel might put the lesser wheels in motion, and turn back the course of nature" (*SEH*, VI: 682, 761). After all, Bacon had written in *Historia vitae et mortis* that age is only a conventional "measure of time" depending on the interrelationship of internal and external factors (*SEH*, II: 115; V: 227).[19] Bacon's faith in the unlimited prospects of medical and technological implementations implied a drastic redefinition of the meaning of the natural order of life. A new understanding of nature progressively dissolved the very concept of natural conditioning. Ageing itself was one of these natural conditionings.

It is certainly no accident that the works in which Bacon expressly dealt with the prolongation of life bear the titles *Historia vitae et mortis* and *De vijs mortis*. Bacon wrote an history of life in which the dominant role is played by death. The permanence of inanimate life, and not the transience of animate lives is Bacon's most genuine meaning of life, and it is within the inanimate life of the spirits that he finds the ultimate ground for successful actions in the domain of living nature and its prolongation. Inanimate life–*demortua natura*–discloses the continuance of the natural appetite of matter, that is, its primordial desire always to restore the original

[19] Discussions of the "ages of man" abounded since the ancient times and the respective stages were considered as an ontological partition of the life-cycle in men and women. See Aristotle, *Rhet.*, II: 1390b.

situation. Can we therefore say that in Bacon the tendency to death, the *via mortis*, is natural? The life of animate beings (living spirits) seems to be, in fact, death; the inanimate beings' tendency to rest, which only for reasons of convenience we call death, is, in fact, life (the persistence of the lifeless spirits), or, if not life, at least a "caricature or shadow of immortality itself."[20] In any case, Bacon points out that the line between life and death is extremely fine. "The fact is that every living thing among us," he explains, "suffers the torment of Mezentius–that the living perish in the embrace of the dead; and although the vital nature enjoys itself and runs things for a while, the influence of the parts nevertheless gets the upper hand not long afterwards, and does so according to the nature of the substance and the lifeless nature and not at all to the nature of the living one."[21] If the most exuberant vitality is an anticipation of death, the slow operations of the lifeless spirits betray the everlasting appetite of matter. Matter is always in a state of appetence: the motion that characterizes it in its deepest level is an annihilable desire to return to the condition prior to any alteration. Bacon closes his analysis of the structural motions of matter in *Novum Organum* with the "motion of rest" (*motus decubitus*), "the motion to which the name of motion scarcely applies, and yet it really is a motion." The whole passage is significant and is worth quoting in full: "By this motion the earth stands in its own mass, while its extremes move towards the center; not towards an imaginary center, but towards union. This is also the appetite by which all highly condensed bodies reject motion; their only appetite is not to move; and though they may be irritated and provoked to move in infinite different ways, they still preserve their nature (so far as they can). And if they are forced to move, they seem to do so simply in order to recover their rest and their position, and not to move again. In this process they show themselves flexible, and they make quite speedy and rapid exertions (as if they were thoroughly fed up, and impatient of any delay). We can only get a partial view of this appetite, because here on earth tangible things are not condensed to the highest degree and are also mixed with some spirit through the warming influence of the heavenly bodies."[22] A passage from the

---

[20] Bacon, *De vijs mortis*, *OFB*, VI: 357: "as far as the lasting and dissolution of things are concerned," says Bacon, there are only two processes that distinguishes the animate from the inanimate: "the one which hastens dissolution much more in animate than in inanimate things, and the other which inhibits it, and is a kind of caricature or shadow of immortality itself." It has been suggested by an anonymous reader that my interpretation of Bacon's notion of appetite and matter echoes the dialectic of life and death outlined by Freud in his *Beyond the Pleasure Principle*. If it is true that in a long-term view Bacon's theorizations can be seen as part of the historic evolution concerning the notion of irritability from the early-modern times to twentieth-century physiology, it is also true that Freud's theory is a distinctly post-Cartesian position that has rejected the metaphysics of living and appetitive matter.

[21] Bacon, *De vijs mortis*, *OFB*, VI: 353. See also *Historia vitae et mortis*, *SEH*, II: 106. Mezentius, the cruel Etruscan tyrant depicted by Virgil in the *Aeneid*, devised a torment whereby a living body was tied to a festering corpse until the infection and putrified exhalations of the dead had killed the living body. See Andrea Alciato's corresponding emblem in *Emblematum libellus* (Venice: Aldus, 1546, 27v). On Mezentius' emblem, see Charles Henebry, "Figures of Speech: The *Emblematum Liber* as a Handbook of Rhetorical Ornaments," *Neophilologus* 87 (2003): 183.

[22] Bacon, *Novum organum*, *SEH*, I: 346; *The New Organon*, eds. Lisa Jardine and Michael Silverthorne (Cambridge: Cambridge Univ. Press, 2000) 205.

Abecedarium novum naturae is even more eloquent: "bodies which are so favourably compounded and placed that they cause annoyance and trouble neither to themselves nor others but are perfectly satisfied in themselves and so are free of all desire for change, evidently have nothing to do except enjoy their own nature. In fact these bodies are of two sorts: ones which rejoice in motion, and others which love quiet and shrink from motion" (*OFB*, XIII: 201). Uninterrupted motion or immobility are simply the two faces of the same appetite, namely, a condition of self-satisfaction caused by freedom from the desire for change. The uneasiness of the interrupted rest and the consequent state of responsiveness, restlessness, and irritability, are the by-products of the same tendency to preserve the primordial state of matter. Vital stir is a reaction to the loss of the original rest.

From what has been said so far, Bacon's insistence on the preservative characteristics of cold and on the connection between cold and long life should come as no surprise. Cold is a clear and concrete example of the natural appetite for rest. The highest degree of rest, condensation and cold, though, is unattainable here on earth. Bacon claimed more than once that the investigation of the nature of heat and cold and their production are characterized by a basic asymmetry. Compared to heat, which with the help of fire man can easily produce in large quantity and intensity, the cold that is available to man or that he can recreate is feeble and weak.[23] And since cold has the power of preserving things through the condensation of the spirits (as is made clear in the use of niter) and through the contraction of their unremitting activity (as is apparent in the use of opiates),[24] a thorough conservation of bodies is hampered by the material and technical difficulties in manufacturing cold. At the time, niter was viewed as a natural source of cold. Bacon shares with many of his contemporary practitioners of science the enthusiasm about the virtues of niter. He replaces the flame with niter as a paragon of life. Niter combines the vitality of the spirits with the refrigerating properties of cold. It is replete with spirits but, unlike any other substance containing spirits, it is cold and capable of producing cold (*SEH*, II: 166). In *New Atlantis*, the members of Salomon's House, in order to better investigate "the secret motions of things," use deep caves under high mountains sufficiently remote from the action of the sun so as to perform all kinds of "coagulations, indurations, refrigerations, and conservations of bodies." These subterranean laboratories called "the Lower Region" are also used to find new therapies for diseases and ways to prolong life by recruiting appropriately chosen hermits (*SEH*, III: 156–7).[25]

---

[23] Bacon, *Historia vitae et mortis*, *SEH*, II: 115. On the connection between cold climate and long life, see *SEH*, II: 148: "Regionibus frigidioribus et hyperboreis diutius homines vivunt plerunque, quam calidioribus; quod necesse est fieri, cum et cutis sit magis astricta, et succi corporis minus dissipabiles; et spiritus ipsi minus acres ad consumendum, et magis fabriles ad reparandum; et aer (utpote modice calefactus a radiis solis) minus praedatorius."

[24] Bacon, *Historia vitae et mortis*, *SEH*, II: 117, 166, 179, 181, 183, 223–4; *idem*, *De vijs mortis*, *OFB*, VI: 299, 321; *Sylva sylvarum* in *SEH*, II: 374.

[25] In *Historia vitae et mortis*, Bacon recalls the Roman physician Antonius Musa's successful treatment of the Emperor Augustus through the use of cold medicines, a therapeutical approach that, says Bacon, in all likelihood contributed to Augustus' length of life (*SEH*, II: 140).

Bacon's lifetime involvement with experiments on prolongation of life, conservation, and freezing of bodies, had a tragicomic finale. John Aubrey tells us the story of Bacon's death as it was reported to him by Thomas Hobbes:

> As he was taking the air in a coach with Dr. Witherborne (a scotchman, physician to the King) towards Highgate, snow lay on the ground, and it came to my lord's thoughts, why flesh might not be preserved in snow, as in salt. They were resolved they would try the experiment at once. They alighted out of the coach, and went into a poor woman's house at the bottom of Highgate Hill, and bought a hen, and made the woman gut it, and then stuffed the body with snow, and my lord did help to do it himself. The snow so chilled him, that he immediately fell so extremely ill, that he could not return to his lodgings (I suppose at Gray's Inn), but went to the earl of Arundel's house at Highgate, where they put him into a good bed warmed with a pan, but it was a damp bed that had not been laid-in about a year before, which gave him such a cold that in two or three days, as I remember Mr. Hobbes told me, he died of suffocation.[26]

However anecdotal the evidence supporting this story may be,[27] the episode encapsulates Bacon's lifetime engagement with the "restitution and renovation of things corruptible" by means of cold. It was Bacon's firm belief that life could be preserved by cold better than by heat. Surely, he would have regarded the fact that he was killed by cold more as a natural accident than a glaring, self-experimental refutation of his theorizing about the properties of life and death. After all, in *De vijs mortis* he had called any process that inhibits dissolution, no matter how powerful, a "caricature of immortality." Seen from the perspective of the life of matter, the successful freezing of a hen would have been more meaningful for him than the temporary prolongation of his own individual life and self.

---

[26] John Aubrey, *Brief Lives*, ed. Oliver Lawson Dick (London: Secker & Warburg, 1960), 16.

[27] On the various accounts of Bacon's death, see Lisa Jardine and Alan Stewart, *Hostage to Fortune: The Troubled Life of Francis Bacon 1561–1626* (London: Orion Books, 1999), 502–11.

Chapter 8

"The Purer Fountains": Bacon and Legal Education[1]

Daniel R. Coquillette

Introduction: an orthodoxy under fire

Today, the classical underpinnings of American legal education are under intense critical review.[2] The dominant pedagogy, the "case book" and the "Socratic method," were established by Christopher Columbus Langdell (1826–1906) at Harvard Law School more than a century ago. Together with Langdell's first-year curriculum, which was exclusively focused on Anglo-American common law doctrine, and his emphasis on a competitive, anonymous graded "meritocracy," this system still exercises an incredible grip on elite American law schools. But Langdell's nineteenth-century model has now been challenged by many rivals, including critical legal studies, "law and economics" empiricism, "global" curriculums, and clinical instruction.[3]

As is so often the case, Bacon anticipated these major forces of change. In his great *De Augmentis Scientiarum* (hereafter, *De Augmentis*), Bacon attacks the narrow parochialism of the common law pedagogy of his day: "For at present there are nothing but schools and institutions for multiplying altercations and controversies on points of law, as if for the display of wit. And this evil is also an old one …" (*SEH*, V: 108). Attacking reliance on decided judicial cases and on the parochial, prevailing common law treatises and pedagogy, Bacon evolved a new system of legal instruction

---

[1] A preliminary version of this article was presented as the Francis Bacon Annual Lecture at the Huntington Library, San Marino, California on May 8, 2003. I am most grateful to the Bacon Foundation and its devoted trustees and to Dr. David Sieburg and his colleagues at the Huntington Library. I am also grateful to my Administrative Assistant, Brendan Farmer, and Deborah H. Goldstein, Boston College Law School. Their intelligence and hard work was invaluable.

[2] See, for example, Harry T. Edwards, "The Growing Disjunction Between Legal Education and the Legal Profession," *Michigan Law Review* 91 (1992): 34, and Randall T. Sheppard, "What the Profession Expects of Law Schools," *Indiana Law Review* 34 (2000): 7. For a different perspective, see Fred C. Zacharias, "Why the Bar Needs Academics–and Vice Versa," *San Diego Law Review* 40 (2003): 701.

[3] The best accounts of Langdell's extraordinary career and influence on all American professional education are by Bruce A. Kimball. See particularly his "'Warn Students that I Entertain Heretical Opinions, Which They Are Not to Take as Law': The Inception of Case Method Teaching in the Classrooms of the Early C. C. Langdell, 1870–1883," *Law and History Review* (Spring, 1999): 57, and "The Langdell Problem: Historicizing the Century of Historiography, 1906–2000s," *Law & History Review* (Winter, 2004), 277.

based on empirical observation, distilled into maxims or aphorisms, one that sought true global significance and universal scientific legitimacy: "[T]here are certain fountains of natural equity from which spring and flow out the infinite variety of laws which individual legal systems have chosen for themselves. And as veins of water acquire diverse flavors … according to the nature of the soil through which they flow … just so in these legal systems natural equity is tinged and stained … according to the site of territories, the disposition of peoples, and the nature of commonwealth. It is worthwhile to open and draw out the purer fountains of equity, for from them all amendment of laws in any commonwealth must be sought."[4] This paper will set out Bacon's philosophy of legal education, analyze its pedagogical and doctrinal elements, and examine its lessons for American legal education today.

In so doing, it will be necessary to traverse a minefield of controversy. As E. O. Wilson has so powerfully described in his book *Consilience: The Unity of Knowledge* (1998), Bacon was "the grand architect" of an Enlightenment dream that "called for the illumination of the moral and political sciences by the 'torch of analysis.'"[5] Bacon was also devoted to a belief in a "unity of knowledge," relying on "the common means of inductive inquiry that might optimally serve all branches of learning." In E. O. Wilson's words, "Bacon envisioned a disciplined and unified learning as the key to improvement of the human condition."[6]

But the "unity" of the modern legal academy has been fragmented into academic specialties and increasingly divorced from the experience of law practice.[7] Equally important, law practice itself has changed. The "three qualities of modern law," described by Max Weber (1864–1920) and articulated in his great *Law and Economy and Society*, seem to be coming true.[8] First, the "legal ignorance of the layman" has increased, as legal rules become more specialized, complex, and technical. Most lawyers in modern firms are divided into such specialties, and usually have little or no idea of what their partners and associates actually do. Second, the "anti-formalistic tendencies of modern legal development" have led courts and tribunals to increasingly depart from "objective" or "universal" rules, and to rely instead on "economical utilitarian meaning." Finally, there is the "lay justice and corporate tendencies in the modern legal profession." Weber adds, "The use of jurors and similar lay judges will

---

4 Francis Bacon, "Aphorisms on the greater Law of nations or the fountains of Justice and Law," edited by Mark S. Neustadt as part of his Ph.D. thesis, "The making of the instauration: science, politics, and law in the career of Francis Bacon" (John Hopkins, 1987, rpt. University Microfilms Dissertation Information Services, 1990), 273, hereafter the "*Aphorismi*." The *Aphorismi* are based on Hardwick Manuscript 51, which was discovered in the Chatsworth House Library and recognized as a unique work of Bacon by Peter Beal. See *Aphorismi*, 241, and Beal, *Index of English Literary Manuscripts, 1450–1625* (London: 1980), 17–52.

5 Edward O. Wilson, *Consilience: The Unity of Knowledge* (New York: Knopf, 1998), 23.

6 Wilson, 27.

7 See Edwards, 34–5; J. Cunyon Gordon, "A Response from the Visitor from Another Planet," *Michigan Law Review* 91 (1993): 1953, 1960; and Sheppard, 11.

8 See Max Weber, *Law in Economy and Society* (trans. Edward Shils), ed. Max Rheinstein, (Cambridge: Harvard Univ. Press, 1954), 301, 302–4, 315–21.

not suffice to stop the continuous growth of the technical element in the law and hence its character as a specialist's domain."[9]

Add to those changes the rapid shrinking of world cultures by improved communications and the welcome, and dramatic, increase in cultural diversity throughout American law schools and American society generally, and it becomes clear that conventional legal pedagogies and curricula will come under great stress. The century-old orthodoxy of American legal education could soon be shattered into a hundred unrelated pieces. Can Bacon help us?

Bacon's university education

Bacon had a low opinion of his formal education, both at Trinity College, Cambridge, and at Gray's Inn of the Inns of Court, where he studied law. But in examining his ideas for education reform, it is logical to start with what happened to him. As was customary for his day, Bacon was matriculated at university at what was, by modern standards, a very early age–thirteen. In June of 1573, he "went up" to Trinity College, Cambridge with his older brother, fifteen-year-old Anthony, whose progress had been delayed by severe health problems. (Anthony, like his father, had severe asthma, which ran in the family.) The two boys roomed together at Trinity College, subject to the constant concern of their mother, whose anxious letters are still a primary source on Bacon's life.[10]

Bacon stayed at Cambridge until March 1576. He left before obtaining a degree. From his later criticisms in *The Advancement of Learning* and *De Augmentis*, it was clear he experienced the typical curriculum and pedagogy of an undergraduate at Trinity College, under its first statutes of 1552. Trinity was no ordinary college. As Arthur Gray observed, under these statutes, "Elaborate provision is made for a teaching staff within the College, and a schedule of the subjects of education is drawn up which conforms to the new curriculum of study which in the reign of Edward VI [1547–53] was substituted for the old routine prescribed for *trivium* and *quadrivium*."[11] This course was rigorous:

> Lectures take place in the morning. Those for bachelors are given in the parlour, for undergraduates in the hall. A "master of the hall" is to note the attendance, diligence and proficiency of the disciples, and to report thereon to the "seniority". If a disciple does not

---

9 See Weber, 301, 302–4, 315–21.

10 See Daniel R. Coquillette, *Francis Bacon* (Stanford: Stanford Univ. Press, 1992), 23–69, 311–15; Catherine Drinker Bowen, *Francis Bacon: The Temper of a Man* (Boston: Little Brown, 1963), 23–37; and Daphne du Maurier's *Golden Lads* (London: Gollancz, 1975) and *The Winding Stair: Francis Bacon, His Rise and Fall* (Garden City, NY: Doubleday 1977).

11 Arthur Gray, *Cambridge University: An Episodical History* (Cambridge: Cambridge Univ. Press, 1926), 93. See also J. M. Fletcher, "The Faculty of Arts," in *The Collegiate University,* ed James McConica, comprised within the *The History of the University of Oxford,* 8 vols., gen. ed. T. H. Ashton (Oxford: Clarendon, 1984–2000), III: 165–211.

> attend regularly he "loses his term". There are certain *quaesitores aulae*, called in the 1560 statutes *lectores*, who are to see that their classes make progress in the special subjects of Plato, Demosthenes and Cicero.[12]

The overall curriculum was dominated by Aristotle and his deductive system:

> The order of study is minutely prescribed. In his first year the disciple reads dialectic and the elements of Euclid: this is considered to be the best introduction to Greek philosophy … In his second year he is engaged on logic; in his third on ethics, politics and rhetoric; in his fourth on physical science. In all these subjects Aristotle is the textbook.[13]

Besides the required lectures, students were required to participate in so-called "repetitions," which were translations of Latin classics into Greek, and reverse. Finally, there were "declamations" in hall three times a week, in Latin, English, and Greek, "the subjects to be discussed … posted in the hall at dinner time".

> In the student's first year he translates Demosthenes into Latin and the orations of Cicero into Greek; in the next year he similarly renders Plato and the philosophical works of Cicero. The third and fourth years are devoted to revision. The disciple must provide himself, within two years, with copies of Aristotle, Plato, Demosthenes, Cicero, a Greek Testament and a Bible in Latin. If he fails to do so he is removed from College.[14]

Unless he chose to observe graduate lectures, Bacon would have had no opportunity to study law, which was taught as a graduate subject. If he had, the course of study would have focused solely on Roman law, primarily the *Institutes of Justinian*, the *Digest* and the *Codex*, all in Latin.[15]

Much of Bacon's mature philosophy was a pent-up rebellion against his university experience. Even after the "reforms" of the Restoration, everything began with the sacred classical texts, and worked down from there. The student, unlike Janus, faced only backward, not forward. "*Plus ultra*" ("even further") was Bacon's battle cry. One had to get past both the contents of the curriculum and the deductive pedagogy to do anything new. Bacon's first major philosophical work, *The twoo Bookes of Francis Bacon, of the proficience and advancement of learning, divine and humane* (London, 1605) [Gibson 81] (hereafter, *The Advancement of Learning*) was a sustained attack on this experience.[16] When, later in his life, Bacon would turn to classical antecedents in both law and philosophy, it was to create a new model, not Aristotle's *Organum*–drummed into his teen-aged head–but a *Novum Organum*, a new "tool," a new way.

---

[12] Gray, *Cambridge*, 93.

[13] Gray, *Cambridge,* 93–4. In 1591, "sons of certain noblemen" at Oxford were "allowed to request the BA degree after three years" (Fletcher, "The Faculty," 165).

[14] Gray, *Cambridge*, 94.

[15] See John Barton, "The Faculty of Law," *The Collegiate University,* ed. James McConica, in *The History of the University of Oxford*, III: 267–9.

[16] See Coquillette, *Bacon*, 77–90.

Bacon and his brother both left Cambridge in March 1575, without taking degrees. Perhaps little should be made of this. They were sons of privilege, and their father was one of the most powerful figures in the country. That privilege soon led to Bacon's appointment to the household of Sir Amias Paulet, ambassador to France. For almost three years, Bacon was able to study continental systems closely, and may well have picked up his taste for comparative analysis. But the death of his father on 20 February 1579 brought him home. Probably by accident, Bacon was ill-provided for by the estate–"I think I had greatest part in his love of all his children, yet in his wisdom served me as last comer."[17] In his own mind, Bacon had little choice but to make his own way. He took advantage of his father's vacant chambers in Gray's Inn, and thus began his legal education.

Bacon's legal education

The evolution of the Inns of Court and of Chancery in London had created a system of legal education that was completely independent of the universities. In John Baker's words: "The fifteenth-century inns of court and chancery together formed a law school not much smaller in size than the University of Cambridge. It would be called in Tudor times the Third University of England and with good reason, for it played at least as full a part as the other two older universities in the education of laymen destined for high or local office."[18] Thanks to John Baker, W. R. Prest, and Samuel Thorne, and their careful study of the readings and moots in the Inns, we now can recreate the world Bacon knew as a law student.[19] (Bacon's own accounts of his time from 1579 to his early admission as a barrister in 1584 consist primarily of letters to his uncle, Lord Burghley, seeking a position that would support his studies.)

In some ways, the process would have reminded Bacon of his university experience–and, indeed, he would later find similar faults with both. Value was placed on two things only, encyclopedic knowledge of the legal past–particularly statutes, pleadings, and some cases–plus oral dexterity in "translating" and using these antiquities for good advantage.[20] The "lectures" on Aristotle were replaced by "readings" on important statutes and processes; the "replications," by pleading exercises and other drills; and the "thrice-weekly declamations in hall," by the "moots," or case arguments. Participation in those exercises was not open to all, but was a function of advancement in rank. As Baker described it:

---

[17] Bowen, *Temper*, 41. Bacon's inheritance came to only £300 a year. By contrast, his eldest step-brother had an income of £6,000 a year (See Coquillette, *Bacon*, 23–4, 311–12).

[18] J. H. Baker, *An Introduction to English Legal History*, 3rd ed. (London: Butterworths, 1990), 184.

[19] See S. E. Thorne and J. H. Baker, *Readings and Moots at the Inns of Court in the Fifteenth Century*, 2 vols. (London: B. Quaritch, 1954–90), 2: 150, 2: xv–lxxvi; 1: 71. J. H. Baker, *The Education of Lawyers* (London: Selden Society, 1978), 94; and W. R. Prest, *The Inns of Court 1590–1640* (London: Longman, 1972). See also J. H. Baker, "Learning Exercises in the Medieval Inns of Court and Chancery," in *The Legal Profession and the Common Law* (London: Hambledon Press, 1986), 7–23.

[20] See Thorne and Baker, "Introduction," *Readings and Moots*, 2: xlix–lii.

> We have already seen how the early lectures and disputations in the common law mirrored those in the university law schools. Now, graduation in those medieval law schools was not a matter of passing an examination and then attending a degree ceremony, but a step (*gradus*) in the educational process. It occurred not by the incantation of a formula of conferment, but by performance of the appropriate exercise by the candidate himself … .[21]

This advancement up the common law ranks by "degree of learning," may well have inspired Bacon's notions of the "*Scala intellectus*," but, as we shall see, Bacon's system was quite different.

Preparation for the "moots" was also divided by experience. "Inner barristers," the lowest rank, helped prepare useful cases for the more senior lawyers, who had passed the bar and thus were "utter barristers." These utter barristers would actually argue the cases in hall, before the "elders of the house," the benchers. The "inner barristers" also had to "recite" the pleadings. Not all "moots" were formal moots in hall. Less formal "library moots" had "students arguing before utter barristers" and informal discussions "at the cupboard." Then there were "board's end cases" and "clerk's common cases" that "evidently involved the putting of specific questions by senior barristers or new benchers, without pleadings, and were therefore not moots in the strict sense of the term." These were found at Gray's Inn as early as the 1530s.[22] Finally, there were "bolts" or "boltings," which were specific doctrinal questions put to the students as a kind of drill, without the case pleadings.[23] According to David Walker and Earl Jowitt, these "bolts" were also private, unlike the public "moots."[24]

Pedagogically, this common law system also worked much like its intellectual rivals in the universities. The students' primary task was to master a vast body of preexisting knowledge in specialist languages (in this case Law French and Latin), with a good deal of highly technical jargon–particularly related to the "art" of pleading. The literature was also highly specialized and, with very few exceptions, there were no conceptual guides through the morass–those were to come more than a century later.[25] *Littleton's Tenures*, *Perkins' Profitable Book*, and *Fitzherbert's Natura Brevium* were essentially outlines or, in the latter case, a form book.[26]

In such a situation, the best "learning aids" were "readings" that efficiently conveyed information and rituals, drills and play-acting to fix facts and forms in your mind. The law was anchored in vast, immutable custom, at least in theory, which must be sorted through, classified, and memorized. No reader of *Coke's Institutes*, particularly *Coke on Littleton*, can fail to admire the vast common law memory that

---

[21] Thorne and Baker, 1: lvi .

[22] Thorne and Baker, 1: lxxiv.

[23] Thorne and Baker, lxxv.

[24] "Unlike moots … they [bolts] were held in private", David H. Walker, *The Oxford Companion to Law* (Oxford: Oxford Univ. Press1980), 140. The term is said to derive from the Saxon "*bolt*, a house." See Earl Jowitt, *The Dictionary of English Law*, ed. C. Walsh (London: Sweet & Maxwell, 1959), 259.

[25] See Baker, *Introduction*, 214–17; Daniel R. Coquillette, *Anglo-American Legal Heritage* (Durham: Carolina Academic Press, 1999), 272–4.

[26] See W. S. Holdsworth, *Sources and Literature of English Law* (Oxford: Clarendon, 1925), 112–61; T. F. T. Plucknett, *A Concise History of the Common Law*, 5th ed. (Boston: Little Brown, 1956), 276–81.

drove those works. Critical analysis was, by definition, worse than useless, it would invariably clog the machine's inherent assumption of legitimacy. Everything was deduced from the preexisting common law, which was studied in a complete vacuum from either social observation or scientific reasoning. Indeed, studies of the student notebooks at Litchfield Law School and Harvard Law School before its dramatic rescue by Joseph Story, show the same patterns. The "lecturer" sets out the law; the "crammer" reinforces it in the memory by questioning and drill; and the "moot court" dramatizes the abstract principles by role acting.[27]

Despite the paucity of records chronicling Bacon's five years as an "inner barrister," we do have some idea of what he thought of all this. First, he saw the system as immensely wasteful, commenting, "I am purposed not to follow the practice of law, it drinketh too much time, which I have dedicated to better purpose." This was, in part, because the system lacked conceptual "short cuts" to guide the efficient mind–which Bacon would later term "Maximes" or "Aphorismi"–and partly because, as he commented on the parallel deductive studies of Aristotle, "being a philosophy strong for disputations and contentions [such as "moots" and "bolts,"] but barren of the production of works for the benefit of the life of man."[28] In other words, this was a system of learning that was not, in any direct sense, an instrument of social good. Finally, in the absence of any direct link between the law–as a body of knowledge–and observation of the law's impact on human affairs, there could be no critical analysis of the system's efficacy or legitimacy. As Bacon observed, "for at present there are nothing but schools and institutions for multiplying altercations and controversies on points of law, as if for the display of wit. And this evil is also an old one. For it was likewise the pride of antiquity, as by sects and factions, to keep alive a number of questions of law, rather than settle them" (*SEH*, V: 108). Bacon was explicitly referring to the Sabinians and the Proculeans of the Roman Republic, but surely was also thinking of the moots and bolts of Gray's Inn! The ancient learning rituals of Bacon's law student days were, like the Aristotelian replications of his university, shackles to the intellect: "The philosophers lay down many precepts fair in argument, but not applicable to use: the lawyers, being subject and addicted to the positive rules … have no freedom of opinion, but as it were talk in bonds" (*SEH*, V: 88).

---

[27] John Langbein at Yale Law School is engaged in major research involving early American legal education in general, and the Litchfield Law School in particular. See John H. Langbein, "Blackstone, Litchfield, and Yale: The Founding of the Yale Law School," *History of the Yale Law School* (A. T. Kronman, ed., New Haven: Yale, 2004), 17–52. In addition, Lynne Templeton Brickley is involved in fundamental research on the Litchfield School. Her exhibition as Project Historian for the Litchfield Historical Society, "The Noblest Study, The Legacy of America's First School of Law," won the Wilbur Cross Award. Karen Beck, Curator of the Coquillette Rare Books Library at Boston College, is engaged in research on early student notebooks from Litchfield and elsewhere. See Karen S. Beck, "One Step at a Time: The Research Value of Law Student Notebooks," *Law Library Journal* 91 (1999): 29 and her exhibit catalogue, "Notable Notes: A Collection of Law Student Notebooks" (1999). See also David Warrington's talk to the 1992 Annual Meeting for the American Society for Legal History, "Under-Utilized Sources for Research in Legal History: Law Student Notebooks" (unpublished ms.).

[28] Bowen, 48, 35.

The "old exercises of [legal] learning" did not long survive the Civil War and the Interregnum. As J. H. Baker has demonstrated, the "moots" in today's law schools owe more to eighteenth-century debating societies than to any timeless rituals of the Inns of Court.[29] But the challenges of formalism and ritual thinking in legal education are very much with us. Out of his own experience, Bacon proposed a better way for his day and, quite curiously, for ours: a new pathway for professional learning.

Bacon's plans for the classification and storage of legal learning: The Advancement of Learning **and** *A Proposition to His Majesty*

Bacon's most important early works focused on the dual topics of education and law reform. In 1605 terrorist conspirators planned to blow up both Houses of Parliament, and James I, on Opening Day. Bacon would have been there. Had he died in that narrowly averted blast, *The Advancement of Learning* (1605), the *Essays* (1597) and the *Maximes of the Law* (written in 1596–97, and published posthumously in 1631), would have ensured his reputation.[30]

Although Bacon was fluent in Latin and professional Law French, he wrote *The Advancement of Learning* in English. This choice was doubtless political, the book was both dedicated to James I and directed at him and the politicians in his court.[31] Perhaps it was a symbolic rebellion as well against the tongue of the Inns of Court, whose rote education systems he challenged: "Here, therefore [is] the first distemper of learning, when men study words and not matter … ."[32] In this challenge he admitted his debt to the Counter-Reformation "college of the Jesuits" who, like himself, have made "a just complaint that states were too busy with their laws and too negligent in the point of education" (*SEH*, III: 277).[33] The ultimate goal would be a legal learning free of both the abstract deductions of the "school" philosopher, bringing forth "cobwebs of learning, admirable for the fineness of thread and work, but of no substance or profit" (*SEH*, III: 475)[34] and the narrow bonds of traditional legal education in the Inns of Court:

---

29 Thorne and Baker, *Readings and Moots,* 2: lxxv–lxxvi.

30 Coquillette, *Bacon*, 77, 325, 332–33.

31 Coquillette, *Bacon,* 78.

32 Coquillette, *Bacon,* 80.

33 See also *SEH*, III: 300, 416 and *SEH*, IV: 494, 496. It is quite appropriate that a Jesuit, Maurice B. McNamara, S. J., has made a major contribution to Bacon scholarship. See Coquillette, *Bacon*, 273, note 200. Bacon admired Jesuit education, continuing: "[W]hich excellent part of ancient discipline hath been in some sort revived of late times by the colleges of the Jesuits; *of whom, although in regard of their superstition I may say, quo meliores, eo deteriores* [*the better the worse*;] yet in regard of this, and some other points concerning human learning and moral matters, I may say, as Agesilaus said to *his enemy* Pharnabazus, *talis quum sis, utinam noster esses* [they are so good that I wish they were on our side" (*SEH*, III: 277).

34 See Coquillette, *Bacon*, 80, 273, note 200.

> [F]or the more public part of government, which is Laws, I think good to note only one deficience; which is, that all those which have written of laws, have written either as philosophers or as lawyers, and none as statesmen. As for the philosophers, they make imaginary laws for imaginary commonwealths; and their discourses are as the stars, which give little light because they are so high. For the lawyers, they write according to the states where they live, what is received law, and not what ought to be law: for a wisdom of a lawmaker is one, and of a lawyer another.

The ideal balance would be both learning and research, freed from rote forms dictated by the past, that focused on "the actual application of laws in practice and their constant improvement."[35] In Bacon's view, as I have noted elsewhere, this would "increase the social value of the legal theorist" while also improving "the quality of the theorist's jurisprudence by constant testing against the truths."[36]

But Bacon was not just concerned with the goals of legal learning, but with pedagogy, the best ways to "deliver" knowledge. Here he turned to his favorite tool, the aphoristic method:

> Another diversity of Method, whereof the consequence is great, is the delivery of knowledge in Aphorisms, or in Methods; wherein we may observe that it hath been too much taken into custom, out of a few Axioms or observations upon any subject to make a solemn and formal art; filling it with some discourses, and illustrating it with examples, and digesting it into a sensible Method; but the writing in Aphorisms hath many excellent virtues, whereto the writing in Method doth not approach. (*SEH*, III: 405)

The aphorism aided the mind by condensing knowledge into memorable "bites," without the seduction of mindless repetition, as in the traditional "replications" of the university or the "exercises" of the Inns of Court. The aphorism stretched the mind, and imposed its own rigor. (*SEH*, III: 405)[37] As we will see, the aphoristic method became a centerpiece of Bacon's expansion of *The Advancement of Learning* (1605) in his mature work, *De Augmentis* (1623), this time in Latin for the ages. But Bacon began using aphorisms much earlier in his first legal work, the *Maximes*, written between 1596 and 1597.

Bacon certainly did not invent the method. Common lawyers as early as 1546 circulated little treatises, such as the *Principia sive Maxima Legum Anglie*, organized around isolated "principal grounds and maxims."[38] But Bacon gave it a new style with his twenty-five maxims, notes written in English, but the maxims in Latin.

---

[35] Coquillette, *Bacon*, 84.

[36] Coquillette, *Bacon*, 84. See *SEH*, III: 475–6.

[37] Latin translation as given by Spedding. See Coquillette, *Bacon*, 83–4.

[38] See discussion in Coquillette, *Bacon*, 37–8. See also John C. Hogan, and Mortimer D. Schwartz, "On Bacon's 'Rules and Maximes' of the Common Law," *Law Library Journal* 76 (1983): 48, 55–6, discussing the *Principia sive Maxima Legum Anglie* (London, 1546) and *Sweet and Maxwell's Legal Bibliography*, 7 vols., 2nd ed., eds. W. H. Maxwell and L. F. Maxwell (London: J. Rees, 1989), I, 244: 9. Another example is William Noy's *A Treatise of the Principall Grounds and Maximes of the Lawes of this Kingdom* (London, 1641). See *Sweet and Maxwell*, I, 243: 8.

Ironically, he claimed to have originally written the book in Law French, although no copies survive (*SEH*, VII: 321–2). Moreover, Bacon was certainly aware that his maximes were student and analytical aids, and no more adequate to store the bulk of the law than Justinian's *Institutes* could replace the *Digest* and the *Codex*. James' accession to the throne in 1603, uniting the Crowns of England and Scotland, not only gave Bacon hope of advancement, but presented the potential problem of uniting the laws of the two kingdoms. Eager to gain favor with James, and also to improve the "storage" of English law in what today we would call a "manageable database," Bacon wrote no less than four tracts on the classification and storage of legal knowledge. The first, and least useful, was his *A Brief Discourse touching the Happy Union of the kingdoms of England and Scotland*, published in 1603, followed by *A Memorial touching the Review of Penal Laws and the Amendment of Common Law* (written in 1614), *A Proposition to His Majesty Touching the Compiling and Amendment of the Laws of England* (written 1616–17) and *An offer to the king of a Digest to be Made of the Laws of England* (written in 1621).[39]

Bacon's first point was that the law was: 1) "stored" in statutes, arranged chronologically with no effort at identifying anachronistic or inconsistent clauses, and with no method to search the statutes by topic; or 2) "stored" in year books and law reports,[40] also arranged chronologically, with no effort to locate overruled cases, anachronistic cases, cases that simply repeated the same proposition many times over, or cases that simply did not make sense as reported.[41] The library at Gray's Inn, doubtless familiar to Bacon, was organized that way, with a modest additional section for the black letter guides to students, like *Littleton's Tenures*, *Abridgements* of the year books, "Registers of Writs," and guide books for pleaders.[42] Indeed, if we were to ignore *Sheppard's Citator*, the various *Westlaw* and *Lexis* computer searchable databases, and the multi-volume treatises, this structure would be familiar to any modern law librarian and student, although on a different, and even more problematic scale.

As a first step to potential unification with Scotland, but also as a first step to a new vision of legal education, Bacon proposed to change all this:

> Wherefore, for the Common Law of England, it appeareth it is no Text law, but the substance of it consisteth in the series and succession of Judicial Acts from time to time which have been set down in books which we term *Year Books* or *Reports* so that as these Reports are more or less perfect, so the Law itself is more or less certain, and indeed better

[39] Despite Bacon's eagnerness for law reform, James I never directed him to this end. See the discussion in Coquillette, *Bacon*, 99–117.

[40] Coquillette, *Bacon*, 111–13.

[41] Coquillette, *Bacon*, 114–15.

[42] This is certainly the picture from the older holdings contained in the 1888 catalogue, assuming that there were no major deaccessions or accessions of old books. See D. R. Douthwaite, *Catalogue of the Books of the Honourable Society of Gray's Inn* (London: C. F. Rowroth, 1888). But, as Douthwaite observed: "There are no means of knowing what, or how many, books the library then contained; [late sixteenth-century] although it is pretty certain that the collection must have been small, even for those days" ("Preface," v).

> or worse. Whereupon a conclusion may be made, that it is hardly possible to confer upon this kingdom a greater benefit than if his Majesty should be pleased that these books also may be purged and reviewed, whereby they may be reduced to fewer volumes and clearer resolutions .... (*SEH*, XII: 85)[43]

Bacon's most complete vision of this massive reform of the fundamental "database" of English law was expressed in his *Proposition to his Majesty ... Touching the Compilation and Amendment of the Laws of England* (1616) (hereafter *A Proposition*). By now Bacon was Attorney General and was elevated to Privy Counsellor, but his criticism of his own legal system was scathing:

> But certain it is, that our laws, as they now stand, are subject to great incertainties, and variety of opinion, delays, and evasions: whereof ensueth,
>
> 1 That the multiplicity and length of suits is great.
> 2 That the contentious person is armed, and the honest subject wearied and oppressed.
> 3 That the judge is more absolute; who, in doubtful cases, hath a greater stroke and liberty.
> 4 That the chancery courts are more filled, the remedy of law being often obscure and doubtful.
> 5 That the ignorant lawyer shroudeth his ignorance of law in that doubts are so frequent and many.
> 6 That men's assurances of their lands and estates by patents, deeds, wills, are often subject to question, and hollow; and many the like inconveniencies (*SEH*, XIII: 64–5).

Bacon's solution was a massive program: first, to sort through the "ancient records" and to select and register those of "most worth and weight" to be used as "reverend precedents, but not for binding authorities"; second, to create a set of scrutinized and approved law reports "the perfect course of law," rejecting those which were anachronistic, over-ruled, repetitious, "idle queries, which are but seminaries of doubts," and "reported with too great proclivity" (*SEH*, XIII: 68–9); third, to create a similarly purged "codex" of all the statutes; and, finally, to produce three authoritative "auxiliary books," and "Institutions," based on Justinian's great textbook of 533 AD, an official law dictionary, *De verborum significationibus*, based on the famous Chapter XVI of Book Fifty of Justinian's *Digest* and–last but not least–a *De regulis juris*, probably in aphoristic form (*SEH*, XIII: 70).[44]

Bacon saw these great projects as the cornerstone of his new system of legal education. Even after his impeachment and disgrace in 1621, he dreamed of completing at least part of his plan. In his sad *An offer to the King of a Digest to be Made of the Laws of England*, sent with his letter of 20 March 1621 thanking James for preventing his continued imprisonment in the Tower, Bacon wrote:

> And because in the beginning of my trouble, when in the midst of the tempest I had a kenning of the harbour which I hope now by your Majesty's favour I am entering into, I

---

43 These proposals were strongly influenced by Roman Law. For discussion of Bacon's relation to the Roman texts see Coquillette, *Anglo-American*, 4–10 and *SEH*, XII: 85–6.

44 For a full discussion, see Coquillette, *Bacon*, 112–13.

> made a tender to your Majesty of two works, *An history of England* and *A digest of your laws*; as I have (by a figure of *pars pro toto*) performed the one, so I have herewith sent your Majesty, by way of an epistle, a new offer of the other. But my desire is further, if it stand with your Majesty's good pleasure, since now my study is my exchange and my pen my factor for the use of my talent, that your Majesty (who is a great master in these things) would be pleased to appoint me some task to write, and that I shall take for an oracle. (*SEH*, XIV: 357–8)

Bacon never gave up his dream of what today we would call a concise, authoritative, systematic, quality-reviewed, fully searchable database of law. The American law student of today crouches in a vast library of hundreds of thousands of chronological law reports, statutes, and impenetrable state and federal "codes," using computer search engines that generate endless unreviewed and unreliable answers. To this student, adrift in a sea of information, Bacon's vision is still a dream.

Bacon's theory of education: the *Scala Intellectus*

Having described Bacon's heroic efforts to reform the vast and desperately untidy attic that was the legal knowledge of his day, we now come to Bacon's theories of education. In general, these are the products of some of Bacon's more mature works, the *Novum Organum* (1620), the *De Augmentis* (1623) and the Chatsworth House Hardwick MS 51, the *Aphorismi* (written *circa* 1614).[45]

In the fifth book of the *De Augmentis*, Bacon observed that "men's labour in rational knowledge is either to invent that which is sought, or to judge that which is invented, or to retain that which is judged, or to deliver over that which is retained" (*SEH*, V: 407). From this observation, he derived the four "logical" or "rational arts": 1) the "Art of Inquiry or Invention"; 2) the "Art of Examination or Judgment"; 3) the "Art of Custody or Memory"; and 4) the "Art of Elocution or Tradition"[46] (*SEH*, V: 407).

Put in modern terms, the "four rational arts" correspond to the basic functions of a modern research university, of the kind described in the *New Atlantis*.[47] The art of "inquiry or invention" is the research function. The "art of examination or judgment" is the function of critical analysis, a function essential to the mature and wise use of research. The "art of custody or memory" is represented by the modern research

---

45 See note 4, *supra*, and the discussion at Coquillette, *Bacon*, 338.

46 The "Art of Elocution or Tradition" is better translated as "transmission." See Harvey Wheeler's "Francis Bacon's 'Verulamium': the Common Law Template of the Modern in English Science and Culture," *Angelaki: Journal of the Theoretical Humanities* 4 (May, 1999): 7 at 20.

47 The utopia of Bensalem had a great research facility called "Salomon's House" or the "College of the Six Days Work" to study the natural world. Coquillette, *Bacon*, 259, notes omitted, citing (*SEH*, III: 159, 163). Wheeler has described this as a "vast mission-oriented 'leading edge' scientific establishment that directed society's development" [Harvey Wheeler, "Francis Bacon's *New Atlantis*: The 'Mould' of a Law Finding Commonwealth" in *Francis Bacon's Legacy of Texts*, ed. W. A. Session, (New York: AMS, 1990), 297].

library and, as Wheeler would put it,"database archival resources."[48] Finally, there is the art of "transmission," the art of transmitting learning from teacher to student, or pedagogy.

As we have seen, Bacon's earlier works focused on the obvious faults of the common law "memory"–a vast, unsearchable, and illogically organized "database archival resource."[49] But, as *The Advancement of Learning* demonstrates, Bacon was also concerned with sophisticated and efficient "transmission" of that knowledge, hence his work on the aphoristic method in the *Maximes of the Law* (written 1596–1597) and elsewhere.[50]

Today, we would all agree that a first-rate university or professional school would require a great knowledge "memory," a "database archive" as intelligently accessible as possible. And we would also agree that first-class "transmission," or teaching, skills were highly desirable, to pass essential knowledge from one generation to the next–or at least the techniques of extracting that knowledge from the great "database archive" or, even better, by direct empirical research. But it is Bacon's latter two "rational" arts that form the most compelling rationale for first-class education: the art of "invention," the creation of new data and insights by research, and the art of "examination or judgment."

Most of us can imagine a fine educational institution handicapped by a poor research library and misguided and inefficient methods for transmitting knowledge. Indeed it clearly resembles the early Harvard Law School, with its grossly defective library–looked after by the janitor–and a "crammer" method of class recitation. But what we cannot imagine is a truly first-class institution that is indifferent to serious research and fails to develop critical judgment in both students and faculty. These latter two "rational" arts, the nature of true research and developing critical judgment, were the focus of Bacon's mature insights in the *Novum Organum* (1620) and *De Augmentis* (1623).

Of course Bacon understood that "invention" and "judgment" require a framework of knowledge. The first rungs of his "ladder of understanding," the *Scala Intellectus sive Filum Labyrinthi* required amassing that critical mass of knowledge (*SEH*, II: 687–92).[51] This is where the "transmission" techniques for conveying knowledge effectively came in, and the abilities necessary to access the knowledge store of the archival "memory." Bacon would not regard the learning rituals of his early days in the university and Gray's Inn as useless or unnecessary–but rather as wasteful, unimaginative, and inefficient in achieving their goals. Nor would he regard the law libraries of the Inns of Court as unnecessary, but rather as pathetically lacking in what today we would term "searchable" databases and, more importantly,

48 Wheeler, "Francis Bacon's 'Verulamium,'" 21.

49 Wheeler "Francis Bacon's 'Verulamium,'" 20–1 and Coquillette, *Bacon*, 99–117.

50 Coquillette, *Bacon*, 325.

51 Bacon used the term "*Scala Intellectus*" to refer to "building tentative axioms from empirical research, seeking practical application, and then attempting to develop even higher truths that, in turn, would suggest new empirical experiments, just as proposed in the *Novum Organum*" (Coquillette, *Bacon*, 260). This, of course, requires a gradual progression of expertise, step by step. See also Coquillette, *Bacon*, 260–1 and 335.

contaminated with poor-quality data, data that was fragmented, unreviewed, anachronistic, misleading, and–as a matter of social utility–profoundly disconnected to any social good (*SEH*, IV: 13–21).[52]

But let us assume that we adopt Bacon's many suggestions on aphoristic pedagogy and on the storage and classification of legal knowledge. Bacon still would be profoundly dissatisfied unless there was similar progress with research and critical judgment. Indeed, once past the first few rungs of the *Scala Intellectus*, all four of the "rational arts" must be integrated. Perhaps this is most obvious with the development of the "invention" or "research" function (*SEH*, IV: 407). Bacon, as we know, believed that theory must be constantly informed and corrected by inductive, empirical research:

> For all those who before me have applied themselves to the invention of arts have but cast a glance or two upon facts and examples and experience, and straightway proceeded, as if invention were nothing more than an exercise of thought, to invoke their own spirits to give them oracles. I, on the contrary, dwelling purely and constantly among the facts of nature, withdraw my intellect from them no further than may suffice to let the images and rays of natural objects meet in a point, as they do in the sense of vision; whence it follows that the strength and excellency of the wit has but little to do in the matter. (*SEH*, IV: 19)

Part of Bacon's genius was that he did not restrict such inductive "invention" to the natural sciences, but foresaw that the empirical study of mankind and human conduct was equally possible and valuable, thus anticipating the social sciences, including linguistics and sociological jurisprudence.[53] Whether it be a study of Aristotle's *Ethics* or Henry VIII's Statute of Uses, the proof was "in the pudding" of actual human affairs.[54] The gap between the university and professional training of his day and the world of inductive observation and actual practice was, to Bacon, like a bad divorce, "the unkind and ill-starved divorce and separation of which has thrown into confusion all the affairs of the human family" (*SEH*, IV: 19). The cure was "a true and lawful marriage between the empirical and the rational faculty" (*SEH*, IV: 19, 50).

But imaginative, empirical research alone was not sufficient. That "rational art" must join with the "Art of Examination or Judgment" (*SEH*, IV: 407). As all true teachers know, instilling critical judgment in students is the greatest task. Bacon saw it as first, overcoming the characteristic impediments to human understanding by self-knowledge and teamwork and second, developing the intellectual courage

---

[52] According to Bacon, "For in like manner the sciences to which we are accustomed have certain general positions which are specious and flattering, but as soon as they come to particulars, which are as the parts of a generation, when they should produce fruit and works, then drive contentions and barking disputation, which are the end of the matter and all the issue they can yield" (*SEH*, IV: 14).

[53] See Coquillette, *Bacon*, 233–4.

[54] See Bacon's extraordinary "Learned Reading … Upon the Statute of Uses," one of the required Double Readings to advance at Gray's Inn. It was delivered in the Lent vacation, 1600 (*SEH*, VII: 391). In this reading, Bacon decried the erosion of the public policy goals of the great Statute of Uses (1536) by the use of blatant legal fictions. See the full discussion at Coquillette, *Bacon*, 48–59.

necessary for true intellectual independence. Only with adequate self-knowledge and intellectual independence could the "judgment" faculty be adequate, in student or teacher, and only with good critical judgment can research, or invention, proceed on a sound footing. The two "faculties" are mutually dependent.

But, of course, there were powerful obstacles to self-knowledge, which Bacon described by his famous "idols":

> The idols and false notions which are now in possession of the human understanding, and have taken deep root therein, not only so beset men's minds that truth can hardly find entrance, but even after entrance obtained, they will again in the very instauration of the sciences meet and trouble us, unless men being forewarned of the danger fortify themselves as far as may be against their assaults. (*SEH*, IV: 53)

I will not here review in depth Bacon's "four idols": the "Idols of the Cave"–the defects of the individual psyche; the "Idols of the Tribe," that "have their foundation in human nature itself, and in the tribe or race of men;" the "Idols of the Market-place," that are "formed by the intercourse and association of men with each other";[55] and finally, the "Idols of the Theatre," the idols which have immigrated into men's minds from the various dogmas of philosophies, and also from wrong laws of demonstration" (*SEH*, IV: 55). It is the latter, the "Idols of the Theatre," that Bacon believed to be the most dangerous to education. A quick review of these "Idols of the Theatre" will show why they are so dangerous, and particularly why they are dangerous to good legal education.

The "Idols of the Theatre" were so named because, in Bacon's view, doctrines and dogmas are inherently artificial. They are over-simplifications and imitations of reality, inherently artificial, although they would have you believe they are real … just like a stage play. In Bacon's words:

> These I call Idols of the Theatre; because in my judgment all the received systems are but so many stage-plays, representing worlds of their own creation after an unreal and scenic fashion. Nor is it only of the systems now in vogue, or only of the ancient sects and philosophies, that I speak; for many more plays of the same kind may yet be composed and in like artificial manner set forth; seeing that errors the most widely different have nevertheless causes for the most part alike. Neither again do I mean this only of entire systems, but also of many principles and axioms in science, which by tradition, credulity, and negligence have come to be received. (*SEH*, IV: 55)

Bacon certainly regarded the classical pedagogy of his university training, including the "repetitions" and "declamations," as this type of dangerous "theater," and the "moots" and "bolts" of the Inns of Court as well. Today, he would be joined by the many postmodern critics of contemporary legal education. Such education remains dominated by the "Socratic" drills and the "case method" of Langdellian formalism.

---

[55] Here Bacon anticipates modern linguistics–"words plainly force and overrule the understanding, and throw all into confusion, and lead men away into numberless empty controversies and idle fancies" (*SEH*, IV: 55).

It is also still concentrated, in the first year at least, on appellate cases which–in their remoteness and intellectual "focus"–can be as detached from reality and as artificial as any stage play. Just ask a trial judge.

So what are Bacon's answers? First, self-conscious awareness of all the "idols" will, in Wheeler's words, be "data cleansing."[56] True, the human mind will always be an "enchanted glass," incapable of purely accurate comprehension (*SEH*, IV: 54). But the inaccuracies of the "false mirror" can be mitigated by making conscious allowance for the defects, in the same way the mirror of a telescope can be repaired. There is a particular trap in the study of law. Bacon observed that "[t]he human understanding is of its own nature prone to suppose the existence of more order and regularity in the world than it finds" (*SEH*, IV: 55). Bacon was prophesying the inherent weaknesses of elegant Newtonian physics, a masterpiece of the Enlightenment. But the seduction is even greater in legal study, where it becomes easy to lose sight of the forest of human needs through the intervening doctrinal trees. Bacon was certainly aware of the peculiar dangers of legal education, as his "Preface" to the *Maximes of the Law* makes clear, and it led to his controversial decision to omit all decided authority–all precedential cases–from that aphoristic text book (*SEH*, VII: 322–3). If his legal proposals did not stand alone, vindicated by direct empirical evidence of utility and common benefit, the trappings of authority were the very idols that the student must self-consciously resist (*SEH*, VII: 322–3).[57]

It was also important how doctrine was taught. Here we come full circle. The rational faculty of "invention," or research, required the rational faculty of "examination" or "judgment" to "cleanse" the data of the ever-present idols. Only in this way could knowledge accurately be taught and stored. Thus good research and teaching were inextricably linked. In Bacon's words, from the Preface of the *Novum Organum*:

> And the same humility which I use in inventing I employ likewise in teaching. For I do not endeavour either by triumphs of confutation, or pleadings of antiquity, or assumption of authority, or even by the veil of obscurity, to invest these inventions of mine with any majesty; which might easily be done by one who sought to give lustre to his own name rather than light to other men's minds. I have not sought (I say) nor do I seek either to force or ensnare men's judgments, but I lead them to things themselves and the concordances of things, that they may see for themselves what they have, what they can dispute, what they can add and contribute to the common stock. (*SEH*, IV: 19)

Bacon was clearly aware of the two great legal pedagogies of his day, the civilian study of the Roman law and the common law study of precedents, authorities, and statutes. As we have seen, he had little sympathy for the latter.[58] But what about the Roman law and the civilian pedagogy of the ancient universities?

---

56 Wheeler, "Francis Bacon's 'Verulamium,'" 27.

57 Bacon observed "although the meaness of mine own person may at first extenuate the authority of this collection … time will … authorize it." (*SEH*, VII, 322–3). When he wrote those words in 1596, Bacon was indeed "mean" in his "own person." Ten years later he was Lord Keeper! (Coquillette, *Bacon*, 313–19).

58 See notes 30–31, *supra* and *SEH*, V: 88, 108.

In Aphorism 93 of the *De Augmentis* Bacon directs that "[T]he lectures and exercises of those who study and labour at the law be so ordered and instituted, as rather to set legal questions and controversies at rest, than to raise and excite them" (*SEH*, V: 108). It could be argued, particularly because of the use of the word "instituted," that this was a reference to the pedagogy symbolized by the classic civilian text, Justinian's *Institutes*. First conceived by the Emperor Justinian himself, as part of his wholesale review of the classical Roman texts, between 528 AD and 535 AD, the *Institutes* were to be a first rung on the "ladder of learning" of Roman law students. Keyed directly to the great *Digest* and *Codex* of the same period, the *Institutes*, published on 21 November 533 AD, were a splendid "nutshell" guide to beginning students, all the more remarkable for being "spoken" directly by the Emperor, and thus law in its own right: "Justinian's design had been to embrace in his three authoritative works every jot and tittle of positive law."[59] All other law books were ordered to be destroyed.

Justinian's teaching philosophy was plain. Not for him the torturous "moots" and "bolts," or the "Socratic" agonies of the modern law student. Directly addressing his law students in "Title I," the Emperor observed:

> Having laid down these general definitions, and our object being the exposition of the law of the Roman people, we think that the most advantageous plan will be to commence with an easy and simple path, and then to proceed to details with a most careful and scrupulous exactness of interpretation. Otherwise, if we begin by burdening the student's memory, as yet weak and untrained, with a multitude and variety of matters, one of two things will happen: either we shall cause him wholly to desert the study of law, or else we shall bring him at last, after great labour, and often, too, distrustful of his own powers (the commonest cause, among the young, of ill-success), to a point which he might have reached earlier, without such labour and confident in himself, had he been led along a smoother path.[60]

How many American law students, frustrated and demoralized by the "hide the ball" teaching techniques of the orthodox first-year "Socratic" method would weep at those words! The civilian approach was not to require a student to reinvent biology by dissecting a frog, but to lay out a simple initial path, rational and clear, that would become increasingly complex once the initial steps were mastered. Was this Bacon's idea of the "*Scala Intellectus*," one rational step at a time?

It is certainly true that Bacon was enamored of the civilian learning, and that his great legal classification schemes, such as *A Proposition to His Majesty*, owed much to the structure of the *Institutes*, the *Digest* and the *Codex*.[61] As Bacon observed, in his letter to George Villiers, Duke of Buckingham and the King's favorite:

> Although I am a professor of the common law, yet am I so much a lover of truth and of learning, and of my native country, that I do heartily persuade that the professors of law,

59 J. B. Moyle, *Imperatoris Iustiniani Institutionum* (Oxford: Clarendon Press, 1883), "General Introduction," 73–81.

60 J. B. Moyle, *The Institutes of Justinian Translated Into English*, 5th edn. (Oxford: Clarendon Press, 1913), 1–3.

61 See Coquillette, *Anglo-American*, 9; Daniel R. Coquillette, *The Civilian Writers of Doctors' Commons* (Berlin: Duncker & Humblot, 1988), 257–9.

> called civilians, because the civil law is their guide, should not be discountenanced or discouraged: else whensoever we shall have aught to do with any foreign king or state, we shall be at a miserable loss, for want of learned men in that profession. (*SEH*, VI: 27, 39)[62]

But Roman law instruction, based on Justinian's "sacred texts," was still a "top down," deductive pedagogy, that assumed the "starting" point of the classical texts. Of course, Bartolist jurists had long discovered how to apply Roman law to improve the emerging legal systems of Renaissance Europe. But the course of instruction–then, as during my Oxford Roman law studies in the 1960s–was based on close analysis of a closed, indeed dead, world.[63]

Not surprisingly, therefore, when Bacon actually attempted to establish an aphoristic guide to the first rungs of the *Scala intellectus*, he refused to blindly adopt civilian models. On the other hand, unlike his famous contemporary, Coke, he equally refused to reject such models outright, particularly when they were useful.[64] In his "Preface" to the *Maximes of the Law*, Bacon noted that some of his maximes, or "rules," were concurrent with Roman models:

> … [W]hereas some of these rules have a concurrence with the civil Roman law, and some others a diversity, and many times an opposition; such grounds as are common to our law and theirs I have not affected to disguise into other words than the civilians use, to the end they might seem invented by me, and not borrowed or translated from them: no, but I took hold of it as matter of great authority and majesty, to see and consider the concordance between the laws penned and as it were dictated *verbatim* by the same reason. (*SEH*, VII: 321)

As will be seen, Bacon rejected the legal nationalism of the common law, and admired the global pretensions of the Roman law and its civilian jurists. But the "global" law he sought to teach was based on experience and reason, not ancient text, or Bartolist adaptations.[65]

Nor was Bacon wedded to a rigorous, civilian curriculum, in which every law was in its exact place in a deductive hierarchy, and carefully related to each other class of laws. Indeed, as Bacon noted below, even Justinian's *Digest* was more "free-form"

---

62 See also Coquillette, *Civilian Writers*, 257–9.

63 Coquillette, *Civilian Writers*, 32–44 and 22–9.

64 In his Preface to 7 *Coke's Reports* (1608), Coke wrote: "And sure I am that no man can either bring over those bookes of late written (which I have seene) from Rome or Romanists, or read them, and justifie them … but they rune into desperate dangers and downefals; for the first offense is a *praemunire* … and they, that offend the second time therein, incurre the heavie danger of high Treason. These bookes have glorious and goodly titles … but there is *mors in olla*: They are like to Apothecaries boxes … whose titles promise remedies, but the boxes themselves containe poyson" [7 *Coke's Reports* (London: 1608), "Preface," 8th unpaginated page].

65 Bacon writes in *De Augmentis*: "Wherefore let it be my present object to go to the fountains of justice and public expediency, and endeavour with reference to the several provinces of law to exhibit a character and idea of justice, in general comparison with which the laws of particular states and kingdoms may be tested and amended" (*SEH*, V: 88).

than that. Thus, Bacon preferred a more open structure for his aphorismic introductory text–to encourage the students to think and, most importantly, to apply the principles to practice. As Bacon observed:

> [W]hereas I could have digested these rules into a certain method or order, which, I know, would have been more admired, as that which would have made every particular rule, through his coherence and relation unto other rules, seem more cunning and deep; yet I have avoided so to do, because this delivering of knowledge in distinct and disjoined aphorisms doth leave the wit of man more free to turn and toss, and to make use of that which is so delivered to more several purposes and applications. (*SEH*, VII: 321)

Not only did Bacon reject the extremes of both the civilian and common law pedagogies, he was one of the first jurists to recognize the importance of comparative analysis of rival legal systems. Indeed, Bacon believed that there was a body of principles common to all legal systems and, like the Roman *ius gentium*, these principles could be discovered by careful observation and comparison of the actual operation of the law in different nations and systems. This was Bacon's theory of the "purer fountains."

Bacon explored this idea most thoroughly in his *Aphorisms on the greater law of nations or the fountains of Justice and Law*, known as the *Aphorismi*. Written about 1614, these *Aphorismi* have only recently been discovered in Hardwick MS 51, a manuscript in a library tended by Thomas Hobbes, Bacon's former secretary, and set out and annotated by Mark Neustadt in 1987.[66] Bacon's belief, as set out in the *Aphorismi*, was that there were universal principles of law, just like that of natural science. While the narrowly educated lawyer learns and appreciates only the positive rules and laws "limited by the boundaries of regions and commonwealths," true legal education must be global in scope, empirical in character, and comparative in technique. Of course, individual states adopt these principles of "natural equity" to their own needs, but this does not detract from their inherent universality:

> There is little doubt, meanwhile, but that there are certain fountains of natural equity from which spring and flow out the infinite variety of laws which individual legal systems have chosen for themselves … It is worthwhile to open and draw out the purer fountains of equity, for from them all amendment of laws in any kingdom or commonwealth must be sought.

And who should teach in the "global law school?" Certainly not "pure academics," detached from empirical study and human reality, nor narrow, nationalistic legal practitioners:

> Certainly it partakes of a higher science to comprehend the force of equity that has suffused and penetrated the very nature of human society. And it should not be concealed that both philosophers and lawyers seem little suited or proper for this work. Philosophers, who wander the pleasant by-ways of contemplation, ornament civil activities more than they

[66] On Bacon and Hobbes see Coquillette, *Bacon*, 237–44; *Aphorismi*, 135–43, 239–46.

> improve them; and lawyers, bound by their rules and formulae, do not employ a free and higher judgment.[67]

What was needed, instead, was a law school by and for "lawyer statesmen," who understood the value of direct observation of human affairs:[68]

> Truly this work and endeavor belongs chiefly to statesmen and those skilled in public affairs who have learned about human dispositions, compulsions and habits by much practice and attentive reading, listening, and observation.[69]

Of course, this could be a description of Bacon himself. Even his choice of the term "equity," rather than "law," for the "purer fountains" anticipated Bacon's elevation to Lord Keeper in 1617, the head of the equitable Court of Chancery. For the term "Law" was traditionally drawn from authority; "equity" was realm of reason and fairness.[70] But lawyer statesmen had a drawback, they were too busy to found law schools:

> But these political men are often made for action and not suited for teaching. Moreover they are not rich in leisure, and are satisfied if they take care for their own times. (Indeed, among them care for posterity commonly does not go beyond the boundaries of their own families.)[71]

Indeed, Bacon himself was too busy, as chief legal officer of the kingdom, and, eventually, Viceroy. He left no detailed plans of any school or curriculum, only the extraordinary vision of a research university–the vast laboratories and "engine-houses" of "Salomon's House" of Bensalem–set out in his haunting utopia, the *New Atlantis* (*circa* 1624).[72] Like his other mature works, the *New Atlantis* was the product of the forced leisure of Bacon's impeachment. And it was incomplete. It cut off, either by accident or deliberately, just before the promised "frame of Laws," or "the best state or mould of a Commonwealth" (*SEH*, III: 127).[73] Thus there was no description of the law school of utopia. So let us do something outrageous. Let us imagine how Bacon would resolve the postmodern crisis facing professional education in general, and American legal education in particular.

---

67 *Aphorismi*, 273, 272.

68 For a current expression of this view, see Anthony T. Kronman, *The Lost Lawyer: Failing Ideals of the Legal Profession* (Cambridge, Mass.: Belknap Press, 1995).

69 *Aphorismi*, 272–3.

70 See Daniel R. Coquillette, "Equity," *Dictionary of the Middle Ages* (1981), set out in Coquillette, *Anglo-American*, 191–5.

71 *Aphorismi*, 273.

72 See note 52, *supra*. See also Coquillette, *Bacon*, 261–2. Salomon's House had a mission statement worthy of any modern university: "The End of our Foundation is the knowledge of Causes, and secret motions of things; and the enlarging of the bounds of Human Empire, to the effecting of all things possible" (*SEH*, III: 156).

73 See Coquillette, *Bacon*, 261–2.

"New Pathways": A Neo-Baconian pedagogy in medicine

In 1985 Harvard Medical School introduced a radical new approach to medical education, the "New Pathway" Program.[74] First conducted as an experiment in its first two years (1985–86 and 1986–87), it aimed at integrating traditional "abstract" learning (anatomy, pathology, pharmacology, etc.) with a more "integrated" medicine, focusing on actual patients and healing. The goal was to graduate doctors with "more positive attitudes toward all behaviors related to humanistic medicine, lifelong learning and issues related to learning in a social environment."[75] In short, the effort was how to train doctors that could exercise all of Bacon's four "rational arts" on a life long basis: to access acquired knowledge ("art of custody or memory"); to test and improve that knowledge by direct experiment and research ("art of inquiry or invention"); to apply critical judgment to such raw experience ("art of examination or judgment"); and to be able to transmit that knowledge effectively–to teach ("art of elocution or tradition and transmission") (*SEH*, IV: 104).[76] All this through a new curriculum, that moved students through the easiest rungs of the *Scala intellectus* by normal classroom pedagogy, but exposed them as quickly as possible to direct experience, in which their doctrinal learning could be quickly tested and improved.[77]

To see how this would work in practice, let us examine the basic four-year American medical school curriculum.[78] It has been traditionally divided in two parts. The first two years is largely classroom learning, taught mostly in lecture form by faculty with Ph.D.s in the sciences. These are mandatory courses (no electives) and are focused on the "building" blocks of the sciences. The goal is to pass Part One of the United States Medical Licensing Examination, administered by the National Board of Medical Examiners ("NBME," hereafter "the Boards"). The examination is multiple choice and tests "whether the student can understand and apply important

74 A. S. Peters, *et al.*, "Long-term Outcomes of the New Pathway Program at Harvard Medical School," *Academic Medicine* 75 (2000): 470, published at http://www.academicmedicine.org/cgi/content/full/75/5/470#REF2.

75 Peters, 470–1.

76 See Wheeler, "Francis Bacon's 'Verulamium,'" 47.

77 See Gordon T. Moore, Susan D. Block, C. Briggs-Style, R. Michell, "The Influence of the New Pathway Curriculum on Harvard Medical Students," *Academic Medicine* 69 (1994): 983–9; Gordon T. Moore, "The First Curriculum: Content and Process," *New Pathways to Medical Education: Learning to Learn at Harvard Medical School*, eds. D. C. Tosteron, S. J. Adelstein, and S. T. Carver (Cambridge: Harvard Univ. Press, 1994); Gordon T. Moore, "The Effect of Compulsory Students in Problem-Based Learning," *Medical Education* 25 (1991): 140–3; Kenneth M. Ludmerer, *Time to Heal: American Medical Education from the Turn of the Century to the Era of Managed Care* (Oxford: Oxford Univ. Press, 1999), 349–99 and *Learning to Heal: the Development of American Medical Education* (New York: Basic Books, 1985); Robert H. Ross, *Innovators in Physician Education: the Process and Pattern of Reform in North American Medical Schools* (New York: Springer Pub., 1996) and *Beyond Flexner: Medical Education in the Twentieth Century*, eds. Barbara Barzansky and Norman Gevitz, (New York: Greenwood, 1992).

78 The English curriculum is very different, usually starting at the undergraduate level. I am indebted to my excellent research assistant, Deborah H. Goldstein, for her research memorandum, "British Medical Training Compared with American Medical Training" (2003, unpublished ms).

concepts of the sciences basic to the study of medicine."[79] It is very hard, with a 20% failure rate. Most medical students regard it as an exercise in memorizing and applying a "given" body of doctrine. Multiple choice tests punish originality, and Part One of "the Boards" is no exception. The next two years are largely clinical in nature, and consist of rotations in cooperation with one or more affiliated hospitals. These usually include a required third-year curriculum including "three months of internal medicine, three months of general surgery, and six weeks of OB/GYN, psychiatry, pediatrics and neurology."[80] Only in the fourth year do students have any serious choices. At the end of the fourth year, students must pass Part Two of "the Boards."[81]

The "New Pathway" was designed to combat the inherent fragmentation and narrowness of this process by focusing on reforming the first two years. As initiated at Harvard in 1988, it introduced students to clinical settings almost at once, replacing the traditional lectures and doctrinal "pigeon holes" with "a problem-based approach emphasizing small-group tutorials" and "self-directed learning complemented by laboratories, conferences and lectures."[82] The students then took Part One of "the Boards" and continued into the third and fourth years of specialty rotation, which remained basically unchanged.

In 1998 a "randomized controlled trial" was conducted "to evaluate the long-term effects of an innovative curriculum, the New Pathway (NP) Program, on behaviors and attitudes related to humanistic medicine, lifelong learning and social learning." This study compared 100 randomly selected 1989 and 1990 Harvard Medical School graduates, 50% who had studied under the "New Pathway" program, and 50% who had "the traditional curriculum." The study did show significant differences–the most striking being the fact that "40% of the NP graduates remained in low-technology, psychosocially oriented careers (primary care and psychiatry)–more than twice the number of traditional graduates (18%)." The study continued "[t]hus, unlike typical residents, the NP graduates were less influenced to leave primary care by the high-technology, hospital based nature of residence education," i.e., the highly fragmented system of specialties described above.[83] Given the fact that the "New Pathway" only made major changes in the first two years of a professional education that, counting internship, residencies, and fellowships, could easily exceed ten years, these results are quite striking.

Bacon would certainly regard the New Pathway as a step in the right direction, but would certainly deride its limitations. To begin, it only applies to the earliest stages in this particular *Scala Intellectus*. In terms of two of his "rational arts," "the Art of Custody or Memory" and "the Art of Elocution or [transmission]," it should be a vast improvement over the lecture hall, by linking the transmission of basic data to actual

[79] See John Langone, *Harvard Med: The Story Behind America's Premier Medical School and the Making of America's Doctors* (New York: Crown Pub., 1995).

[80] Goldstein, "British Medical Training," note 118. See also her research memorandum "The Medical School Experience" (2003, unpublished ms.), 1–3.

[81] Goldstein, "The Medical School Experience," 1.

[82] See Peters, 470–1.

[83] Peters, 470–1, 477.

experience as soon as possible (*SEH*, V: 407).[84] But pure research and critical judgment, the "Art of Inquiry or Invention" and the "Art of Examination or Judgment," are developed over time, later down the long path of medical training. Increasingly, medical experts concerned with fundamental cancer research have returned to Neo-Baconian principles. In particular, research developed within the traditional hierarchies of the medical profession, and driven by the goals of funding agencies, has been disappointing. Bacon would have predicted the failure of any program that was not based on "pure," i.e., purely empirical research, free from such agendas and preconceptions (*SEH*, V: 53–5).[85] But the inherent narrowness of the medical certification system and the culture of hierarchies it promotes would make a "New Pathway" at this level–still dedicated to "lifelong learning, integrated knowledge, and a humanistic approach"–a challenge indeed.

"New Pathways": A Neo-Baconian pedagogy in law

Can legal education learn from the medical "New Pathway," in particular, and Neo-Baconian reform initiatives, in general? The first year of law school approximates the first two years of the traditional medical program in that it is largely mandatory and focuses on time-honored doctrinal "pigeon-holes," "contracts," "torts," "civil procedure," etc. Although the conventional pedagogy is now less rigid than Langdell's so-called "Socratic Method," the data analyzed is still artificially structured appellate cases–quite far from the empirical trenches of the law. Equally important, the "critical judgment" inspired by this pedagogy is often transparently phony. Extremely large classes, and professors with a carefully programmed "theater" of exposition, are closer to the "replications" and "bolts" of Bacon's day, than a system that truly rewards originality and inquiry.[86] Finally, the importance of first-year grades, and the rigor of that competition, is at least comparable to Part One of the medical "Boards."

---

84 See text at note 51, *supra* and *SEH*, V: 407.

85 See also *SEH*, V: 108.

86 See, for example, Duncan Kennedy's famous attacks on conventional legal pedagogy: *Legal Education and the Reproduction of Hierarchy: A Polemic Against the System* (Cambridge, Mass.: Afar, 1984); and "Legal Education as Training for Hierarchy," in *The Politics of Law: A Progressive Critique*, ed. David Kairys (New York: Pantheon, 1982). The "Socratic method," the dynamics of the traditional classroom, and individual competition for grades engendered by the formal nature of law school examinations has been accused of discouraging cooperative skills, encouraging confrontation rather than mediation, and creating disadvantages for women, among other charges. See Lani Guinier, Michelle Fine, Jane Balin, Ann Barton and Deborah Lee Stachel, "Becoming Gentlemen: Women's Experiences at One Ivy League Law School," *University of Pennsylvania Law Review* 143 (1994): 45–8; Jenny Morgan, "The Socratic Method: Silencing Cooperation," *Legal Education Review* 1 (1989): 151–64; J. Rifkin, "Teaching Mediation: A Feminist Perspective on the Study of Law," *Gendered Subjects: The Dynamics of Feminist Teaching*, eds. Margo Culley and Catherine Portuges (Boston: Routledge, 1985), 96; and T. L. Banks, "Gender Bias in the Classroom," *Journal of Legal Education* 38 (1988): 137.

The second two years in the conventional medical school curriculum really have no equivalency in the legal academy except in the clinical programs. But these programs are certainly not universally required, and there are no compulsory rotations. In addition, at many law schools clinicians do not hold tenure-track positions, and the programs they teach do not have the prestige of the regular classroom. At Harvard Law School, only half the students take clinical programs of any kind.[87] The second year is still quite necessary for learning required "building blocks," such as taxation, professional ethics, corporations, constitutional law (in many schools), and legal accounting. These courses–with the exception of professional ethics–are rarely required, but they so much form part of the language and culture of American practice that they are rarely skipped over.

It is, of course, the third year which is the problem. Many law students complain bitterly that it would be better spent in a legal form of "internship," i.e., "apprenticeship." As it stands, only an emphasis in some law schools on supervised research and writing gives the third year any structure or focus.[88] Ironically, it has been the required first-year curriculum that is seen to be the most successful by graduates. Equally ironic, most serious efforts at reform, such as Todd Rakoff's "Experimental First Year" at Harvard, or Judith Ahreen's "Optional First Year" or "Curriculum B" at Georgetown, have focused on the first year.[89] At the close of the third year, many law graduates take one or more state bar examinations. These form the "minimal" qualifications to practice law and are a function equivalent of the far more difficult Part Three of the medical "Boards." Most of these examinations are, at least in part, multiple choice, and they stick closely to an unimaginative memorization of doctrine. "Lifelong learning" skills and critical judgment are not tested at all.

Only at this point does true apprenticeship begin, as graduates take junior positions in large firms or organizations. But recent developments in law firm economics have led to early specialization, and widespread abandonment of required rotations–even though these rotations filled the same needs as in medical training, i.e., providing enough generality to at least know where to go to solve problems of a patient/client that are outside your specialty. And, of course, some law students just

---

[87] The Office of Clinical Programs, Harvard Law School, gave the figure of "just over 50%" for the year 2002–03.

[88] See Kristine Strachan, "Curriculum Reform in the Second and Third Years: Structure, Progression and Integration," *Journal of Legal Education* 39 (1989): 523–33.

[89] See Todd D. Rakoff, "The Harvard First-Year Experiment," *Journal of Legal Education* 39 (1989): 491–9. Despite the valiant efforts of Professor Rakoff, the Harvard experiment proved so costly in faculty time that it was discontinued. Georgetown's "optional" first year "curriculum B" continues as "an innovative and integrated approach to the study of law" (www.law.georgetown.edu/curriculum/jdprog.CFM#first). For the view that the rigor of the first year represents the best of professional education, see Mary Ann Glendon, *A Nation Under Lawyers: How the Crisis in the Legal Profession is Transforming American Society* (New York: Farrar, Straus, Giroux, 1994), 239–53. For an uncomplimentary and anecdotal description of an "experimental" first-year section, see Richard D. Kahlenberg, *Broken Contract* (New York: Hill and Wang, 1992), 11–20.

"hang out a shingle" and practice law.[90] Bacon would find this existing system reminiscent of the flawed professional training of his day. Only the specific nature of the "idols" have changed–with Langdell's pedagogy and curriculum creating the "theater" of today's first-year classroom, rather than the received doctrine of the common law moots, bolts, and readings of his day. But these rituals, isolated from experience, make for poor learning.

Here is a modest proposal. Let us assume that the first year–if modestly enriched by the kinds of interdisciplinary research and writing and clinical experiences proposed by Rakoff, Ahreen, and others–does an acceptable job of transmitting "building blocks" of information or, at least, of teaching students where such information is stored, i.e., Bacon's "Transmission" and "Memory" functions. Let us further assume that the next two years do a poor job of inculcating Bacon's most crucial "rational arts": 1) the "Art of Inquiry or Invention," i.e., the creative skills required to "think outside the box" and learn from life experience; and 2) the "Art of Examination or Judgment," i.e., the art of critical analysis and wisdom. In Bacon's view, these two years could hardly be expected to be successful, fragmented into scores of electives, unrelated to each other by any overlapping concept, and divorced from actual experience–except for optional clinical programs, ignored by many students and, shamefully, often the tenure track faculty.

So why not design a "New Pathway" for the second and third years, drawing from the Harvard Medical School experience? Assuming that first-year doctrinal, lawyering, and research and writing programs are adequate to transmit the basic legal vocabulary and date access skills (i.e., Bacon's "Transmission" and "Memory"), the second year could offer a choice of "concentrations," but–as with the third and fourth years of medical school–some rotation being mandatory. These "concentrations" could integrate classroom instruction, supervised research, and clinical experience through faculty "institutes," that had both academic and clinical components.

For example, assume a second-year student decides–as many do–that litigation is not for him or her. "Transactional" or "corporate" law is what attracts. A "corporate" institute could combine academic instruction in business planning, law and economics, secured transactions, corporate taxation, securities law, and legal accounting, with a program of actually representing small businesses and neighborhood

---

[90] According to the American Bar Association's *"McCrate Report"* (*Legal Education and Professional Development–An Educational Continuum*, Chicago, 1992), the gap between university legal education and practice was, at least in 1992, growing dangerously (284–99). If so, the result could be Bacon's worst nightmare. In Mary Ann Glendon's words: "It would be as though medical faculties decided to devote themselves entirely to basic research in chemistry, biology, and genetics, but continued to send their graduates out to treat patients and to design systems for the delivery of health services. The more law professors give themselves over to high theory, the closer we come in a full circle to a stripped-down version of the old apprenticeship system. Law school provides advanced general education on top of a liberal arts degree, and new graduates, it is assumed, will pick up what they need to know on the job as they go along. The flaw in that assumption, however, is that students are being sent from law schools without law into law firms where senior lawyers are too busy to provide much supervision" (*A Nation*, 245).

development schemes, as is done now at Harvard's Hale and Dorr Center.[91] Such an "Institute" would not just be a vehicle to transmit research and judgment skills to students, but would be a vehicle for faculty research as well, and would encourage collaboration among faculty and students alike–the "teamwork" Bacon found essential for genuine creativity. Another student, interested in litigation, would be taught a different package of academic, research, and clinical skills, while a third, interested in a career in government legal service, could be supervised by an Institute that actually provided counsel to government agencies. A mandatory "rotation" among at least two or three of these "concentrations" would provide a conclusion of the universal legal or jurisprudential principles represented by a rule of law.[92]

It might be objected that these "Institutes," to the extent they actually practice law, would compete unfairly with the practicing bar, or provide uneven services. Exactly the same charges could be leveled at the medical internships and residencies, but they provide invaluable services for the poor and middle class that regular doctors cannot or will not provide. Existing legal clinics, applying strict means tests, have far too much work. Or the "Institutes" could work in support of government projects, reducing the cost to the taxpayer.

The real objections are likely to be by legal educators themselves. Such "Institutes" would raise the cost of legal education by requiring smaller student/faculty ratios and replacing classroom instruction by more costly clinics. Simply integrating existing clinical programs would not be enough–these programs would have to have the added prestige of full equality with classroom and tenure-track instruction, and that, in turn, would require a change in faculty culture.[93] All to the good, Bacon would say. If better education of students breaks down traditional law

---

91 The Community Enterprise Project (CEP) of the Hale and Dorr Legal Service Center "works toward facilitating access to capital and equity participation for neighborhood residents, as well as stimulating home ownership, business development and overall economic stability." See www.law.harvard.edu/academics/clinical/lsc/cep.htm.

92 These are hardly new ideas, but the case for reform grows more compelling. See, for example, Dean Kristine Strachan's implemented programs for integrated "capstone" and "cornerstone" programs in the third year, implemented at the University of San Diego School of Law, Kristine Strachan, "Curriculum Reform in the Second and Third Years: Structure, Progression, and Integration," *Journal of Legal Education* 39 (1989): 523–33. She candidly acknowledged that "[T]he direction of our reform runs against the present mainstream of legal education" ("Curriculum Reform," 530).

93 For one thing, most tenure-track law faculties are not qualified to teach such integrated courses. In 1998 Dean Patrick J. Schlitz took the Harvard Law School faculty as an example: "Although most Harvard graduates go on to practice in private law firms, most Harvard professors have little or no such experience. Only three of the seventy-five members of the Harvard Law faculty have more than five years experience in private practice. The youngest of those three is sixty-six years old, last practiced law in 1964, and was appointed to the Harvard faculty in 1966. In other words, not a single member of the (huge) Harvard Law School faculty has practiced in a law firm at even a senior associate level in the last thirty years. If current trends hold, the Harvard faculties of the future will have even less collective experience in private practice" ["Legal Ethics In Decline: the Elite Law Firm, the Elite Law School, and the Moral Formation of the Novice Attorney," *Minnesota Law Review* 82 (1998): 760–2]. Of course, Dean Schlitz is concerned about the effect on the students. Bacon would find the disconnect between theory and experience of the profession damaging to research and the intellectual life of the faculty as well.

faculty formalism, develops new bonds between faculty formerly working in isolation, and encourages learning by direct experience as a primary goal, rather than an optional afterthought, the results will not just be better legal education, but better legal research and critical judgment from students and faculty alike. The "New Pathway" would lead to "New Tools" of analysis and invention! And what about the cost? As Bacon emphasized, the concern of society should not be the modest cost of better education, but the extraordinary social cost and danger of a blinkered formalism, indifferent to human welfare.[94]

Bacon, for all his intense frustration with contemporary legal education, was devoted to his Inn of Court, Gray's Inn, serving as Treasurer for the exceptionally long term of eight years, 1608–1616:[95] "[F]ew men," he observed, "are so bound to their societies, by obligations ancestral and personal … ."[96] Indeed, for all the attacks of his contemporary rivals, and of Whig historians long after his death, Bacon was loyal to his profession: "I hold every man a debtor to his profession," reads his Preface to the *Maximes of Law*, "from the which as men of course do seek to receive countenance and profit, so ought they of duty to endeavor themselves, by way of amends, to be a help and ornament thereunto" (*SEH*, VII: 319).[97] How it would please him if the great educational insights of *The Maximes, The Advancement of Learning*, the *Novum Organum*, and the *De Augmentis* could serve professional education, particularly legal education, today! And so it should, for the Harvard Medical School's "New Pathway" and its hope for a better education for tens of thousands of American medical students, is a Baconian idea, and its application to the serious plight of our law students offers even more promise. Bacon's faith in the power of true research, in the ability of the human mind to overcome the idols of doctrinal theaters, in the value of integrating knowledge by true experience, in the focus of learning on human need, and in the need to pursue "global" perspectives that unite all study–these speak powerfully to the condition of all American professional schools today, and particularly to law schools.

As William Twining observed in his great Maccabean Lecture of 1989, despite the "structuralists, post-structuralists, deconstructionists, semioticians and even narratologists" it is not bad history to "converse" with great historical figures, and he used the example of Bentham.[98] Indeed, "a failure to equip students with a basic method for conducting reasonably disciplined conversations" is a fault.[99] Bacon was a man of

---

94 See *SEH*, IV: 13–21.

95 See Coquillette, *Bacon*, 316–17.

96 See Sir Dunbar P. Barton, *The Story of the Inns of Court* (Boston: Houghton Mifflin, 1926), 198–9.

97 The feeling was mutual. Gray's Inn religiously observes Bacon's anniversaries, and unveiled a statue to his honor in the front quadrangle on 27 June 1912 to mark his tercentenary at Gray's Inn. See *Francis Bacon: The Commemoration of his Tercentenary at Gray's Inn* (London: Chiswick, 1912) and *The Unveiling of the Statue of Francis Bacon* (London: Chiswick, 1912). Of course, Bacon's alleged "totalitarianism," as promulgated by Whig historians like Macaulay, has not endeared him to common lawyers ideologically. See Coquillette, *Bacon*, 14–17, 293.

98 Coquillette, *Bacon*, 292–3.

99 See William Twining, "Reading Bentham," Maccabean Lecture in Jurisprudence, published in *Proceedings of the British Academy* 75 (1989): 97–141.

his day, but, in his own words, his works were "dedicated only to use, and submitted only to the censure of the learned, and chiefly of time" (*SEH*, VII: 323).[100] One of his goals was to converse with future generations about the nature of good education so that "there may spring helps to man, and a line and race of inventions that may in some degree subdue and overcome the necessities and miseries of humanity" (*SEH*, IV: 27). We should listen carefully, for Bacon has much to say to us.

100 See also *SEH*, IV: 13–33.

Chapter 9

A Society of Baconians?: The Collective Development of Bacon's Method in the Royal Society of London[1]

William T. Lynch

Tracing how methodological thinkers shaped science has gone out of fashion. Historians of science have challenged the view that scientific practice can be neatly tied to rules of scientific procedure. An emphasis on rhetoric and social context leads many to see method as a *post hoc* rationalization of embedded scientific practice, the details of which are better explicated by local features of a scientist's culture. Historians now tend to view collective reference to a shared methodology in terms of a pragmatic need to legitimate new approaches to science. The situation is only exacerbated when the source is Francis Bacon, a figure seen as the epitome of the naive empiricism debunked by recent history and sociology of science.

But why would Bacon–a courtier convicted of corruption whose name had accreted dangerous Puritan associations–serve the Royal Society's collective need for legitimation during the Restoration if the Society had no genuine concern to implement his method? What are we to make of systematic reference to Bacon's philosophical program in the charter of the Society, its meeting minutes, and its earliest published works? In what follows, I demonstrate the seriousness with which the early Fellows of the Royal Society attempted to implement Bacon's program for the reform of knowledge in their collective activity and the influence this had on their scientific activity and publications. I argue that tracing the influence of methodological doctrines requires teasing out productive tensions in methodological writings of Royal Society members. Key to this process is the productive use of metaphors for understanding the nature of the world observed and the proper stance of the investigator. Bacon articulated his call to attend to "things themselves" by referring to three different metaphors: the passive observer of nature, the active manipulator of nature, and the uncoverer of hidden powers. These metaphors correspond to three views of how the object of natural philosophy is known–what I call specular, manual, and

[1] This paper draws upon William T. Lynch, *Solomon's Child: Method in the Early Royal Society of London* (Stanford: Stanford Univ. Press, 2001). I am grateful for comments by Julie Robin Solomon, Catherine Gimelli Martin and an anonymous reviewer.

generative objects–which in turn underwrites three views of how objective knowledge is attained–empiricism, constructivism, and theoretical realism. In the first two decades of its life, the Royal Society developed each of these interpretations of Bacon's philosophy before their associated criteria of truth diverged to shape different segments of the scientific community in the eighteenth century. Each approach drew sustenance from methodological traditions besides Baconianism, yet Bacon's influence was the primary touchstone for the early Royal Society.

Ambivalence about the early Royal Society

Historians have seen the founding of the Royal Society as a crucial step in the development of modern science and have focused a lot of attention on identifying the institutional origins of the Royal Society in a variety of informal meetings and associations in London, Oxford, and elsewhere. The focus on the birth of the modern scientific society accounts for the interest in questions about the origins of the Royal Society. The work of certain key figures in the Royal Society has also garnered much attention, most significantly that of Robert Boyle and Robert Hooke, but also the contributions of other early figures like Christopher Wren, John Wallis, John Wilkins, and William Petty. Another strand of work looks at the often strained relationship between Isaac Newton and those early Fellows interested in and qualified to discuss physical and mathematical questions, following Newton's 1672 paper on optics submitted to the society. What captures attention in this literature is the significance of a carefully selected figure or group from the early society that turned out to be significant in some larger narrative about the history of science. But when looking at the early *Philosophical Transactions*, the unofficial journal, or meeting minutes, of the society, historians discovered a wide-ranging, eclectic fascination with odd facts and an apparent confusion as to the purposes of mature science.

This waning of the reputation of the very early Royal Society reflects a similar change in the reputation of Francis Bacon as a philosopher of science. When inductivism was taken for granted in describing scientific method, Bacon and the Royal Society were seen as highly significant players in the history of science.[2] Related to this denigration of Baconian inductivism is the commonplace that the Royal Society was filled with dilettantes, amateurs who were not "serious" scientists.[3] The recent trend among historians of science to downsize the role of method in science began in part with the denigration of Bacon's significance for understanding the Royal Society. Charles Webster launched this process by drawing attention to the way historians took for granted that historical actors' self-descriptions could be taken at face

[2] Richard Yeo, "An Idol of the Market-Place: Baconianism in Nineteenth Century Britain," *History of Science* 23 (1985): 251–98, 266–7, 277–83; Antonio Perez-Ramos, *Francis Bacon's Idea of Science and the Maker's Knowledge Tradition* (Oxford: Clarendon Press, 1988), chapter two.

[3] Richard S. Westfall, *The Construction of Modern Science: Mechanisms and Mechanics* (Cambridge: Cambridge Univ. Press, 1977), 112; Michael Hunter, *Science and Society in Restoration England* (Cambridge: Cambridge Univ. Press, 1981), 67.

value. Not only might self-described Baconians not actually practice Baconian method as advertised; they might be engaged in tactical dissimulation about the origins of the Royal Society in order to make the Royal Society look better in the political context of the Restoration. Wood continued this line of analysis in rereading Sprat's polemic defense of the Royal Society in terms of tactical positioning in a hostile environment rather than a straightforward advertisement of the kind of work carried on by its Fellows. Webster and Wood share the view that any effort by the Royal Society to present a united front behind Bacon is designed for public relations and obscures a greater diversity of influences and approaches within the Royal Society. This development of a complex and contextual historiography of the Royal Society finds its fullest expression in the work of Michael Hunter, who effectively demolishes most simple accounts of either the methodological or institutional coherence of the Royal Society by demonstrating the variety of influences and practices within the Royal Society. At the same time, Hunter tells us a story of early enthusiasm followed by institutional failure as the Royal Society proves unable to develop a sustainable collective research program.[4]

These contextual approaches partially reinforce similar ideas put forth by the sociology of scientific knowledge that assert the practical impotence of methodological ideas. Applying the insight that science is made not found, constructivists argue, in effect, that words and deeds do not connect. Descriptions of a practice may have a practical purpose (that is, they may be actions themselves), but they can not be "applied" to science in a manner that will direct scientists in the details of their practice. Thus, we are led to look for one set of reasons for why scientists practice their craft as they do and a different set of reasons for why they describe that craft as they do. As a result, constructivists deny the possibility of explicit methodological discourse shaping research practice in any substantive sense.[5] Together, contextualists and constructivists reject the idea that scientists actually follow the rules of scientific method in carrying out their work.

Severe and indulgent method and the question of hypothesis

I grant the value and significance of contextualist and constructivist approaches to the history of science. There is little doubt that members of the Royal Society who claimed to be following Bacon's plan did not in fact follow strictly the steps of inductive ascent outlined in *Novum Organum,* nor can the prevalence of Bacon-tinged

---

[4] C. Webster, "The Origins of the Royal Society," *History of Science* 6 (1976): 106–28; P. B. Wood, "Methodology and Apologetics: Thomas Sprat's *History of the Royal Society,*" *British Journal for the History of Science* 13 (1980): 1–26; Michael Hunter, *Science and Society,* 11–21; Michael Hunter, *Establishing the New Science: The Experience of the Early Royal Society* (Woodbridge, Eng.: Boydell Press, 1989), 6, 11–12.

[5] John A. Schuster, "Methodologies as Mythic Structures: A Preface to the Future Historiography of Method," *Metascience* 1 (1984): 15–36, 17; Steven Shapin and Simon Schaffer, *Leviathan and the Air-Pump: Hobbes, Boyle, and the Experimental Life* (Princeton: Princeton Univ. Press, 1994), 14.

rhetoric be used as an excuse to ignore the diversity of methods and practices within the Royal Society. Nevertheless, I think it is important to understand what happens when scientists claim to follow a particular account of scientific method. How did the Royal Society's collective effort to implement Bacon's program impact the work of its Fellows? Particular attention must be given to the collective effort to carry out Bacon's program which can be seen in its official activities–meetings, publications, and committees.

In order to trace this less direct approach, it is helpful to pay attention to crucial metaphors within Bacon's writings that appear to resonate with his followers. Key metaphors in Bacon's writings illuminate his meaning in two ways. First, Bacon spoke of being either "strict" or "indulgent" in following the steps outlined in *Novum Organum,* depending upon whether the groundwork had been fully established or not. Second, Bacon uses three different images of what it means to pay attention to "things themselves"–as objects of vision, as objects to be manipulated, and as hidden causes of observed phenomena. Focus upon these metaphors helps us to understand productive tensions in Bacon's own thought. These metaphors, in turn, were exploited by the Royal Society, both collectively in its institutional efforts to implement a new instauration, and individually, in the way Fellows shaped their work to make it fit these images.

Bacon's induction was broader and more theoretical than the enumerative induction of modern textbooks. In theory, there was nothing hypothetical about it, since proper execution of his method was to guarantee the certainty of his outcome.[6] In practice, however, there were two ways in which hypotheses crept back in, one in the articulation of the method itself and the other in actually carrying out preliminary research. First, Bacon's inductive machinery outlined in Book II of *Novum Organum* inevitably broke down and required additional helps to arrive at a conclusion. After running through his elaborate inductive method applied to an examination of the form of heat, we are led to believe that a true induction of the nature of heat would be forthcoming. Collecting tables of instances possessing heat, similar phenomena lacking heat, and tables comparing degrees of heat, Bacon is able to exclude a number of potential candidate explanations for heat, a process he describes as "the foundations of true Induction, which however is not completed till it arrives at an Affirmative" (*SEH*, IV: 149). Bacon observes that the satisfactory completion of exclusion of simple natures depends upon an understanding of what simple natures are like–precisely what is at issue and unknown prior to successful induction. This seeming circularity (and breakdown) of method is rescued by practical expedience,

6 There is disagreement about whether Bacon endorsed a corpuscular hypothesis or a Paracelsian chemical theory. See Graham Rees, "Francis Bacon's Semi-Paracelsian Cosmology and the Great Instauration," *Ambix* 22 (1975): 161–73; *idem*, "The Fate of Bacon's Cosmology in the Seventeenth Century," *Ambix 24* (1977): 27–38. Yet it is clear that his discussion of atoms and the latent configurations of matter lent themselves to adaptation by the Royal Society's mechanical philosophers. John A. Schuster argues that in the scientific revolution a proliferation of ontologies is narrowed down eventually to some version of the mechanical philosophy ["The Scientific Revolution," *Companion to the History of Modern Science,* ed. R. C. Olby *et al.* (London: Routledge, 1990), 217–42].

which requires a special permission to assert affirmatively, with the promissory note that additional "helps of the understanding in the Interpretation of Nature and true and perfect Induction" will be forthcoming (*SEH*, IV: 155):

> And yet since truth will sooner come out from error than from confusion, I think it expedient that the understanding should have permission … to make an essay of the Interpretation of Nature in the affirmative way; on the strength both of the instances given in the tables, and of any others it may meet with elsewhere. Which kind of essay I call the *Indulgence of the Understanding,* or the *Commencement of Interpretation,* or the *First Vintage* (*SEH*, IV: 149).

Bacon's apparent oscillation between a "legitimate, chaste, and severe" limitation on inference and an affirmative "indulgence" or "liberty" of the understanding is fundamental to the dynamic quality of Bacon's methodological discourse.[7] The further promise that examination of significant observations or "prerogative instances" will aid the inductive process at this point end up introducing explicitly analogical reasoning and heuristic guides for further inquiry.[8]

A second way in which hypothetical reasoning could survive Bacon's inductive machinery was through his own concrete examples of the kinds of investigation he favored. Bacon realized that he could not follow his method strictly while at the same time actually beginning the project of which he spoke. Without examples of the kind of enterprise he had in mind, he felt he could not convince others to contribute to his task. Yet his method was too large for any one person or any one generation to carry out alone. While in principle, the method could be self-warranting and automatic, in practice, it had to be begun on imperfect terms since "in the present condition of things and men's minds," his great instauration "cannot easily be conceived or imagined" (*SEH*, IV: 32). Like "interest payable from time to time until the principal be forthcoming," Bacon saw a need to exemplify his discoveries, even those not made "according to the rules and methods of interpretation, but by the ordinary use of the understanding in inquiring and discovering."[9]

As a system of rules intended to direct inquiry, Bacon's method is subject to the normal interpretive flexibility attending any rule-following activity.[10] In addition,

---

[7] *SEH*, IV: 32, 149; Ronald Levao, "Francis Bacon and the Mobility of Science," *Representations* 40 (1992): 1–32.

[8] *SEH*, IV: 155–246; L. Jonathan Cohen, "Some Remarks on the Baconian Conception of Probability," *Journal of the History of Ideas,* 41 (1980), 219–31; William Eamon, *Science and the Secrets of Nature: Books of Secrets in Medieval and Early Modern Culture* (Princeton: Princeton Univ. Press, 1994), 288–90; Perez-Ramos, "Bacon's Forms," 108. For a similar analogy among techniques of problem-solving, see Lisa Jardine's discussion of learned experience or *experientia literata* (*Francis Bacon: Discovery and the Art of Discourse* (Cambridge: Cambridge Univ. Press, 1974), 144–7).

[9] *SEH*, IV: 31. For a discussion of Bacon's speculative natural philosophy, see Graham Rees, "Bacon's Speculative Philosophy" in *The Cambridge Companion to Bacon,* ed. Markku Peltonen (Cambridge: Cambridge Univ. Press, 1996), 121–45.

[10] H. M. Collins, *Changing Order: Replication and Induction in Scientific Practice* (London: Sage, 1985); David Bloor, *Wittgenstein: A Social Theory of Knowledge* (New York: Columbia Univ. Press, 1983).

however, Bacon allowed himself the luxury of opting out of strict conformity to his precepts whenever they interfered with the preliminary work of establishing the method on a secure practical foundation. Speaking of his own contributions to substantive natural philosophy as "wayside inns" for the weary traveler, he adopted a fallibilist attitude which he did not allow for the outcome of the method itself:

> Nevertheless I wish it to be understood in the meantime that they are conclusions by which (as not being discovered and proved by the true form of interpretation) I do not at all mean to bind myself. Nor need any one be alarmed at such suspension of judgment, in one who maintains not simply that nothing can be known, but only that nothing can be known except in a certain course and way; and yet establishes provisionally certain degrees of assurance, for use and relief until the mind shall arrive at a knowledge of causes in which it can rest. (*SEH*, IV: 32)

We can see here the roots of Boyle's probabilism and yet Bacon is clear that a more certain induction is possible, a view shared by many Royal Society fellows, as we will see in discussing Hooke below. Bacon's followers would often distinguish their incomplete and unsatisfactory efforts to reform knowledge from what would be possible when the project was firmly established and supported.

Baconian Background to the Royal Society

While Bacon's essays and political writings attracted a wide audience during his lifetime, interest in his scientific and methodological writings grew only by the late 1630s.[11] Aubrey went so far as to claim that "the searching into Naturall knowledge began but since or about the death of King Charles the first" (1649).[12] Beginning in the mid-1640s–and continuing throughout the 1650s–a number of distinct, but loosely interacting and partially overlapping, groups drew on Bacon's writings to help motivate their work. Relative emphasis varied but most groups evidenced at least some interest in the broad range of topics associated with Bacon's natural writings: practical knowledge, the history of trades, natural history, state support of knowledge and technology, utopian reform, inductive or experimental approaches to knowledge, antischolasticism, and the reform of education. Despite their differences in philosophy, social background, and primary orientation, these groups recognized each other as engaged in similar enterprises.

---

[11] M. L. Donnelly demonstrates an early neglect of Bacon's methodological writings at the universities through analysis of the printing history of Bacon's writings, the personal library of John Harvard bequeathed to Harvard University in 1638 and a booklist attributed to Richard Holdsworth ["Francis Bacon's Early Reputation in England and the Question of John Milton's Alleged 'Baconianism'," *Prose Studies* 14 (1991): 1–20].

[12] Quoted in Donnelly, 12. Alvin Snider believes that Bacon became a figure of origin for natural philosophy mostly clearly after the Restoration, but cites evidence that natural philosophers had begun to appeal to him as the source of their various anti-scholastic programs in the 1640s and 1650s ["Bacon, Legitimation, and the 'Origin' of Restoration Science," *The Eighteenth Century* 32 (1991): 123–5].

Indeed, an important reason for the growing coalescing of Baconian fervor in the nation was the existence of individuals who moved back and forth between groups, reconstituting each group's picture of a diverse reform movement in the process.[13] Robert Boyle moved from a strong association with Prussian émigré Samuel Hartlib's web of correspondence, with its Puritan emphasis on reform and utility, to John Wilkins' "experimental philosophical club" at Oxford in 1655–56, with its less practical and more philosophical interests. His experiences with artisans–and the similar experiences of other Hartlib correspondents like John Evelyn and William Petty–shaped the ongoing interests of the Oxford group, and later, of the Royal Society.[14] Similarly, Evelyn moved between the Hartlib circle, the Oxford group, and a primary affiliation with London Royalist virtuosi, including later Royal Society Fellow Thomas Henshaw. As Hunter has shown, his interest in a Baconian history of trades differed from the populist approach of Hartlib as the result of his circle's commitment to virtuous and literate public service by an elite, resulting in a gentlemanly reading of Bacon that also shaped the early Royal Society.[15]

Before 1640, interest in Bacon's natural writings was sporadic. Occasional references by instructors to Bacon's natural philosophy in university settings can be found, but typically as an afterthought or supplement to traditional course work.[16] Samuel Hartlib had a long-standing interest in Bacon's ideas dating from his education at Cambridge in the 1620s. Hartlib's brand of Baconianism, infused with Puritan and millenarian strains by his association with John Dury and Jan Amos Comenius, shaped the ideas of reform of the country party of gentry opposed to the domination of court over Parliament in the 1630s.[17] In the correspondence and notes of Hartlib, we find occasional mention of Bacon's *Novum Organum* and *The Advancement of Learning* by the mid-1630s.[18] However, it is only in 1639 and 1640 that the scattered, latent interest in Bacon began to coalesce as the wider Hartlib circle began to wed Bacon's philosophy more explicitly to their utopian, religious,

---

13 David S. Lux and Harold J. Cook, "Closed Circles or Open Networks?: Communicating at a Distance during the Scientific Revolution," *History of Science* 36 (1998): 202.

14 Michael Hunter and Edward B. Davis, "General Introduction," *The Works of Robert Boyle,* eds. Hunter and Davis, 7 vols. (London: Pickering & Chatto, 1999), I, xxxii–xxxiii. Hunter, "How Boyle Became a Scientist," *History of Science* 33 (1995): 59–103.

15 Michael Hunter, *Science and the Shape of Orthodoxy: Intellectual Change in Late Seventeenth-Century Britain* (Woodbridge, Eng.: Boydell Press, 1995), ch. 3.

16 Charles Webster, *The Great Instauration: Science, Medicine and Reform, 1626–1660* (London: Duckworth, 1975), 134; Donnelly, "Reputation."

17 Hugh Trevor-Roper, *Religion, Reformation, and Social Change,* ch. 5; Charles Webster, ed., *Samuel Hartlib and the Advancement of Learning* (Cambridge: Cambridge Univ. Press, 1970); *idem,* "Macaria: Samuel Hartlib and the Great Reformation," *Acta Comeniana* 2 (1970): 147–64; G. H. Turnbull, *Hartlib, Dury and Comenius: Gleanings from Hartlib's Papers* (London: Univ. Press of Liverpool, 1947); *idem,* "Samuel Hartlib's Influence on the Early History of the Royal Society," *Notes and Records of the Royal Society of London* 10 (1953): 101–30.

18 *The Hartlib Papers: A Complete Text and Image Database of the Papers of Samuel Hartlib (1600–1662),* 2 CD-ROMS (Ann Arbor, Mich.: UMI, 1995) 11/1/41A, 30/4/4B, 21/27/1A, 46/8/3A, 36/4/15B.

economic, and educational programs.[19] By 1640, interest in Bacon's work had begun to gain footholds inside and outside the academy. An edition of *De Augmentis Scientiarum* was put out by Oxford Fellow Gilbert Wats in 1640. The purges of faculty beginning at Cambridge in 1644 and Oxford in 1648 provided for a new influx of faculty enthusiastic about new approaches to natural philosophy. These changes helped reshape pedagogy to some extent and informal association to a much greater extent.[20]

At Gresham College in London, a group began to meet regularly around 1645 to discuss natural philosophy, at the instigation of Theodore Haak, a Hartlib associate who was an enthusiastic proponent of Bacon and Comenius. Early on the group included primarily physicians and mathematicians interested in experiment, and meetings continued on and off with changing membership until the Restoration. When John Wilkins left London in 1648 to become Warden of Wadham College, Oxford, he was followed by members of the London group, including later Royal Society Fellows Jonathan Goddard and John Wallis. Other later Fellows, including William Petty and Robert Boyle, added new energy to the group.[21] Boyle had developed an interest in Baconian utility and experimentation as part of his self-styled "Invisible College" that Webster has shown to have emerged around 1646 among a circle of Anglo-Irish politicians and intellectuals centered around Benjamin Worsley and Boyle's sister, Katherine, Lady Ranelagh.[22] Canny has argued that the Invisible College had deeper roots among English settlers in Ireland who saw the reconquest of Ireland as an opportunity to carry out many of the same reforms Hartlib sought in England.[23]

Following the death of Oliver Cromwell in 1658, a sense of renewed hope for political settlement animated the landed elite in particular, whatever their political or religious differences. Accompanying that sense of hope was a concern to establish the Gresham meetings in London on a more regular basis. These meetings were revived in 1657 with much of the active core of what would become the Royal Society following the Restoration, a core that had roots in the diversity of Baconian groups of the 1640s and 1650s. The interest in formal establishment may have drawn sustenance from the growing hope for political stability, but the concern with institutionalization was long-standing among the groups we have surveyed.

Following an astronomical lecture by Christopher Wren at Gresham College on 28 November 1660, a group interested in promoting experimental philosophy met at the lodgings of Lawrence Rooke, a professor of geometry there. Together they

19 Stephen Clucas, "In Search of 'The True Logick': Methodological Eclecticism among the 'Baconian Reformers,'" *Samuel Hartlib and Universal Reformation: Studies in Intellectual Communication,* eds. Mark Greengrass, Michael Leslie, and Timothy Raylor (Cambridge: Cambridge Univ. Press, 1994), 51–74.

20 Webster, *Instauration,* 129–78.

21 For discussion of the Oxford group, see Webster, *Instauration,* 153–78.

22 Webster, *Instauration,* 57–67; Charles Webster, "New Light on the Invisible College: The Social Relations of English Science in the Mid-Seventeenth Century," *Transactions of the Royal Historical Society* 24 (1974): 19–42.

23 Nicholas Canny, *The Upstart Earl: A Study of the Social and Mental World of Richard Boyle, first Earl of Cork, 1566–1643* (Cambridge: Cambridge Univ. Press, 1982), ch. 7.

resolved to formalize weekly meetings and drew up a list of suitable candidates for the Society, appointing Wilkins as chair. At the next meeting, Sir Robert Moray, a Royalist courtier close to the King, brought word of Charles' approval of their endeavor and his readiness to support it.[24] What Evelyn soon dubbed the Royal Society (in published efforts to win support from the Crown) was chartered as the Royal Society of London in 1662. Its statutes called for the explicit recording of observational and experimental matters of fact in a manner clearly influenced by Bacon's method of induction from tables of facts.[25] In addition, the Society was licensed to print books under its own imprimatur and to conduct philosophical correspondence with foreign nations. Henry Oldenburg, appointed Secretary along with Wilkins, carried out extensive correspondence on behalf of the Society, and his journal, *Philosophical Transactions,* begun in 1665 and widely associated with the Society, reported observations and experiments brought to the Society's notice. *Ad hoc* and permanent committees were formed to carry out different parts of Bacon's program.[26]

Specular, manual, and generative objects

The most significant shared commitment employed by Royal Society Fellows was a determination to attend to "things" rather than "words," following Bacon's observation in *The Advancement of Learning* that the "first distemper of learning" is that "men study words and not matter" (*SEH*, III: 284). This rhetoric has usually been seen as part of the Society's atheoretical empiricism, wherein matters of fact–consensual accounts of the behavior of "things"–replace endlessly divisive theoretical disputes–"words." Bacon does distinguish between facts and hypothetical anticipations of nature, but his distinction between things and words has a different connotation. "Matters of fact" or "facts" served as shorthand for observed claims about nature that avoid the traps of Bacon's idols.[27] Bacon's view of "things" served as a metaphorical ontology enabling nature to speak once the interference of the idols had been checked. The injunction to focus on "things" themselves has at least three different connotations: specular, manual, and generative (see Table 9.1).

First, in closest conformity with emphasis on matters of fact is the *specular* conception of objects. Things are the ordinary objects that we can observe with our senses. Once we have set aside the contamination of the idols and cast our net widely, the objects act on us without any further activity on our part:

---

[24] Thomas Birch, *The History of the Royal Society of London for improving of natural knowledge, from its first rise,* 4 vols. (London, 1756–57), I: 3–5.

[25] John Evelyn, *Instructions Concerning Erecting of a Library* (London, 1661); idem, *A Panegyric to Charles the Second* (London, 1661); *The Record of the Royal Society of London,* 3rd ed. (London: Oxford Univ. Press, 1912), 85, 82, 59; Hunter, *Establishing,* 4–5.

[26] Hunter, *Establishing,* 91–3, 104–5.

[27] Shapin and Schaffer, *Leviathan*; Barbara Shapiro, "The Concept 'Fact': Legal Origins and Cultural Diffusion," *Albion* (1994): 236–48.

Table 9.1 Objects and corresponding criteria of truth in Bacon's philosophy and the early Royal Society

Objects of Knowledge: How Known	Corresponding Criterion of Truth	Represented in Royal Society's Work	Conflict over Relative Emphasis
Specular: Vision, passive, let nature speak, avoid corruption by idols: "Let the images and rays of natural objects meet in a point, as they do in vision; whence it follows that the strength and excellency of the wit has little to do in the matter."	**Empiricist:** Do not "seek either to force or ensnare men's judgements, but … lead them to things themselves and the concordances of things."	Moray's rules for recording observations; natural history; microscope; matters of fact; collective observation of experimental demonstrations; Graunt and Petty's statistical observations	Concern over Hooke's bold hypotheses in *Micrographia*; suspicion of Digby's account of sympathetic medicine; appeal to microscope to ground mechanical philosophy
Manual: Active, constructed by art, corrects sense, recipe for producing effects, systematic experiment: "For the end which this science of mine proposes is the invention not of arguments but of arts … and directions for works."	**Constructivist:** To make is to know: "Truth … and utility are here the very same things: and works themselves are of greater value as pledges of truth than as contributing to the comforts of life."	History of trades; Hooke's experimental program and inventions	Hooke's experimental investigations into causes versus bracketing of causes in meeting minutes and Boyle's writings on airpump; Hooke's impatience with demonstration experiments; actual construction of new instruments versus untested proposals
Generative: Alphabet of hidden forms, theoretical understanding that allows us to combine powers in new ways. Letters of alphabet: "to enquire the form of a lion … is a vain pursuit; but to enquire the form of dense, rare, hot, cold … and the like … which (like the letters of the alphabet) are not many and yet make up and sustain the essences and forms of all substances" is not.	**Theoretical realism:** "The nearer [investigation] approaches to simple natures, the easier and plainer will everything become; the business being transferred from the complicated to the simple; from the incommensurable to the commensurable; from surds to rational quantities; from the infinite and vague to the finite and certain; as in the case of the letters of the alphabet and the notes of music."	Wilkins' philosophical language; Hooke's philosophical algebra and similitude in the nature of cause; mechanical philosophy; Petty's political arithmetic	Wilkins caught between specular classification of nature and method of discovery, language of communication versus transcendental denomination; John Ray's objections to distortion of empirical classification of plants and animals to fit Wilkins' system; Hooke's appeal to underlying theory of congruity after experimental confirmation and manipulation of capillary action

> For all those who before me have applied themselves to the invention of arts have but cast a glance or two upon facts and examples and experience, and straightway proceeded, as if invention were nothing more than an exercise of thought, to invoke their own spirits to give them oracles. I, on the contrary, dwelling purely and constantly among the facts of nature, withdraw my intellect from them no further than may suffice to let the images and rays of natural objects meet in a point, as they do in vision; whence it follows that the strength and excellency of the wit has little to do in the matter (*SEH*, IV: 19).

The basis for the *empiricist* interpretation of Bacon and the Royal Society depends upon this view of objects. Philosophers ought to see how nature behaves in all its particulars and not rely upon selective observation or upon prior testimony. Objectivity consists in consensus organized around observation rather than judgments:

> I have not sought (I say) nor do I seek either to force or ensnare men's judgments, but I lead them to things themselves and the concordances of things, that they may see for themselves what they have, what they can dispute, what they can add and contribute to the common stock. (*SEH*, IV: 19)

The specular account of objects and the accompanying empiricist view of objectivity motivated the Society's interest in natural history, the development of instruments to extend the senses like the telescope and microscope, and the cultivation of virtual and real witnesses for experimental demonstrations performed at the weekly meetings.

Second, the *manual* conception of objects considers things themselves as objects handled and constructed by art. We have already noted that Bacon believes that nature reveals itself more clearly through art than detached observation. Technological improvements come about since artisans and mechanics are forced to manipulate and control the material world. In invoking this image of active manipulation, Bacon questions the passive reliance upon the senses that is privileged in the specular account of objects:

> For the end which this science of mine proposes is the invention not of arguments but of arts; not of things in accordance with principles, but of principles themselves; not of probable reasons, but of designations and directions for works. And as the intention is different, so accordingly is the effect; the effect of the one being to overcome an opponent in argument, of the other to command nature in action. (*SEH*, IV: 24)

Nature is only truly understood when we operate upon the world and learn to produce a variety of effects at will rather than rely upon the accidental arrangement of qualities existing in unaltered nature. Like the artisan, the experimentalist operates upon the world and does not merely observe it. Experimentation according to proper method can more closely approximate the subtlety of nature than can detached observation, however augmented by instruments it may be:

> I have sought on all sides diligently and faithfully to provide helps for the senses–substitutes to supply its failures, rectifications to correct its errors; and this I endeavour to accomplish not so much by instruments as by experiments. For the subtlety of experiments is far greater

> than that of sense itself, even when assisted by exquisite instruments; such experiments, I mean, as are skilfully and artificially devised for the express purpose of determining the point in question. To the immediate and proper perception of the sense therefore I do not give much weight; but I contrive that the office of the sense shall be only to judge of the experiment, and that the experiment itself shall judge of the thing. (*SEH*, IV: 26)

Experimentation ends when the operative form of the nature under investigation is uncovered, allowing the production of the phenomenon by the experimenter. The ability to reproduce natural effects through disciplined art is the ultimate test of the validity of knowledge for Bacon.

Corresponding to manual objects is a *constructivist* definition of objectivity: to make is to know. Bacon suggests that the only way to understand the natural world is through imitating the act of creation in nature by the act of creation in manufacturing useful items:

> Be it known then how vast a difference there is ... between the Idols of the human mind and the Ideas of the divine. The former are nothing more than arbitrary abstractions; the latter are the creator's own stamp upon creation, impressed and defined in matter by true and equisite lines. Truth therefore and utility are here the very same things: and works themselves are of greater value as pledges of truth than as contributing to the comforts of life. (*SEH*, IV: 110)

After describing the form of heat in his first vintage, Bacon provides an operative definition that gives instructions for producing heat from motion:

> Viewed with reference to operation it is the same thing. For the direction is this: *If in any natural body you can excite a dilating or expanding motion, and can so repress this motion and turn it back upon itself, that the dilation shall not proceed equably, but have its way in one part and be counteracted in another, you will undoubtedly generate heat.*[28]

Utility, in the sense of the production of effects, is not a secondary benefit of induction but the crucial proof that one has attained knowledge. The Royal Society's interest in the history of trades, practical mechanics, and systematic experimentation derives from a focus on manual objects and a belief in constructivist objectivity.[29] According to a constructivist epistemology, the Royal Society's experimental demonstrations before witnesses are less significant than the directed series of experiments to answer particular questions, written up as recipes for the reader to perform,

---

[28] *SEH*, IV: 155; Perez-Ramos, *Maker's Knowledge,* chs. 10, 13; *idem*, "Bacon's Forms and the Maker's Knowledge Tradition," *The Cambridge Companion to Bacon,* ed. Markku Peltonen (Cambridge: Cambridge Univ. Press, 1996), 109. In England, a constructivist interpretation of mathematics converged with experimental philosophy. See Peter Dear, *Discipline and Experience: The Mathematical Way in the Scientific Revolution* (Chicago: Univ. of Chicago Press, 1995), ch. 8.

[29] Kathleen H. Ochs, "The Royal Society of London's History of Trades Programme: An Early Episode in Applied Science," *Notes and Records of the Royal Society of London* 39 (1985): 129–58; J. A. Bennett, "The Mechanics' Philosophy and the Mechanical Philosophy," *History of Science* 24 (1986): 1–28.

as found in Hooke's *Micrographia*. Bacon believed that experimental manipulations of matter would suggest further experiments. Embodied experimentation would define a dynamic and unending research program, just as the arts continually improved by their nature. Symptomatic of this constructivist tendency in the Royal Society is Hooke's impatience with the performative and specular quality of the demonstrations he was required to perform as curator of experiments.[30]

Third, a *generative* conception of objects reaches beyond ordinary objects familiar to our senses and subject to direct manipulation to describe an alphabet of hidden powers that can combine in numerous ways to produce the ordinary objects of our everyday experience. Bacon's simple natures, to be put together in new combinations, are like

> the letters of the alphabet in themselves and apart have no use or meaning, yet they are the subject-matter for the composition and apparatus of all discourse. So again the seeds of things are of much latent virtue, and yet of no use except in their development. (*SEH*, IV: 107)

Bacon compared ordinary objects to words, while things in themselves are compared to letters. It is from the endless combination of letters into words that we must look to understand nature's works:

> For as it would be neither easy nor of any use to inquire the form of the sound which makes any word, since words, by composition and transposition of letters, are infinite; whereas to inquire the form of the sound which makes any simple letter … is comprehensible, nay easy; and yet these forms of letters once known will lead us directly to the form of words; so in like manner to enquire the form of a lion, of an oak, of gold, nay even of water of air, is a vain pursuit; but to enquire the form of dense, rare, hot, cold, heavy, light, tangible, pneumatic, volatile, fixed, and the like, as well configurations as motions … and which (like the letters of the alphabet) are not many and yet make up and sustain the essences and forms of all substances;–this, I say, it is which I am attempting. (*SEH*, III: 355–6)

True forms are the "very thing itself"; in short, they are more real than everyday objects since everyday objects are composed out of them.[31] The combination of generative powers entails a comprehensive *theoretical realist* understanding of objectivity: the world has depth that escapes our spontaneous experience of it and a true induction leads to a parsimonious explanation of the world's diversity. In order to discover the "Latent Configuration" underlying appearances, we are guided by the realization that a true understanding of the simple natures at the root of things would simplify our picture of the world and would conform to real matter rather than some verbal doctrine:

> Nor shall we thus be led to the *doctrine* of atoms, which implies the hypothesis of a vacuum and that of the unchangeableness of matter (both false assumptions); we shall be led only to

---

[30] Michael Hunter and Paul B. Wood, "Towards Solomon's House: Rival Strategies for Reforming the Early Royal Society," *History of Science* 24 (1986): 49–108, reprinted in Hunter, *Establishing*, 199–204.

[31] *SEH*, IV: 137. For discussion, see Perez-Ramos, "Bacon's Forms," 102–4.

> *real particles,* such as really exist. Nor again is there any reason to be alarmed at the subtlety of the investigation, as if it could not be disentangled; on the contrary, the nearer it approaches to simple natures, the easier and plainer will everything become; the business being transferred from the complicated to the simple; from the incommensurable to the commensurable; from surds to rational quantities; from the infinite and vague to the finite and certain; as in the case of the letters of the alphabet and the notes of music. (*SEH*, IV: 126, emphasis added)

For Bacon, the identification of the simple nature of heat led immediately to the delineation of this nature's similarities and differences from other simple natures, like cold. Ultimately, a small number of primitive powers and a grammar of their possible combinations were to be found. The Royal Society pursued this idea most explicitly through Wilkins' development of a real character and a philosophical language, which was not only to be a convenient classification of nature, but a method for generating new knowledge of nature. Hooke's reflections on the "similitude of the nature of cause" and his development of a "philosophical algebra" reflect a similar idea that knowledge of basic forms, few in number, would makes possible endless new discoveries and inventions. The widespread acceptance of a flexible, mechanical philosophy within the Royal Society reflects a similar concern to identify the basic building blocks of nature.

Competing interpretations of the statutes

It is possible to trace divergence of emphasis over which interpretation of Bacon's *res/verba* distinction should predominate. Nevertheless, early Fellows writing under the Society's imprimatur combined empiricism, constructivism and theoretical realism. In the remainder of the paper, I will draw out the relative emphasis of three early authors. In Evelyn's horticultural work, a specular emphasis predominated, though with a generative interest in mechanical philosophy and an "Alphabet of Earths and Composts."[32] Hooke's manual conception of experimentation sought causal inference within a work that nevertheless insisted on its specular credentials via his observations with the microscope. In turn, Wilkins' work on philosophical language attracted Hooke's attention since it promised a generative view of hidden forms constituting ordinary objects that made manipulation of nature possible. To show tensions existed within an overall commitment to all three interpretations, I first examine the Society's statutes, where Hooke followed Bacon's "indulgence of the understanding," while Sir Robert Moray and the committee overseeing Hooke's *Micrographia* objected to his illegitimate "anticipation of nature."

The 1663 statutes provide explicit guidance on the observation and reporting of experiments, ordering that the Secretary

---

[32] Evelyn, *Philosophical Discourse,* 48.

> shall jointly draw up the Report of the matter of fact, in every such Experiment or Observation … In all Reports of Experiments to be brought into the Society, the matter of fact shall be barely stated, without any prefaces, apologies, or rhetorical flourishes; and entered so in the Register-book, by order of the Society. And if any Fellow shall think fit to suggest any conjecture, concerning the causes of the *phaenomena* in such Experiments, the same shall be done apart; and so entered into the Register-book, if the Society shall order the entry thereof.[33]

The specular view of objects predominates here: matters of fact are to be "barely stated" and should avoid "rhetorical flourishes." These in turn are to be segregated from conjecture about causes. This injunction found echoes throughout the works produced by the Royal Society and serves as the foundation of Shapin and Schaffer's argument that Boyle and the Royal Society promoted consensus about facts over disputes about causes.

The statutes do not disallow conjecture, yet the divide between conjectures and matters of fact could certainly lead to an empiricist suspicion of hypothetical explanations. A manuscript elaborating upon the statutes' distinction between facts and conjectures (probably by Sir Robert Moray) calls for an even greater suspicion of hypotheses:

> [T]his Society will not own any Hypothesis, systeme, or doctrine of the principles of Naturall philosophy, proposed or maintained by any Philosopher Auncient or Moderne, nor the explication of any phaenomenon, where recourse must be had to Originall causes, (as not being explicable by heat, cold, weight, figure, consistence & the lyke, as effects produced thereby,) Nor dogmatically define, nor fixe Axiomes of Scientificall things, but will question and canvas all opinions, adopting nor adhering to none, till by mature debate & clear arguments, chiefly such as are deduced from legittimate experiments, the trueth of such positions be demonstrated invincibly.[34]

A key issue in the institutional life of the Royal Society involved the issue of ownership: what did it mean to say that the society would not "own" any hypothesis? What obligations would this impose upon individual fellows pursuing their own ideas? Moray exempts certain kinds of entities from the taint of illegitimate "Originall causes"–"heat, cold, weight, figure, consistence"–that bear similarity to Bacon's list of primitive natures quoted earlier. Moray's emphasis on the need for explanatory

[33] *The Record of the Royal Society of London for the Promotion of Natural Knowledge*, 4th ed. (London: Morrison & Gibb, 1940), 290. On the Society's use of register books and record keeping see Adrian Johns, *The Nature of the Book: Print and Knowledge in the Making* (Chicago: Univ. of Chicago Press, 1998), 480–91; Hunter, *Establishing,* 4–6.

[34] This document was attributed to Hooke when published in Charles Richard Weld, *A History of the Royal Society, with Memoirs of the Presidents,* 2 vols. (London: John W. Parker, 1848), I: 147–8. Hunter questioned the attribution of this document to Hooke (*Establishing,* 242, n. 211). In *Orthodoxy,* Hunter attributes the document to Moray and reproduces the version cited here (172–3). Hunter's view of the document suggests that Moray's views were widely shared among serious Fellows within the Royal Society.

claims to be "demonstrated invincibly" by "legitimate experiments" enacts the Royal Society's institutional suspicion of hypothesis. For Moray, the time was not ripe for the careful, inductive assent Bacon calls for, but remains at the stage of the collection of facts:

> And till there be a sufficient collection made, of Experiments, Histories, and observations, there are no debates to be held at the weekely meetings of the Society, concerning any Hypothesis or principle of philosophy, nor any discourses made for explicating any phenomena, except by speciall appointment of the Society, or allowance of the president: But the time of the Assemblyes is to be employed, in proposing and making Experiments, discoursing of the trueth, manner, grounds & use therof; Reading & discoursing upon Letters, reports, and other papers concerning philosophicall & mechanicall matters; Viewing and discoursing of curiosities of Nature and Art; and doing such other things as the Council, or the president alone shall appoint.[35]

In effect, Moray provides one translation of Bacon's philosophy into institutional rules, a "chaste, and severe" interpretation, which we shall contrast with the "indulgence" and "liberty" of Hooke's view (*SEH*, IV: 32, 149). Conjectures are not just to be segregated from facts, but are to be banned, at least for a period of time. No "discourses made for explicating any phenomena" are allowed without special permission. The passage does call for opinions to be canvassed and eventually those "deduced from legitimate experiments" can be adopted. Thus, even this strident call for avoiding premature speculation does not indicate that hypothesis and conjecture should not play a role in natural philosophy.

The issue of interpretation of the timeliness or relative strictness of the suspicion of hypothesis came to a head in the discussion of Hooke's *Micrographia* (1665), the second work officially published by the Royal Society. *Micrographia* was the outcome of efforts by the Royal Society to determine the proper relationship between the individual views of its members and its official representation in those works explicitly offered to the public under its imprimatur.[36] In order to understand what it meant for the Royal Society to implement a Baconian program *collectively*, attention must be paid to those early works that were published under their newly won authority. The books published, significantly with none by Boyle, included John Evelyn's *Sylva, or a Discourse of Forest Trees* (1664), a horticultural work that was part of a broader contribution to the history of trades to which Evelyn aimed, John Graunt's *Natural and Political Observations ... upon the Bills of Mortality* (1665), which showed the hand of William Petty's interest in statistical questions, and John Wilkins' *An Essay towards a Real Character, and a Philosophical* Language (1668), developing a philosophical language of things. All explicitly invoked a common rhetoric in calling attention to things themselves rather than words, but they also reflected an interest in producing knowledge that would satisfy

[35] Hunter, *Orthodoxy*, 173.

[36] John T. Harwood, "Rhetoric and Graphics in *Micrographia,*" in Hunter and Schaffer, *Hooke*, 121.

each Baconian image of objectivity that I have discussed, albeit with differences in emphasis on the part of each.[37] In Hooke's case, his work demonstrated the possibilities that a constructivist focus on knowing as doing offered, though his often bold use of hypotheses appeared to violate the Society's statutory concern to segregate matters of fact from conjecture about causes.

The manual interpretation of Bacon: Hooke's constructivism

Though the Society was appreciative of Hooke's observations, they were also suspicious of his tendency to propose speculative hypotheses. Hooke agreed with Moray's injunction that explanations ought to be "deduced from legittimate experiments";[38] he believed that he had in fact deduced some of his explanations from experiments. They were particularly concerned that the Royal Society not be seen as "owning" Hooke's hypotheses, so it was necessary to indicate their hypothetical status.[39] When the license to publish under the Society's name was given, Hooke was cautioned that he

> give notice in the dedication of that work to the Society, that though they have licensed it, yet they own no theory, nor will they be thought to do so: and that the several hypotheses and theories laid down by him therein, are not delivered as certainties, but as conjectures; and that he intends not at all to obtrude or expose them to the world as the opinion of the Society.[40]

Hooke complied, in the process observing that hypothetical statements were not completely banned by the Royal Society's method:

> The Rules YOU have prescribd YOUR selves in YOUR Philosophical Progress do seem the best that have ever yet been practisd. And particularly that of avoiding Dogmatizing, and the espousal of any Hypothesis not sufficiently grounded and confirm'd by Experiments. This way seems the most excellent, and may preserve both Philosophy and Natural History from its former Corruptions. In saying which, I may seem to condemn my own Course in this Treatise; in which there may perhaps be some Expressions, which may seem more positive then YOUR Prescriptions will permit: And though I desire to have them understood

[37] These works by key figures active in the society (counting Petty as the driving force behind Graunt's work), along with Sprat's *History of the Royal Society* (1667), are five of the first six books printed by the Royal Society and discussed in Lynch, *Solomon's Child* (omitting discussion of Walter Charleton, *Inquisitiones II Anatomico-physicae* [1665]). Publications under the Royal Society's imprimatur are listed in Charles A. Rivington, "Early Printers to the Royal Society," *Notes and Records of the Royal Society of London* 39 (1984): 1–27; *idem*, "Addendum: Early Printers to the Royal Society, 1667–1708," *Notes and Records of the Royal Society of London* 40 (1986): 219–20.

[38] Hunter, *Orthodoxy*, 173.

[39] Hooke to Boyle, 24 November 1664, in R. T. Gunther, *Early Science in Oxford*, 15 vols. (Oxford, 1923–67), VI: 223.

[40] 23 November 1664 in Gunther, VI: 219; Birch, *History*, I: 491.

> only as Conjectures and Quaeries (which YOUR Method does not altogether disallow) yet if even in those I have exceeded, 'tis fit that I should declare, that it was not done by YOUR Directions.[41]

The key question became whether Hooke's hypotheses were "sufficiently grounded and confirm'd by Experiments." In addition, an unanswered question remained what determined whether an author (or the whole Society) "espoused" or "owned" a hypothesis. What did it mean to treat possible causes in a "more positive" fashion than allowed? Should the line be drawn at entertaining speculations in a non-committal fashion, at pursuing them as tentative working hypotheses, or at accepting them as facts?

The point I would like to make about the conflict between Hooke and the Royal Society is not that Hooke's appeal to hypothesis was more advanced than the society's suspicion of hypothesis. Rather, I want to emphasize the parties' *agreement* that ungrounded theorizing should be avoided and the necessity, eventually, of arriving at a true theory by induction. The common commitment of the Royal Society and Hooke to Baconian method explains both their agreement on the rules of method and their conflict over its interpretation. Hooke himself developed the most explicit attempt to follow Bacon's lead in *Novum Organum,* so it makes little sense to see him as adopting a Cartesian hypotheticalism in place of the Society's Baconianism.[42] Hooke's combination of a Baconian aversion to premature theorizing and highly speculative explanations is extreme, but not unique.

The relationship between experiment and theory (or between manual and generative objects) in instrumental practice remained problematic in the larger philosophical community. While Hooke held that theory was to be *deduced* from experiment, in practice he treated plausible explanations of experimental findings as firm bases for further theoretical speculation and even for empirical descriptions of as-yet unconstructed instruments. Hooke's description of a lens-grinding tool in *Micrographia* elicited criticism from the French astronomer Adrian Auzout, soon to be a Royal Society fellow, who was amazed that the instrument had been advertised without practical confirmation of its working. Hooke turned the tables, wishing that Auzout's criticisms had proceeded "not by speculation, but by experiments" and suggesting that despite Auzout's claim to distinguish sharply theory from fact, he treated his theory of apertures "very positive, not at all doubting to relie upon it." At the same time, he insisted that "it was not meer Theory I propounded, but somewhat of History

[41] Robert Hooke, *Micrographia: Or Some Physiological Descriptions of Minute Bodies Made by Magnifying Glasses with Observations and Inquiries thereupon* (London, 1665; New York: Dover Publications, 1961), "To the Royal Society."

[42] Robert Hooke, "A General Scheme, or Idea of the Present State of Natural Philosophy, and How its Defects may be Remedied by a Methodical Proceeding in the making Experiments and Collecting Observations" in *The Posthumous Works of Robert Hooke,* ed. Richard Waller (London, 1705; New York: Johnson Reprint Corporation, 1969), 1–70. Larry Laudan, "The Clock Metaphor and Hypotheses: The Impact of Descartes on English Methodological Thought" in *Science and Hypothesis: Historical Essays on Scientific Methodology* (Dordrecht: D. Reidel, 1981); Rose-Mary Sargent, "Robert Boyle's Baconian Inheritance: A Response to Laudan's Cartesian Thesis," *Studies in History and Philosophy of Science* 17 (1986): 469–86.

and matter of Fact," having made tentative trials "not without some good success."[43] More importantly, Hooke took offense that the Royal Society had been implicated despite his efforts "even in the Beginning of my Book, to prevent such a misconception," a reference to the warning about his use of hypotheses which the Royal Society had asked him to include.[44]

Auzout admitted that he had not read the preface, but believed that the Royal Society "would allow nothing to be published on scientific matters or concerning machines unless the former were based upon observations and the latter upon practice," criticizing Hooke for having "made public under their auspices so important a machine without having tested it." For Auzout, an explicit warning of its status would have been acceptable, thereby preventing workmen from wasting their time and money and "stop them from making fun of theorists when they perceive that their machines do not work."[45] In this case, Auzout and Hooke were agreed that hypotheses were necessary but disagreed in practice about the proper discursive framing of hypotheses within instrumental practice, as well as the extent to which it was proper to rely upon them in further work.

If knowing was doing, Hooke also sought an understanding that drew analogical connections between different kinds of phenomena, a feature also found in Bacon's approach. He developed his theory by first confirming a proposed explanation for capillary action experimentally, improving the effect through manufacturing thin glass tubes, then using his explanation as a springboard for constructing analogous explanations of a variety of phenomena, including combustion, gravity, and biological activity. Specifically, Hooke argued for a non-homogeneous conception of matter, in which any form of matter would tend to cohere or repel with other forms, depending upon whether the two forms were "congruous" or "incongruous." Congruity and incongruity were intended to be mechanically based concepts, although they could operate as explanations without requiring the specification of detailed micro-mechanisms and hence expanded the mechanical philosophy to include "active powers."[46]

This example brings out quite clearly what neglecting the constraining effect of the Royal Society's methodological commitments would leave unexplained. It would

---

43 "Considerations of Monsieur Auzout upon Mr. Hook's New Instrument for Grinding of Optick-Glasses," *Philosophical Transactions,* I: 57–63, 5 June 1665; "Mr. Hook's Answer to Monsieur Auzout's Considerations, in a Letter to the Publisher of these Transactions," *Philosophical Transactions,* I: 63–9, 5 June 1665, reprinted in Henry Oldenburg, *Correspondence,* eds. A. Rupert Hall and Marie Boas Hall, 13 vols. (Madison: Univ. of Wisconsin Press, 1965–86), II: 383, 387.

44 Oldenburg, *Correspondence,* II: 384.

45 Auzout to Oldenburg in Oldenburg, *Correspondence*, II: 420.

46 John Henry, "Robert Hooke, the Incongruous Mechanist" in *Robert Hooke: New Studies,* eds. Michael Hunter and Simon Schaffer (Woodbridge, Eng.: Boydell Press, 1989), 168; *idem*, "Occult Qualities and the Experimental Philosophy: Active Principles in Pre-Newtonian Matter Theory," *History of Science* 24 (1986): 335–81; Keith Hutchison, "What Happened to Occult Qualities in the Scientific Revolution?," *Isis* 73 (1982): 233–53; *idem,* "Supernaturalism and the Mechanical Philosophy," *History of Science,* 21 (1983): 297–333; Simon Schaffer, "Godly Men and Mechanical Philosophers: Souls and Spirits in Restoration Natural Philosophy," *Science in Context* 1 (1987): 55–85; Patrick Curry, "Revisions of Science and Magic," *History of Science* 23 (1985): 299–325.

be possible to treat the construction of hypotheses as an unproblematic process. It could be argued that Hooke employed a hypothetico-deductive approach.[47] According to this interpretation, in confronting particular problem areas, Hooke proposes the best explanation he can arrive at based upon contextually available intellectual resources. This interpretation ignores the particular manner in which Hooke's process of constructing hypotheses was *constrained.* Hooke did not propose hypotheses at just any particular point nor did he treat as equal the status of any conjecture not directly confirmed by experimentation. Congruity can only become a central concept for Hooke following the experimental confirmation of its role in capillary action. If Hooke had been an ordinary hypothetico-deductivist, he could have proposed congruity-based explanations before this point. Moreover, there is no reason why he could not have proposed different kinds of explanations for the variety of problem areas addressed in *Micrographia.* Explanations derived by analogy with one confirmed explanation are justified in Hooke's mind since good explanations should satisfactorily address a wide variety of phenomena rather than adapt themselves narrowly to a particular phenomenon, just as Bacon had sought to discipline the inductive ascent to the form of heat by a sufficiently wide experience of types of heat.[48] Hooke developed explicit reflections upon this dynamic via the concept of the "Similitude of the nature of Cause."[49]

Hooke's concepts of the similitude of the nature of cause and of congruity were developed by Newton in the form of the methodological rule called the "analogy of nature" and the concept of sociability, suggesting that historiographical approaches that sharply distinguish Newton's theory building from the early Royal Society's positivist focus on accumulating facts are misguided.[50] The old idea, popularized in the eighteenth century, that Newton followed a Baconian method in rejecting hypotheses has been rejected by historians of science who have shown that he employed hypotheses and that hypotheses had a legitimate role in his method.[51] However, as we have seen, Bacon himself allowed for something like hypotheses both in assessing candidate explanations of underlying forms before excluding explanations incompatible with his factual tables and in his ultimate "indulgence of the understanding" that gave him permission to assert affirmatively. Where he differed from modern hypothetico-deductive views was by his insistence that hypotheses

---

47 Carl G. Hempel, *Philosophy of Natural Science* (Englewood Cliffs, N.J.: Prentice-Hall, 1966), ch. 2. Despite the widespread rejection of this account of scientific method, for philosophers like Bacon or Descartes to be taken seriously, they must be rescued from the accusation of deductivism or inductivism, obliterating the differences in their philosophies for a shared recognition of the role of hypothesis. See John Losee, *A Historical Introduction to the Philosophy of Science,* 2nd ed. (Oxford: Oxford Univ. Press, 1980); David Oldroyd, *The Arch of Knowledge: An Introductory Study of the History of the Philosophy and Methodology of Science* (New York: Methuen, 1986).

48 R. H[ooke], *An Attempt for the Explication of the Phaenomena, Observable in an Experiment Published by ... Robert Boyle* (London, 1661), 40–2; *SEH*, IV: 126–49.

49 Hooke, *Posthumous Works,* 165.

50 J. E. McGuire, "Atoms and the 'Analogy of Nature': Newton's Third Rule of Philosophizing," *Studies in History and Philosophy of Science* 1 (1970): 3–58; Schaffer, "Godly Men," 64.

51 Alexandre Koyre, *Newtonian Studies* (Chicago: Univ. of Chicago Press, 1965), ch. 2.

must be established by induction from a wide base of experience–he did not give free reign to the creative imagination–and by his requirement that certainty was the goal of science. Hypotheses were at best temporary scaffolding on the way to truths established by experience. Bacon's instance of the fingerpost or crucial instance (*instantia crucis*) sought to distinguish a true explanation from a false one by a carefully selected observation. This concept was adapted by Boyle and Hooke into a crucial experiment (*experimentum crucis*) designed to test a hypothesis by a single, well-chosen experiment or act "as a Guide or Land-mark, by which to direct our course in the search after the true cause" of a phenomenon. The concept was in turn adopted by Newton to establish decisively that white light is constituted by a mixture of colored rays in his famous 1672 optical paper addressed to the Royal Society.[52]

Just as the Royal Society failed to agree with Hooke on the status of his explanations, so too did Hooke fail to endorse Newton's optical claims even while largely sharing a sophisticated Baconian reading of experimentation. Thus, Newton could instruct Hooke that "You know the proper Method for inquiring after the properties of things is to deduce them from Experiment."[53] To be sure, Newton's method drew more heavily upon developments in mixed mathematics linking mathematical explanations to natural philosophy and he endorsed a deductive, mathematical ideal of science. Still, in his disputes with Hooke on the status of his optical explanations and in his frequent insistence on distinguishing hypotheses from true, confirmed theories, Newton could be seen as partially sharing in the Royal Society's Baconianism.

Empiricism with a generative twist: Analogy and natural powers in Evelyn

While it is easy to credit Hooke with carrying forward Bacon's program for induction to natural forms underlying appearances, John Evelyn's work on horticulture looks more trivial, at best instantiating the specular, empiricist side of Bacon. In fact, however, Evelyn's contribution in *Sylva* was part of a systematic effort by the Royal Society to address itself to agricultural questions and borrowed Hooke's use of disciplined analogy to illuminate the growth of trees, the nature of composts and mists, and the study of natural processes like calcification. While the collaborative and practical context of *Sylva* support the view that Evelyn participated in the Royal Society's Baconian methodological program, his reliance upon classical and contemporary literary sources have been taken as evidence of his orientation to texts rather than

[52] In *Micrographia,* Hooke speaks of the cause of colors (54). Brian Vickers, *English Science, Bacon to Newton* (Cambridge: Cambridge Univ. Press, 1987), 198–211. Vickers observes that Boyle first coined the phrase *experimentum crucis* in referring to Pascal's Puy-de-Dome experiment in 1662 (237, n. 8). See also J. A. Lohne, "Experimentum Crucis," *Notes and Record of the Royal Society of London* 23 (1968): 169–99. On the difficulties Newton had establishing (retrospectively) the experiment as crucial, see Simon Schaffer, "Glass Works: Newton's Prisms and the Uses of Experiment" in *The Uses of Experiment: Studies in the Natural Sciences,* eds. David Gooding, Trevor Pinch, and Simon Schaffer (Cambridge: Cambridge Univ. Press, 1989), 67–104.

[53] Quoted in Richard S. Westfall, *Never at Rest: A Biography of Isaac Newton* (Cambridge: Cambridge Univ. Press, 1980), 247–8.

things themselves. William Petty fits the traditional picture of a Baconian when he declares that he avoids reading books, even going so far as to criticize Boyle for wasting his time by "continual reading" when his own "stock of experience" exceeds what he could find from books.[54]

In fact, however, Evelyn's reading practice was heavily shaped by Bacon's interest in commonplace books as a good way to prepare for induction. Distrust of written sources can take two forms: avoiding them or carefully using a method to weed out unreliable claims. Bacon preferred the later approach, since his method was supposed to weed out false information as the form of a nature slowly became evident following the collection of comprehensive natural and experimental histories. The Royal Society followed the approach outlined in Bacon's *Parasceve ad historiam naturalen et experimentalem* in the formation of a committee for collecting natural phenomena and accounts of natural and preternatural phenomena, leaving the history of art for a separate committee.[55]

Evelyn's reliance upon commonplace books where information was digested into units and systematically compared amounted to an artificial augmentation of memory, just as instruments augmented senses and Bacon's inductive logic amplified reason. Hooke's "memoranda" similarly relied upon Bacon's account of artificial aids for remedying the limitations of sense, memory, and reason.[56] In one of his commonplace books, Evelyn quotes Bacon's endorsement of commonplacing in *The Advancement of Learning*:

> A substantiall and Learned digest of *Common-Places* is a solid, and a good aide to memory; And, because it is a counterfeit thing in knowledge, to be forward and pregnant, unlesse you be withall deep and full; I hold that the dilligence, and paines in collecting Commonplaces, is of great use.[57]

As learned as Evelyn's approach may be, rooted as it is in practice in humanist methods, he insisted in *Sylva* upon the need to focus on useful facts without rhetorical flourishes, "abstracting things Practicable, of solid use, and material, from the Ostentation and impertinences of Writers; who receiving all that came to hand on trust, to swell their monstrous volumes, have hitherto impos'd upon the credulous World, without conscience or honesty."

While he credits classical authorities as the source of much factual information, he considers effort spent pursuing their speculations to be wasted. Following "what our Verulam has justly pronounc'd concerning their Rhapsodies," he "must deplore

[54] Petty to Boyle in Boyle, *Works,* VI: 138.

[55] The report of the committee is given in Hunter, *Establishing,* 104–5 (for commentary, see 91–3). *SEH,* I: 369–411, IV: 249–63.

[56] Lotte Mulligan, "Hooke's 'Memoranda': Memory and Natural History," *Annals of Science* 49 (1992): 47–61; Johns, *Nature,* 433.

[57] Quotation from Bacon's *Advancement of Learning,* as recorded in Evelyn, *Locorum Comm: Tomus IIdus,* Evelyn papers, Brit. Lib., JE.C3, f. 1. See also Evelyn's *A Booke of Promiscuous Notes & Observations concerning Husbandry, Building &c,* JE.D6.

the time which is (for the most part) so miserably lost in pursuit of their Speculations." Evelyn considers the Society's motto, *Nullius in verba,* to testify less to a rejection of all reliance upon books than a "base and servile submission" to the words of some author.[58] If Evelyn felt that his gentlemanly status and his attachment to Bacon made possible a better empirical understanding of success in horticulture, he also believed that the work of the Royal Society would uncover the true causes of natural phenomena, in particular the investigations with the microscope by Hooke.

Evelyn mentions the microscope at two places in *Sylva;* in both cases, it functions as a potential empirical foundation for underlying corpuscular explanations, in effect "deduced from legittimate experiments."[59] In the first case, Evelyn suggests that the microscope could make evident the "Secret Schematismes"–Bacon's "latent configurations"–accounting for the quality of seeds.[60] In the second case, he quotes extensively from Hooke's account of a piece of petrified wood, which microscopical observation reveals to result from calcification by minerals in the water.[61] The presence of pores allowing silt to accumulate, thereby turning wood into stone, served as a prerogative instance revealing how the motion of sap brings nutrients to all parts of the plant in ordinary growth, a subject of extensive discussion in the Royal Society.[62] If Bacon's initial stages of induction avoided privileging certain observations over others, in later stages "prerogative instances" played an important role. In general, prerogative instances revealed nature in a clearer fashion than ordinary examples. An observation might exhibit similarity to other instances only in the nature under consideration, and not in any other qualities, as prisms isolate color apart from the ordinary color fixed in objects (*SEH*, I: 180, 156).

Petrified wood is a good example of this kind of anomalous object, a puzzle Evelyn had already discussed in corpuscular terms in his 1656 translation of Lucretius' *De rerum natura.* Petrification attracted interest among natural philosophers both for its practical relevance to the cure of the stone and its theoretical significance.[63] John Webster's encyclopedia on metals considered petrification to be a kind of transmutation in his history of metals.[64] Within the alchemical tradition that Digby carried forward into the Royal Society, minerals grew in the ground just as plants did; the Society debated this question at the same time they were considering the cause of

58 John Evelyn, *Sylva, or a Discourse of Forest-Trees, And the Propagation of Timber in His Majesties Dominions* (London, 1664), "To The Reader."

59 Hunter, *Orthodoxy,* 173.

60 Evelyn, *Sylva,* 5.

61 Evelyn, *Sylva*, ch. 30. Birch, *History*, I: 260–2, 17 June 1663.

62 See, for example, "Queries Concerning Vegetation, especially the Motion of the Juyces of Vegetables," *Philosophical Transactions* 40 (19 October 1668): 797–9; [Francis] Willugby and [John] Wray, "Concerning the motion of the sap in trees," *Philosophical Transactions* 48 (21 June 1669): 963–5, draft in Royal Society Classified Papers X(1).12.

63 Hartlib suffered from the stone and Digby possessed a cure; see Beale to Evelyn, 30 July 1664 and 3 October 1664, Evelyn Papers, Brit. Lib., JE.A12, ff. 42, 45.

64 John Webster, *Metallographia: or, An History of Metals* (London, 1671), 358–64.

petrification.[65] The analogy between plant growth and the origin of minerals gave plausibility to the alchemist's attempt to transform minerals into gold, while endowing petrification with theoretical significance for understanding the transformation of matter.

From the Baconian perspective espoused by Evelyn, Beale, and Boyle, petrification attracted attention as a prerogative instance that carried greater epistemic weight than ordinary vegetables or minerals.[66] What kind of process could turn a formerly living tree into a stone? If a plant could turn into a mineral, the motion of inanimate objects might in turn make up living objects; the application of the mechanical philosophy to living things depended upon an examination of prerogative instances like this. Appearing as it does as a process linking the animate and inanimate worlds, petrification was a matter of fact that served as the basis for analogical theory development. Theoretical insight would build upon surreptitious analogies between unusual natural effects like petrification and the ordinary observed behavior of plant growth.

Among Royal Society Fellows, Digby had earlier adapted alchemy to mechanical explanations and operational definitions, but his unregulated metaphorical thinking typified by his defense of Paracelsian sympathetic medicine was treated suspiciously by them. For Evelyn, the development of theory was to take place within the context of experimental work; experiment was not to be interpreted by a pre-existing theoretical tradition. This view was broadly shared within the Royal Society and accounts for the skeptical view taken of Digby's writings.[67] Beale observed to Evelyn that Sir Kenelm Digby's book on plants ought to have "every paragraph confirmed, (as far as it could be) with experimts. or such demonstrations, as ye maker will beare." Lacking this, he preferred that Digby's work not be associated with the Royal Society (the work was not published by the Royal Society; in addition, by mistake, the book omitted mention that Digby was a member of the Royal Society, to the relief of Beale).[68]

To say that Evelyn, Boyle, and Beale were suspicious of the role of sympathetic analogy does not mean that similar analogies found no place in their work. Indeed, in his *Philosophical Discourse on Earth*, Evelyn explicitly invoked Digby's account of the sympathetic powder to explain the recovery of a soil's fertility after it attracts vital spirits in the air, an account he confirmed through his own observation.[69]

[65] Betty Jo T. Dobbs, "Studies in the Natural Philosophy of Kenelm Digby. Part I," *Ambix* 18 (1971): 10. For differing opinions on whether minerals grew, discussed at the same meeting at which Hooke examined the piece of petrified wood with a microscope, see Birch, *History*, I: 247 (27 May 1663).

[66] See Beale's discussion of Dr. Huret's experimental study of petrification in a letter to Evelyn, 3 August 1664, Evelyn papers, Brit. Lib., JE.A12.

[67] Oldenburg to Boyle, 20 March 1660, in Boyle, *Works,* VI: 145.

[68] Sir Kenelm Digby, *A Discourse concerning the Vegetation of Plants,* an early version of which was read to the Society on 28 May 1661; see Oldenburg *Correspondence,* I: 25; Beale to Evelyn, 28 April 1666 and 2 January 1668/69, Evelyn Papers, Brit. Lib., JE.A12. Sorry to see "Sr K Digbyes closet opened" following the publication of *Of Bodies, and of Man's Soul* (1669), Beale remarked that he is "not sorry yt his Title to R S" was omitted from the title page by a translator's mistake (Beale to Evelyn, 2 January 1668/69, JE.A12).

[69] J. Evelyn, *A Philosophical Discourse of Earth* (London, 1676), 67–8. For background on the search for vital spirit, see Allen G. Debus, "Chemistry and the Quest for a Material Spirit of Life in the Seventeenth Century" in *Spiritus,* eds. M. Fattori and M. Bianchi (Edizioni dell'Ateneo, 1984), 245–63.

Nevertheless, in order to arrive at an understanding of vegetative activity through induction, Evelyn cannot remain satisfied with empirical confirmation of the existence of a vital power. Rather, he must seek an underlying explanation for the power that would allow him to produce the phenomenon at will. For this, Evelyn turned to Boyle's mechanical philosophy, which he understood through the lens of his Baconian pursuit of generative knowledge.[70] He considered petrification caused by water to indicate that the element water could take on solid form. If solids can be produced from fluids, the process can be reversed to create "artificial *Dews* and *Mists* impregnated with several qualities" to nourish exotic plants that otherwise cannot survive in British soil.[71]

Hermetic theories on the possible role of May-Dew in carrying a vital principle underlying plant and animal growth and spontaneous generation were taken seriously enough to launch a series of experiments in the early Royal Society. Hooke's mechanical philosophy incorporated an account of how the "congruity" between air and small particles allowed dews to impregnate plants with specific nutrients. This mechanical translation of Digby's occult power was matched by the Society's determination to carry out careful experimentation. Carried out by Thomas Henshaw, with input from John Pell and Robert Boyle, these experiments included the collection and weighing of dews from different environments, the isolation of salts to be used as fertilizer, and the study of their putrefaction leading to spontaneous generation.[72] Evelyn considered mists impregnated with specified qualities to allow for the cultivation of plants from different climates.

While often considering chemical and Aristotelian explanations in a non-committal fashion throughout this work, only the mechanical philosophy could shed light on the *generation* of qualities, much like Bacon's alphabet of natural powers:

> And if that be true, that there is but one *Catholic, homogeneous*, fluid matter, (diversified only by *shape, size, motion, repose*, and various *texture* of the minute Particles it consists of; and from which affections of matter, the divers qualities result of particular bodies;) what may not mixture, and an attent inspection into the anatomical parts of the vegetable family in time produce, for our composing of all sorts of Moulds and Soils almost imaginable … ?

For Evelyn, the mechanical philosophy is a language of causes suitable for a Baconian program since it allows one to envision recombining and rearranging the fundamental elements of matter to produce desired qualities. Theoretical understanding requires "attent inspection" to be sure, but only if it aspires to a generative language that explains nature's productions, so that in turn art may generate

---

[70] See Brian P. Copenhaver's "A Tale of Two Fishes: Magical Objects in Natural History from Antiquity through the Scientific Revolution," *Journal of the History of Ideas* 52 (1991): 373–98 for a discussion of how the mechanical philosophy was used to explain the workings of magical objects.

[71] Evelyn, *Philosophical Discourse,* 45 and 47. See also the discussion of Beale's analysis on petrified wood, 171.

[72] Alan B. H. Taylor, "An Episode with May-Dew," *History of Science* 32 (1994): 163–84.

variety through its own combination of a natural "Alphabet of Earths and Composts."[73]

The idea that the microscope would directly reveal the hidden, corpuscular causes of natural phenomena was widespread in the Royal Society. Beale, himself skilled in the construction of optical instruments, notes that the Royal Society's interest in "philosophical glasses" included uncovering the causes of generation among animals.[74] The link that Evelyn draws explicitly between his own work and the promise that Hooke's work offers to carry it further shows how these dramatically different philosophers nevertheless saw themselves as part of the same collective activity to carry out a broadly Baconian program for the reform of knowledge.

Generative objects/generating knowledge: Wilkins' language of things

It would be easy to see John Wilkins' *An Essay Towards a Real Character, and a Philosophical Language* as out of step with the Royal Society's Baconianism, reflecting the residual "taxonomic" interests" of a university-trained scholar. Alternatively, Wilkins could be seen as part of a group of "physico-mathematicians" within the Royal Society who pointed the way from a Continental emphasis on mixed mathematics towards Newton's mathematical physics.[75] And yet no work better exemplifies the Royal Society's collective effort to fulfill Bacon's charge to focus upon things themselves than Wilkins' language of things. This seems paradoxical, since Wilkins has attracted the attention primarily of historians of linguistics to the extent to which his efforts to create an artificial language illuminate emerging understanding (or misunderstanding) of natural language. Wilkins, however, reflecting the Royal Society's Baconian rhetoric of *res/verba,* insisted that his undertaking was superior to a focus on natural languages, "it being as much to be preferred before that, as things are better than words, as real knowledge is beyond the elegancy of speech."

Unless we are willing to understand the significance that Wilkins' project had for the Royal Society in terms of Bacon's *res/verba* distinction, we will fail to see how the value attributed to Wilkins' work reflects the collective project of Baconian reform of the early years of the Royal Society. If we focus attention of the relationship between Wilkins' project and other language schemes of the period, we will fail to understand why Hooke, in particular, took so much interest in the language. Of particular interest to Hooke was the promise that the language might have to *uncover* true knowledge of nature, not just lay it bare after discovery. Language could thus

---

73 Evelyn, *Philosophical Discourse,* 45–6, 48. For the production of "very unexpected *Phaenomenas*" by this method, see 47. Beale answered queries on composting that Evelyn had put to him between the publication of *Sylva* and *A Philosophical Discourse on Earth* (letter from 11 December 1668 inserted in *Rude Collections to be Inserted into Elysium Britannicum, Referring to the several Chapters of what is begun,* Evelyn papers, Brit. Lib., JE.D9, ff. 79–80).

74 Beale to Evelyn, 26 April 1665, Evelyn papers, Brit. Lib., JE.A12.

75 M. M. Slaughter, *Universal Languages and Scientific Taxonomy in the Seventeenth Century* (Cambridge: Cambridge Univ. Press, 1982); Dear, *Discipline,* 227.

possibly provide a generative method and not just a static, specular taxonomy. To this extent, Wilkins' language bears similarities to Hooke's Bacon-inspired scheme to construct an automatic machine for constructing new natural knowledge, his "philosophical algebra." Wilkins insisted that a method of discovery was the language's most important virtue:

> That the reducing of all things and notions, to such kind of Tables, as are here proposed (were it as completely done as it might be) would prove the shortest and plainest way for the attainment of real Knowledge, that hath been yet offered to the World. And I shall add further, that these very Tables (as now they are) do seem to me a much better and readier course, for the entring and training up of men in the knowledge of things, then any other way of Institution that I know of.[76]

Like Bacon, Wilkins offered his system as a legitimate stand-in for a more perfected system that would actually *generate* knowledge. Even as it stood, he insisted that familiarity with the language will prepare men to attain true "knowledge of things." Consequently, he believed that his language met Bacon's requirement for a true understanding of nature, spelled out in *Novum Organum,* that it would actually direct new work and generate new inventions:

> [T]he works already known are due to chance and experiment rather than to sciences; for the sciences we now possess are merely systems for the nice ordering and setting forth of things already invented; not methods of invention or directions for new works. (*SEH*, IV: 48)

Wilkins' efforts to describe a natural grammar of powers reflects the theoretical realist side of Bacon's focus on things, rather than just the empiricist classification of visible objects that it seems to be at first sight.

Like Bacon's method for the discovery of a phenomenon's operative form or Hooke's philosophical algebra, the philosophical grammar of Wilkins' language was to map the forms of nature directly and in principle generate alternative ways of producing any complex object. Hooke shared this belief; in a letter discussing Leibniz's interest in philosophical languages, Hooke suggested that philosophical language can assist natural philosophy just as algebra assists geometry and arithmetic. A follow-up letter refers to a philosophical language as "the Algebra of Algebras or the Science of Methods."[77] Just as Bacon believed that a correct classification of heat as a particular variety of motion would allow heat to be produced at will by bringing about the motion in question, Wilkins believed that a correct categorization of natural phenomena would allow for their manipulation and control.

Wilkins' *Essay* compiles tables mapping out all simple notions organized taxonomically. The difficulty in setting out a real character, referring directly to things

---

[76] John Wilkins, *An Essay Towards a Real Character, and a Philosophical Language* (London, 1668), epistle dedicatory.

[77] Hooke to Leibniz, 12 July 1680, Royal Society Letterbooks H3.64; Hooke to Leibniz, 15 May 1681, H3.64.

themselves rather than notions becomes apparent when one considers how one is to organize a system of simple notions in such a way that allows more complex notions to be easily and reliably combined from the simple notions. One strategy, promoted by Royal Society Fellow Seth Ward in *Vindiciae Academiarum* (1654),[78] is to search for a small number of primitive notions that, as is the case with existing vowels and consonants, could be conjoined in such a way as to produce every conceivable word. Like vowels and consonants, the natural character would represent a small core of primitive notions from which every other notion could be constructed by combination. A name would then be a definition and would specify the true nature of the thing named.[79]

The difficulty with such an approach is that one would have to know already the true nature of a thing before one could refer to it. If there was disagreement about how to define a thing, each party would employ their own name based upon their own definition. Communication would break down failing prior consensus as to the true essence of things. Moreover, one would have to be able to organize a taxonomy such that a small number of primitive ideas would serve not only to mark out general kinds, but to discriminate varieties within each kind. One would in effect have to be able to reduce the variety of objects in the world to a few fundamental concepts that could combine to define the nature of any referent.[80] While the "algebraic," generative approach seemed out of reach, Wilkins had difficulty in achieving a specular, empirical adequacy, as well. The tensions between Wilkins' efforts to construct a language convenient to memorize and his claim to identify a natural classification was a source of exasperation for John Ray, who was engaged by Francis Willoughby on behalf of Wilkins to construct the tables of plants and animals, and complained of being "required to fit the plants to the author's own system."[81] Wilkins' attempt to forge a workable compromise between the ease of use of the language and classification based upon nature was unsuccessful.

Wilkins was aware of the imperfections of his language and called upon the Royal Society to appoint a committee to examine the language further.[82] Although the committee did not formally report back to the Society, interest in perfecting Wilkins' work continued after his death in 1672, particularly by Andrew Paschall, Francis Lodwick, Thomas Pigott, John Aubrey, and original committee members Robert Hooke and John Ray. Their interests seemed to have reflected the range of Wilkins'

---

[78] Reprinted in Allen G. Debus, *Science and Education in the Seventeenth Century: The Webster-Ward Debate* (London, Macdonald, 1970), 193–259.

[79] Debus, *Science*, 214–17.

[80] Slaughter, *Universal Languages*, 38–9; David Cram, "Language Universals and 17th-Century Universal Language Schemes" in *John Wilkins and 17th-Century British Linguistics*, ed. Joseph L. Subbiondo (Amsterdam: John Benjamins,1992), 191–203.

[81] Ray to Lister, 7 May 1669 in John Ray, *The Correspondence of John Ray: Selections from the Philosophical Letters Published by Dr. Derham and Original Letters of John Ray ...*, ed. Edwin Lankester (London, 1848), 39–42; Benjamin DeMott, trans., "Science versus Mnemonics: Notes on John Ray and on John Wilkins' *Essay toward a Real Character, and a Philosophical Language*," *Isis* 48 (1957): 5.

[82] Wilkins, *Essay*, epistle dedicatory. The committee was formed on 14 May 1668 (Birch, *History*, II: 283).

own motivations, including method for discovery, ideal communication system, improved pedagogy, and guarantor of religious harmony. Hooke believed that it might be possible to cut the number of radicals in half, with the result that it would be "applicable and usefull not onely for common Discourse and keeping correspondence but for the strict & philosophycall way of Reasoning and enquiry" allowing the identification of "the true proprietys of bodys" and "the power of Causes."[83] Aubrey thought the real character would be useful in educating children and Frances Lodwick developed a character or "universal alphabet" to improve the teaching of pronunciation.[84] Paschall wrote to Aubrey in the character, Wilkins and Wallis exchanged letters with it, and a letter from Cave Beck to Hooke in the character can be found in the Royal Society papers.[85] Hooke thought the language might serve for scientific communication that would preserve priority claims while confining the details to "true Lovers of Art, and they only." Consequently, he articulated the details of his spring balance watch in the real character. Richard Towneley, Andrew Paschall, and Thomas Pigott each translated the passage.[86] Paschall believed that religious controversy would be ended by perfecting the system, although he also despaired of the technical difficulties.[87]

These technical difficulties were the same ones Wilkins confronted: how to arrive at a transcendental classification of all natural powers and their combination. Based on correspondence beginning in 1676 reconstructed by Vivian Salmon, we now know that debate continued over the possibility of arriving at a transcendental determination for the entire classification scheme–Ward's ideal, although they could not determine how such a language would be practical for communication.[88] Thomas Piggott was skeptical, suggesting that only angels could "compound words just as things are compounded" and that "we justly expect a second confusion when we strive to build such a babel."[89]

---

[83] Hooke to Leibniz, 12 July 1680, Royal Society Letterbooks H3.63.

[84] John Aubrey, *Aubrey on Education: A Hitherto Unpublished Manuscript by the author of "Brief Lives"*, ed. J. E. Stephens (London: Routledge & Kegan Paul, 1972), 40, 81, 126. Francis Lodwick, "An Essay Towards An Universal Alphabet," *Philosophical Transactions* 16 (26 June 1686): 126–33; *idem*, "A Second Essay concerning the Universal Primer," *Philosophical Transactions* 16 (26 June1686): 134–7, both reprinted in Vivian Salmon, *The Works of Francis Lodwick: A Study of His Writings in the Intellectual Context of the Seventeenth Century* (London: Longman, 1972), 235–46. The drafts of Lodwick's papers in the Royal Society archives draw on Wilkins' *Essay* (Royal Society Classified Papers, XVI.4–5).

[85] Paschall to Aubrey, 13 February 1676/7 in A. J. Turner, "Learning and Language in the Somerset Levels: Andrew Paschall of Chedsey" in *Learning, Language and Invention: Essays Presented to Francis Maddison,* eds. W. D. Hackmann and A. J. Turner (Aldershot, Eng.: Variorum, 1994), 306–7; John Wallis, *A Defence of the Royal Society, and the Philosophical Transactions, Particularly those of July, 1670. In Answer to the Cavils of Dr. William Holder* (London, 1670), 17; Beck to Hooke, 17 January 1675/6, Royal Society Classified Papers XVI.1.

[86] Hooke, *A Description of Helioscopes* (London, 1676), reprinted in Gunther, *Early Science,* 150; Vivian Salmon, *The Study of Language in 17th-Century England* (Amsterdam: John Benjamins, 1979), 195; E. N. da C. Andrade, "The Real Character of Bishop Wilkins" in *British Linguistics,* ed. Subbiondo, 261.

[87] Turner, "Learning," 305.

[88] Salmon, *Study of Language,* 191–206.

[89] Quoted in Turner, "Learning," 305.

The method that sought to overcome Babel merely reinstated it. Yet to classify Wilkins' *Essay* as a dead end would be to miss its most important effects. As an early expression of the idea that natural science could overcome religious and ideological disagreement by displaying the fact of the matter, Wilkins codified the general feeling of the early Royal Society and influenced epistemological reflections of philosophers like Locke and Leibniz.[90] Perhaps Wilkins' most important influence was felt through his pupil, Thomas Sprat, in his *History of the Royal Society,* who tried to show that the Royal Society's focus on things themselves demonstrated how to control the divisiveness of verbal disputes throughout English society.[91]

Conclusion

I have argued that Bacon's philosophy of science incorporated three different conceptions of knowledge: empiricism, constructivism, and theoretical realism, corresponding to three different metaphors of "things themselves." These objects of knowledge were specular (objects of passive vision), manual (objects actively manipulated to achieve some end), and generative (underlying forms of ordinary objects capable of being recombined like letters of the alphabet). The early publications of the Royal Society and its committee deliberations demonstrate that each component of Bacon's philosophy was incorporated into their collective work, despite public rhetoric that seemed wedded to the specular, empiricist interpretation. Evelyn's collection of horticultural observations sought an underlying theoretical understanding of causes in an "Alphabet of Earths and Composts." Hooke's ostensibly specular recording of observations with the microscope included efforts to better manipulate material phenomena, using the microscope as a tool, and experiments directed at elucidating underlying causes. Wilkins' seemingly specular (or even non-Baconian) taxonomical focus obscured the real thrust and significance of his work: the uncovering of a hidden alphabet of forms that would generate new knowledge and powers. These efforts to implement collectively Bacon's reform of knowledge characterize the early years of the Royal Society.

---

90 Leibniz to Oldenburg, 23 August 1670; Leibniz to Oldenburg, 29 April 1671; Leibniz to Oldenburg, 16 April 1673 (Oldenburg, *Correspondence,* VII: 64–8; VIII: 22–9; IX: 593–601). Wilkins' scheme remained influential for future English grammarians as well. See Salmon, *Study of Language*, 203.

91 Lynch, *Solomon's Child*, ch. 5.

Chapter 10

Francis Bacon, *The Advancement of Learning*, and Historical Thought

Fritz Levy

In 1621, with his career in ruins, Francis Bacon began an effort to repair the damage by at last writing one portion of the history of Britain that he had been promising for more than two decades. What resulted–*The History of the Reign of King Henry VII*–was dedicated to Prince Charles and was intended to serve him as a pattern, one parallel to that other royal pattern, the Prince's father, King James. For Henry, like James, was "a wise man, and an excellent King," despite the fact that "the times were rough, and full of mutations, and rare accidents," and his reign was of the sort that pleased the biographer more than it had pleased its subject.[1] In its effort to link the present to the past, to see the past, properly interpreted, as providing explicit guidance for living in the present, Bacon's *History* was conventional enough. That Bacon's account of Henry's reign was more successful than most such endeavors was due in large measure to the author's style and to the intelligence of his analysis.

Yet that is not quite the whole story, for it is possible to argue that the acuity of Bacon's biography is at least in part a result of his theoretical analysis of historical writing and the way in which that subject might be divided between history of times and biography on the one hand and analytical discourses upon historical topics on the other. These categories were not intended to be exclusive, and Bacon's view of the interplay among them not only makes up a significant portion of *The Advancement of Learning* but also helps illuminate the novelties of his life of Henry VII. That said, the complexity and grandeur of the structure of *The Advancement* should not blind us to the fact that what Bacon has to say about history and about how historians work has roots in contemporary debates, and that his placement of varieties of historical thought within his system reflects those debates. It may also be argued that the relatively limited influence of Bacon's theory of history resulted from his neglect of some of the discussions going on around him, and that his own practice, in *The History of the Reign of King Henry VII,* also served to lessen the impact of his theoretical pronouncements. The complete story would involve an analysis of the critical

[1] Francis Bacon, *The History of the Reign of King Henry VII*, ed. Brian Vickers, Cambridge Texts in the History of Political Thought (Cambridge: Cambridge Univ. Press, 1998), 3.

thought, literary and historical, of late fifteenth- and sixteenth-century Europe which, while worth doing, would take us too far afield. Limiting our attention to England in the years just before and after the publication of *The Advancement of Learning*, should, however, provide us with at least an outline of what was happening.[2]

The classic Elizabethan discussion of the interrelationship among philosophy, history, and poetry was Sir Philip Sidney's *Defence of Poesy*; it has been suggested, however, that Sidney wrote the *Defence* in part as an answer to Jacques Amyot's learned preface to his translation of Plutarch's *Lives*.[3] Amyot defined history "as an orderly register of notable things said, done, or happened in time past, to mainteyne the continuall remembrance of them, and to serue for the instruction of them to come." The utility of history, according to Amyot, lay in the way historical examples allow us to judge present events and foresee those to come, "so as we may knowe what to like of, & what to follow, what to mislike, and what to eschew."[4] Such historically based judgment might be applied to public affairs as well as private, it might enable a man to give good advice and–perhaps most important–it might provide the materials necessary to teach him how to live well. Amyot acknowledged that philosophy could offer precepts, but insisted that examples were the more effective arguments because they offered more detail and were accompanied with reasons. And history exceeded poetry in effectiveness because truth was more persuasive than fiction. While reading history did not displace experience in preparing a man for action, at worst it made a useful supplement and, in certain circumstances, could stand in place of actual participation. Many a good general learned by fighting in the ranks, but–as preface after preface reminded us–the Roman general Lucullus had successfully learned his business through a careful study of the records of the past.

Even as Sir Thomas North translated Amyot's preface, Thomas Blundeville was preparing his brief treatise on *The true order and methode of wryting and reading hystories*, a summary translation and amalgamation of works by the theorists Francisco Patrizzi and Jacopo Aconcio. Blundeville dedicated his work to the Earl of Leicester–for whom Aconcio had written the original Italian–to allow the Earl "to gather thereof such iudgement and knowledge as you may therby be the more able, as well to direct your priuate actions, as to giue Counsell lyke a most prudent Counseller in publyke causes."[5] The sentiment was much like Amyot's. However, Blundeville's view of history was much more political. For him, an historical narration had not only to give an account of events in order, but had as well to allow the reader to see the rise, continuation, and fall of a city, or country, or empire. In addition, the reader should be able to learn about the public revenue, the military force, the manner of

[2] Some of the same issues have been discussed in a different context in Barbara Shapiro, *A Culture of Fact: England, 1550–1720* (Ithaca: Cornell Univ. Press, 2000), particularly chapter two.

[3] Elizabeth Story Donno, "Old Mouse-Eaten Records: History in Sidney's *Apology*," *Studies in Philology* 72 (1975): 275–98.

[4] Plutarch, *The Lives of the Noble Grecians and Romanes*, trans. Sir Thomas North [from Amyot's French] (London: Thomas Vautrollier, 1579), "To the Reader," *iij^{v}.

[5] Thomas Blundeville, *The True Order and Methode of Wryting and Reading Hystories* (London: William Seres, 1574), A.ij^{r}.

government, even the economic foundations of the country whose history was being studied. This is not to say that Blundeville ignored matters of morality; on the contrary, the good historian also examined carefully whoever had excelled either in virtue or in vice. Underlying this was the idea that the reader should be able to enter into the minds of the historical actors, to discern their intent. If the reader were to learn how to live well and to give good advice, the historiographer had to speak the unvarnished truth: he could not follow the poets by making much of little or nothing, nor yet follow the orators who enlarged small things and abased large as suited their case. Thus, the historian "ought not to fayne anye Orations nor any other thing, but truely to reporte euery such speach, and deede, euen as it was spoken, or done."[6] While not disagreeing in substance, Amyot emphasized history's potential for moral teaching while Blundeville emphasized the political. The result, however, was that Blundeville insisted that the historiographers concentrate their attention on history as a process, in which every action had to be seen in terms of its time and place, along with the motives of the principal actors. In this scheme, there was no place for the fictional.

Yet the fictional was hard to resist. When Sidney came to defend poetry for its moral value–after all, in real life virtue did not always triumph, nor was evil always punished–he also noted that, even with the best of intentions, the historian was reduced to fictional invention. In part, he had in mind the imaginary speeches that Blundeville and others were already attacking, and this despite the fact that such speeches had good classical antecedents. Thucydides indeed had claimed to have sought out the best information available for the content of the speeches in his history, but had had to admit that all too often, details were sparse. The *lacunae* were filled with speculation. The speeches in Livy's history were lauded as models of rhetoric, but most lacked verisimilitude; as later skeptics put it, there was something inherently improbable about generals addressing their troops at such length and in such elegant language. But the speeches *did* provide causal explanations. So, when an historian such as Sir John Hayward invented speeches, as he did in his *Historie of King Henrie IIII* (1599), he did so to provide an explanation for the actions of his characters; there was, of course, plenty of evidence for the actions themselves. Thus, for example, when a speech had to be put in the mouth of the Bishop of Carlisle, who was guilty of treason, his speech had at least to sound treasonable–and Hayward succeeded well enough to find himself in the uncomfortable position of being thrown into the Tower and then having to explain to the justices why he was inventing such plausibly seditious talk.[7]

It was an extension of this problem that led Sidney to remark that historians were guilty of fictionalizing when they assigned causes to events. Only rarely was there any direct evidence for the intentions of historical actors; good documentation for the causes of historical events was perhaps only slightly less uncommon. In practice, historians worked backward from the event to the cause, but in doing so were highly

---

6 Blundeville, E.iv^{v}.

7 Margaret Dowling, "Sir John Hayward's Troubles over his *Life of Henry IV*," *The Library*, 4th ser., 11 (1930): 212–24; *Calendar of State Papers Domestic, Elizabeth*, 1598–1601, 404–5, 539–41.

speculative. There is no question that Sidney took seriously both the moral and political utility of reading history,[8] but in the intermittently playful *Defence* he made very clear that the historian's pretensions rested on sand. The mouse-eaten records upon which he built his case were hard to authenticate. Oral evidence was often no more than hearsay. Causation was built on fictive imaginings–which may help explain why Bacon, amongst many others, insisted that the best histories were written by statesmen whose analyses possessed a greater degree of verisimilitude than those of mere sedentary scholars immured in an ivory tower. In the end, Sidney was convinced that fiction was inherently part of the historian's work. It was not only that accurate history provided very dubious moral guidance, and that poetry did the job better, it was also that

> in painting forth the effects, the motions, the whisperings of the people, which though in disputation one might say were true, yet who will marke them well shall finde them taste of a poeticall vaine, and in that kinde are gallantly to be marked, for though perchance they were not so, yet it is enough they might be so.[9]

Even here, in a letter written to his brother praising history, Sidney admitted that often verisimilitude was the most the historian can hope for.

The potentially fictive nature of history did as much to help undermine its utility for moral discrimination as the perverse stubbornness of historical facts had done. Following (or eschewing) the moral example of some historical protagonist made sense if the story was true; it was less persuasive if the story were doubtful. That said, the historian might still retain his role as a kind of hanging judge. Amyot had threatened evildoers with the judgment of history. Oddly enough, Machiavelli made much the same point in his discussion of the Sicilian tyrant, Agathocles, whose climb from the bottom of society to be ruler of Syracuse had been accomplished through a steady parade of crimes. There was no moral end to the story, no retribution: Agathocles died peacefully in bed. Nevertheless, his career did not receive a Machiavellian accolade. Successful or not, he deserved no glory.[10] Bacon's judgment of Henry VII, though more balanced, followed the same path.

The prudential uses of history were not necessarily open to the same objections. Much depended on just what was at issue. Frequently, an historian picked his subject because it appeared to echo the problems of the present. Hayward wrote about the accession of Henry IV, and more particularly about the fall of Richard II, because the circumstances surrounding those events reminded him of what was going wrong in the Elizabethan government of the 1590s. Samuel Daniel chose to write his play, *Philotas*, because that story of how evil counselors persuaded Alexander the Great to kill one of his best generals was reminiscent of what had just

---

[8] This point is made in detail by Donno, "Old Mouse-eaten Records," 278–87.

[9] *The Prose Works of Sir Philip Sidney*, ed. Albert Feuillerat, reprint, 4 vols. (Cambridge, 1968), III: 131.

[10] Niccolò Machiavelli, *The Prince*, trans. Robert M. Adams, 2nd ed. (New York and London: W. W. Norton, 1992), 25.

happened to the Earl of Essex.[11] This sort of thing happened often enough for even those being pilloried to recognize themselves and, occasionally, take a measure of revenge. But, as my admixture of play scripts with histories indicates, no one expected complete accuracy or a complete correspondence from such accounts: some overlap, some similarity was enough to make the point–and, perhaps, to evade the censor.

This sort of history-writing depended on a partnership between author and reader. The reader of histories who sought only examples of morality might write them down in a commonplace book, each under the appropriate heading–though some considered this unnecessary as the philosophical precepts were valid even without the examples. Nevertheless, someone engaged in "courtly reading" might find the historical instances useful in the never-ending process of self-fashioning.[12] As the examples followed upon the precepts, their detailed accuracy mattered little. The same was not true of the political reader. Such a reader compared the ages and places he read about with his own, and then, "when occasion is giuen to make in a mans minde application of things past to the present, and to consider whether, and why, they hold, or hold not, is a kind of imaginary practice, to confirme, and make a man the readier for reall action."[13] Reading of this sort might, of course, be engaged in silently and alone, but many felt it was even better to read or study with someone else. Conference with the learned made the obscurities of the historical narrative clearer, and helped reveal the connections between past and present. So, for instance, the Rome described by Tacitus could be seen to resemble late Elizabethan England–a point made by Sir Henry Savile, by Ben Jonson, and, perhaps, by Bacon himself.[14] And it was this kind of reading-by-analogy that made the treasonous speeches that Hayward put in the mouth of the Bishop of Carlisle appear so dangerous to the authorities: if the events of 1599 so closely resembled those around 1400, might not some readers find Carlisle's speech persuasive?

There was yet another way that history might be useful. From a careful study of the past, a prince, an adviser to princes, a politician, even a young man seeking to make his way in the world could learn what to do and what to avoid. Provided the narrative were detailed enough, the reader could work out for himself the causes of

---

[11] Most of these instances have been examined in my "The Theatre and the Court in the 1590s," in *The Reign of Elizabeth I. Court and culture in the last decade,* ed. John Guy (Cambridge: Cambridge Univ. Press, 1995), 274–300. When Bacon himself wrote a brief "direccion for the readeinge of histories with profitt," he also took his examples from the reign of Alexander. David M. Bergeron, "Francis Bacon: An Unpublished Manuscript," *Papers of the Bibliographical Society of America* 84 (1990): 402–3.

[12] Julie Robin Solomon, "'To Know, To Fly, To Conjure': Situating Baconian Science at the Juncture of Early Modern Modes of Reading," *Renaissance Quarterly* 44 (1991): 513–58.

[13] Anon., *Horae subsecivae* [probably by William Cavendish, 2nd Earl of Devonshire, and Thomas Hobbes] (London: Edward Blount, 1620), 219. The essay is entitled "Of reading History."

[14] J. H. M. Salmon, "Stoicism and Roman Example: Seneca and Tacitus in Jacobean England," *Journal of the History of Ideas* 50 (1989): 199–225. On reading, see Anthony Grafton and Lisa Jardine, "'Studied for Action': How Gabriel Harvey Read his Livy," *Past and Present* no. 129 (1990): 30–78.

events; indeed, doing so was good practice.[15] Of the ancients, Thucydides' writings were among the best, "as having in them profitable instruction for Noblemen, and such as may come to haue the mannaging of great and waighty actions."[16] Of the moderns, the best choice was probably Guicciardini, whose enormous *History of Italy* provided more than enough detail. The difficulty here, however, was that the sheer length of the book inhibited those who wished to learn from it. Individual episodes might be instructive, but a busy prince or harried counselor rarely had the time to seek them out. Scholars rushed to the rescue, producing epitomes and books of aphorisms drawn from the larger work.[17] These might allow their readers to appear learned but, in stripping away the detail, they also stripped away much of the history's usefulness. Little besides psychological generalizations remained. In addition, Guicciardini himself constantly warned against the dangers of drawing conclusions rashly. Piero de' Medici lost his place in Florence by setting off to negotiate with the King of France in imitation of his father's superficially similar trip to Naples. But, Guicciardini noted in his *Ricordi*, "To judge by example is very misleading," and then continued: "unless they are similar in every respect, examples are useless, since every tiny difference in the case may be a cause of great variations in the effects. And to detect these tiny differences takes a good and perspicacious eye."[18] That last sentence, however, pointed to the contradiction inherent in all these arguments, for anyone who had an eye perspicacious enough to detect the relevant, if tiny, differences that marked off one example from another presumably had little need to read the history in the first place.

Such ideas about the moral and prudential uses of history were part of the humanist legacy, derived ultimately from the writings of Cicero and Quintilian. Another strain of humanist thought, deriving from the medical works of Hippocrates and the natural history of the elder Pliny, eventuated in the descriptive writings of the antiquaries. Unlike the chronological narratives of writers such as Leonardo Bruni or Polydore Vergil, the antiquaries organized their work geographically or by subject. Flavio Biondo's *Italia illustrata* described the Italy of the ancients, and did so by following the remnants of the Roman roads, stopping at each way station and investigating what had happened at that spot from Roman times to the present. William Lambarde's *Perambulation of Kent* looked at one county; William Camden's

---

[15] Something akin to this was used to test the suitability of young travelers for government service: they were encouraged to send letters describing, in full detail, the countries through which they passed. Such descriptions included geography, economics, military preparedness (insofar as it could be discerned), and, of course, politics. A young man who gave a good account of himself through his letters might find a patron, and advancement.

[16] Thucydides, *Eight Bookes of the Peloponnesian Warre*, trans. Thomas Hobbes (London: Henry Seile, 1628), A.1^{v}.

[17] Among those translated into English were Robert Hitchcock, *The Quintesence of Wit* (London: Edward Allde, 1590), derived from Francesco Sansovino; Remigio Nannini, *Civill Considerations*, trans. W. T. (London: F. K. for Matthew Lownes, 1601); and, most considerably, Robert Dallington, *Aphorismes Civill and Militarie* (London: Edward Blount, 1613).

[18] Francesco Guicciardini, *Maxims and Reflections of a Renaissance Statesman*, trans. Mario Domandi (New York: Harper Torchbooks, 1965), 71. He makes the same point at greater length in the *History*.

Britannia examined the whole country. The underlying assumption was that geography and climate determined the way peoples behaved. More than that, most of the antiquaries were driven by a kind of patriotism. Camden began with the history of the Roman province, and thus linked Britain to the glories of Rome; nevertheless, the sub-divisions of his book were based on the old British kingdoms that preceded Caesar's invasions, then (finally) on the counties. By way of introduction, he worked his way through the historical layers represented by the Angles and Saxons, the Danes and the Normans. In the central portion of his book, Camden found room for antiquities, inscriptions, coins as well as notices of the most important families of his own day. This too was a kind of "useful" history, proto-nationalist yet localized, allowing its readers to find their own places within the broader scheme of a conception of England that shared the intellectual legacy of the ancients while insisting on national particularity.

The antiquaries also raised questions about the nature and use of historical evidence. In Sidney's jaundiced view, the historian authorized "himself (for the most part) upon other histories, whose greatest authorities are built upon the notable foundations of hearsay."[19] While not true of writers like Guicciardini, it came embarrassingly close to describing the practice of altogether too many. The antiquaries, for their part, had no hesitation in using old chronicles, but added to them documents searched out of the repositories as well as the "hard facts" of physical remains. They also developed the habit of authenticating their statements. It was as part of a battle among heralds–whose concerns overlapped considerably those of the antiquaries–that Augustine Vincent challenged his opponent to "lay them [his Records] downe vpon the Margent," while John Selden praised Vincent for seeking out manuscripts, and insisted that "For your so often taking helpe from the *Records*, or publique Acts of the State; you haue therein … taught the right way for such as shall collect a store for the new writing of any part of our Storie."[20] If carried through, this would be to put the writing of English history on an altogether new footing.

So, were a gentleman of scholarly tastes to look at the works of history published in the years around 1600, he would have been struck by the sheer variety of what was available. He would have had a choice of straightforward narratives, ranging from the various if slightly old-fashioned chronicles of John Stow, through the second edition of Holinshed, to the more modern assemblage of John Speed.[21] Taking a very different approach was John Hayward's account of the deposition of Richard II and accession of Henry IV, a work heavily charged with the politics of the day.[22] In addition,

---

19 Sir Philip Sidney, *A Defence of Poetry,* in *Miscellaneous Prose of Sir Philip Sidney*, edited by Katherine Duncan-Jones and Jan van Dorsten (Oxford: Clarendon, 1973), 83.

20 Augustine Vincent, *A Discoverie of Errours* (London: William Jaggard, 1622), "To Raphe Brooke, Yorke Herald," 93^{v} and "To my singular good Friend, M^{r}. Augustine Vincent," a^{v}.

21 John Stow, *The Chronicles of England* (London: R. Newberie, 1580), with later editions called *Annales*; Raphael Holinshed, *The first and second volumes of chronicles* (London: H. Denman, 1587); John Speed, *The History of Great Britaine* (London: W. Hall and J. Beale for J. Sudbury and G. Humble, 1611).

22 John Hayward, *The first part of the life and raigne of king Henrie the IIII* (London: J. Wolfe, 1599).

and rather outside the focus of our discussion, he would have found a number of historical poems, some (like Michael Drayton's) concerned with particular incidents, others (like Samuel Daniel's) exploring the English civil wars, still others (like William Warner's *Albion's England*) covering the whole range of English history.[23] Moreover, such poeticized history had also spilled over onto the stage, leading for example not only to Shakespeare's account of all the English monarchs from Richard II to Richard III, with King John as an outrider and King Henry VIII still to come, but also to a series of plays on Roman themes, by (among others) Shakespeare, Chapman and Ben Jonson. Yet even here some of the fissures we have noted were beginning to appear. Samuel Daniel had tried to write history in verse, intending to demonstrate how his *Civile Warres*

> Reueale to me how all the strife begunne
> Twixt *Lancaster* and *Yorke* in ages past.
> How causes, counsels, and euents did runne
> So long as these vnhappie times did last,
> Vnintermixt with fictions, fantasies;
> I versifie the troth, not poetize.[24]

As Daniel continued to work on his historical verse, he came increasingly to the conclusion that this was not a viable compromise, that versifying the truth without poetizing was impossible. Still interested in how politics played out in the past, he set out to write a prose history of England, beginning with the Norman Conquest because earlier periods were too poorly documented. By then, Daniel had come to the same point of view as Selden, that documents were crucial and should be included with the history. Like Augustine Vincent, he was prepared to give a true account of his documents in the margin, "so that the Reader shall be sure to be payd with no counterfeyt Coyne, but such as shall haue the Stampe of Antiquitie, the approbation of Testimony, and the allowance of Authority."[25] The documents themselves–treaties, letters, articles, charters, ordinances, entertainments, passages of state–he intended to publish separately. This idea seems to have been too far ahead of its time and, in the end, the documents were never published, though the manuscript survives as testimony to Daniel's intentions.[26] At the same time that Daniel was thinking his way

---

[23] Michael Drayton, *The Barrons Wars* (London: J. Roberts for N. Ling, 1603), which incorporates a number of "heroical epistles;" Samuel Daniel, *The first fowre bookes of the Ciuile Warres betweene the two houses of Lancaster and Yorke* (London: P. Short for J. Waterson, 1595); William Warner, *Albions England* (London: G. Robinson for T. Cadman, 1586), enlarged in later editions.

[24] Samuel Daniel, *The ciuile wars* (London: P. Short for Simon Waterson, 1595), 67. The poem is quoted and discussed in this context by Jean R. Brink, "Framing Philosophy: Sidney and Daniel," in *Perspective as a Problem in the Art, History and Literature of Early Modern England*, eds. Mark Lussier and S. K. Heninger (Lewiston: Edwin Mellen, 1992), 81–93.

[25] Samuel Daniel, *The Collection of the Historie of England* (London: Nicholas Okes, 1618), A3^{v}.

[26] John Pitcher, *Samuel Daniel. The Brotherton Manuscript*, Leeds Texts and Monographs, n. s. 7 (Leeds, 1981), 178–84, describes a MS. in the National Library of Scotland which appears to be a draft of the projected appendix.

through the historical maze, Camden, already known for his antiquarian writings, published the first installment of his history of the reign of Queen Elizabeth. He called his work *Annales*, a hint to readers that this account owed a debt to Tacitus and, like his Roman predecessor, Camden claimed to write *sine ira et studio*, "without spleene, and partialitie."[27] This work too was of a kind never yet seen in England, a spacious book, written about the recent past, full of documents quoted at length within the body of the text.

How much of this activity was known to Bacon in 1604–05, when he was writing his *Advancement of Learning*? Any answer must remain speculative, but there is a certain amount of suggestive evidence. The key to much of this is the Earl of Essex. Bacon lived in Essex House for much of the 1590s. Most likely, it was in that way that he came into contact with Sir Henry Savile (whom he also visited at Eton after Savile became Provost) and Henry Cuffe, both advisers to the Earl, and both confirmed admirers of Tacitus. Bacon himself came to know Tacitus well enough to be able to salt *The Advancement* with references to him, and to identify Hayward's borrowings from his Roman model.[28] One of Samuel Daniel's chief patrons was Fulke Greville and for a considerable time Greville also lived at Essex House: it is not unreasonable to believe that Greville brought Bacon and Daniel into contact. Something similar may be said about Camden who, like Daniel, had also been the recipient of Greville's patronage. In addition, Camden and Henry Cuffe had known each other since their Oxford days. Once again, it is likely that Bacon and Camden met, and that Bacon was aware of what Camden was doing–and, in the case of the *Annales*, long before the publication of the first part of the book in 1615.[29] And, as Bacon contributed to Speed's *History of Great Britaine*, he surely knew about that chronicle well before it appeared. The arguments concerning history in *The Advancement of Learning* were not put forward in ignorance of current activity in the field.

It is not always easy to discern Bacon's meaning in *The Advancement of Learning*, not only because the language is occasionally elliptical, but because many of the details of the argument are missing. Though he touches on all aspects of learning, here Bacon concentrates his attention on human learning, not divine, on matters to do with the activities of people rather than the works of nature. Moreover, the object of the book was not to record what had already been accomplished but to persuade King James to help advance learning in the future; thus the emphasis lay on the unfinished, on what had not yet been done. Inevitably, this led to a kind of imbalance, in that the amount of attention paid to any given topic did not necessarily reflect its importance. Nevertheless, Bacon was very clear on the utility of learning, either in terms of the individual or in those of society. Book One of *The Advancement* in particular adopted the humanist ethos that learning, all learning–including historical learning–could be

[27] The translation of Tacitus is from *Horae subsecivae*, 248 (in "A Discourse vpon the beginning of Tacitus").

[28] Bacon later claimed that he had saved Hayward from execution for treason by showing him guilty of theft.

[29] Bacon examined Camden's text before 7 April 1610, when he returned the manuscript to Sir Robert Cotton (*LL*, IV: 212). Camden began working on his *Annales* around 1596.

used in pursuit of the active life, and that (one way or another) it might be shown that it *was* useful.[30] If King James could be persuaded to help the pursuit of learning, by lending his name to it, perhaps even by funding it, then rapid progress would be assured.

The writing of history needed to be advanced along with most other sorts of human learning, and Bacon meant more by this than merely recommending a new history of England. In what segment of historiography was such an advance to take place? One problem was to define the varieties of history and determine which of them needed reclamation the most. According to the Baconian schema, all learning was divided into three parts, those associated respectively with memory, imagination, and reason.[31] History depended on memory, poesy on the imagination, philosophy on reason. Bacon drew no logical distinction between religious learning and secular; thus, sacred history paralleled human history, parables matched up with poesy. That said, Bacon was quite clear in arguing that, despite similarities, poesy was separate from history. He agreed with Sidney that history and poesy shared subject matter: indeed, Bacon defined poesy as "FAINED HISTORIE," by which he meant that it could improve on nature (*OFB*, IV: 73).[32] The poet could make events greater or more heroic; he could more closely match success with virtue or failure with vice. The poet might even abandon the commonplaces of history and produce a narrative that was more unexpected, rarer, more interesting. Bacon was ready to accord the poet some measure of the divine, because he could bend the things of this world to the power of the mind, while in the world of reason, the mind had to adjust to the nature of things. Poesy could give insights into the mind of man, as could history, but in the harsh world of politics, reason–and reality–had finally to rule.

Then, when Bacon contemplated the realm of history at its most abstract, he saw in it four divisions: natural history, civil history, ecclesiastical history, and what he called literary history. All the forms of history had in common a very broad definition of the word: history here meant a record of happenstances, and it did not matter whether these were the result of human, natural, or divine activity. Of these, literary history–which we would call the history of learning–he deemed altogether deficient, though he was convinced it had enormous potential in that "it will make learned men

30 See Brian Vickers, "The Myth of Francis Bacon's 'Anti-Humanism,'" in *Humanism and Early Modern Philosophy*, eds. Jill Kraye and M. W. F. Stone (New York: Routledge, 2000), 135–58. See also Stephen Gaukroger, *Francis Bacon and the Transformation of Early-Modern Philosophy* (Cambridge: Cambridge Univ. Press, 2001), 48ff.

31 According to Grazia Tonelli Olivieri, "Galen and Francis Bacon: Faculties of the Soul and the Classification of Knowledge," in *The Shapes of Knowledge from the Renaissance to the Enlightenment*, eds. Donald R. Kelley and Richard H. Popkin (Dordrecht: Kluwer Academic Pub., 1991), 61–81, Bacon's locating the three faculties of the soul–memory, imagination and reason–in the three ventricles of the brain may have been a unique contribution to the otherwise well-worn theory of faculty psychology.

32 There is an interesting comparison of Bacon and Sidney in Judith Anderson's *Biographical Truth* (New Haven: Yale Univ. Press, 1984), 124–5, 164ff. On Bacon's theory of poetry, still useful is Murray W. Bundy, "Bacon's True Opinion of Poetry," *Studies in Philology* 27 (1930): 244–64. John L. Harrison, "Bacon's View of Rhetoric, Poetry, and the Imagination," *Huntington Library Quarterly* 29 (1957): 157–25 takes the view that Bacon is not unfavorable to poetry.

wise, in the vse and administration of learning" (*OFB*, IV: 62).[33] In other words, the history of learning would provide a record of the advances Bacon was proposing.

Civil history itself might be further subdivided into memorials, antiquities, and perfect history, the first defined as unfinished history, the second as history defaced, and only the last in any sense complete. Memorials provided the materials for a more carefully analyzed history: here could be found simple records of events (without motives, counsels, or speeches), the sort of thing now called, rather derisively, a chronicle; included here too were registers of public acts, very like state papers.[34] Such memorials resembled the documents that Augustine Vincent and John Selden insisted should be referred to in the margin of a serious history, the sort Samuel Daniel planned to publish in the supplement to his book. At the other extreme, but equally incomplete, might be found antiquities, collected by learned and industrious persons, "out of Monuments, Names, Wordes, Prouerbes, Traditions, Priuate Recordes, and Euidences, Fragments of stories, Passages of Bookes, that concerne not storie, and the like" (*OFB*, IV: 66). That list looks rather like the table of contents of William Camden's *Remaines of a Greater Worke concerning Britaine*, published the same year as *The Advancement.*[35] Bacon took no notice of Camden's *Britannia*, but instead insisted that the deficiencies in this sort of imperfect history were part of their very nature, and in so saying missed the importance of the labors of the antiquaries in building up a new kind of history based on material objects and documents rather than on documents alone.[36]

Naturally, Bacon was most interested in the sorts of histories he described as perfect. These were divided into three kinds: chronicles, that is, the histories of a time, large, circumstantial, full of great actions and great persons, but omitting smaller actions as beneath notice; lives, the biographies of great men, which were the most readily applicable because they included small actions as well as large; and last, the narrations or relations of an action such as the Peloponnesian War, smaller and more comprehensible than histories of times and as a result more accurate and less susceptible to conjecture. Of the large histories of times, only those of ancient Greece and Rome provided enough material for such matters as law, learning, moral virtue, policy, and military valor. Their modern equivalents were, in Bacon's judgment, in large part below the level of mediocrity, and as an example he noted that what might be the very interesting and instructive history of Britain since the end of the Wars of

---

33 On Bacon's non-political history, see A. B. Ferguson, "The Non-Political Past in Bacon's Theory of History," *Journal of British Studies* 14 (1974): 4–20.

34 Three other sorts of sources–orations, letters and brief speeches or sayings–are included in a separate, later section, "appendices to history."

35 William Camden, *Remaines* ... (London: G. E. for Simon Waterson, 1605). The title continued, "Concerning Britaine, the inhabitants thereof, their Languages, Names, Surnames, Empreses, Wise speeches, Poësies, and Epitaphes."

36 When he returned to this issue in *De augmentis*, Bacon had reconsidered, and described the work of antiquaries as an activity "laborious indeed, but agreeable to men, and joined with a kind of reverence; and well worthy to supersede the fabulous accounts of the origins of nations, and to be substituted for fictions of that kind" (*SEH*, IV: 303–4).

the Roses remained to be written. Good modern histories of lives were also somewhat deficient, as were histories of actions. The absence of the latter was especially to be lamented, for were these events only tolerably narrated as they occurred, "the compiling of a complete HISTORIE of TIMES mought be the better expected, when a Writer should arise that were fit for it: for the collection of such relations mought be as a Nursery gardein, whereby to plant a faire and stately gardein, when time should serue" (*OFB*, IV: 69). Ironically, Bacon's dream was not to be realized for another century, and then only as a composite history of Britain, with Bacon's own *History of the Reign of Henry VII* as one of the centerpieces, one author's work following another with no real unifying principle except elegance of style.[37]

Having established a carefully worked out and complex scheme encompassing all the varieties of history, Bacon halted and noted an alternative way of looking at things, derived not from faculty psychology but from Tacitus, surely a reflection of the current interest in that author, which Bacon shared.[38] This was altogether simpler. Annals recorded matters of state, journals affairs of lesser importance such as triumphs, ceremonies, and novelties. This contrasted the distant perspective of an annual view of events with the much more intimate if narrower view of what passed by day by day. Tacitus himself wrote annals; the journalist was, one supposes, Suetonius. As a matter of pride, some princes had indeed had journals kept of what transpired in their time, as had generals and navigators, but Bacon leaves us in no doubt that such accounts might be the sources of history without achieving the dignity of being themselves histories in the Tacitean mode. The distinction is also one that separates Camden's *Annales* of Queen Elizabeth's reign from a chronicle such as Stow's.

More confusing still–and for our purposes, more interesting–is what Bacon called ruminated history, "a forme of Writing, which some graue and wise men haue vsed, containing a scattered History of those actions, which they haue thought worthy of memorie, with politique discourse and obseruation thereupon" (*OFB*, IV: 70).[39] Bacon differentiated ruminated history from the more ordinary sort in that, in the latter, authors laced their analysis into the text, while in the former–he had in mind, of course, Machiavelli's discourses on the first ten books of Livy–they separated out the commentary and indeed made it their focus. The distinction was based on a notion of history that defined its "true office" as "to represent the euents themselues, together with the counsels, and to leaue the obseruations, and conclusions thereupon, to the liberty and facultie of euery mans iudgement" (*OFB*, IV: 70). But this is to argue that the "counsels"–for which we may here read "causes"[40]–are somehow inherent in the

[37] See Joseph M. Levine, *Humanism and History* (Ithaca: Cornell Univ. Press, 1987), 164–71.

[38] See also Salmon, "Stoicism and Roman Example,"199–225 and Edwin B. Benjamin, "Bacon and Tacitus," *Classical Philology* 60 (1965): 102–10.

[39] The same point was made by Jean Bodin, *Method for the Easy Comprehension of History*, trans. and ed. Beatrice R. Reynolds, Columbia Records of Civilization (New York: Columbia Univ. Press, 1945), 51–4.

[40] "Counsel" usually means political advice, often of a political nature: this is the meaning borne by the word in the title of the 1625 edition of the *Essayes* (*OFB*, XV: xxii–xxiii). "Causes" fits the context better here, though the notion of advice may remain as a distant echo.

narrative of events, and not at least in part the product of the historian's imagination. Sidney had, of course, argued the opposite: many times the historian must describe events "whereof he can yield no cause; or, if he do, it must be poetically."[41] In Sidney's view, the "counsels" were no different from the "obseruations and conclusions" in being the product of the historian's memory informed by his imagination, the whole then ordered by his reason. The purpose of Bacon's maneuvering was to render history more useful, to give it "much greater life for practise." The historical narrative, done properly, was "set down with al circumstances," and this level of detail would at least "controul the discourse thereupon made and some times supply it; as a verie patterne for action" (*OFB*, IV: 162–3). The brief examples that often were used to illustrate discourses had not the same practical usefulness. Thus Bacon moved discourses from the province of memory into that of civil knowledge, within the realm of philosophy.

Besides civil knowledge, philosophy also included natural philosophy. Natural philosophy, in turn, was divided into the search for causes and the production of effects. For Bacon, the operative was as (or more) important than the speculative. Men needed to learn in order to act. And here, as so often in this section of *The Advancement*, Bacon used an analogy from civil knowledge to explain natural philosophy: "For as in Ciuile matters there is a wisedome of discourse, and a wisedome of direction: So is it in Naturall" (*OFB*, IV: 80). In both civil and natural philosophy, "knowledges are as PYRAMIDES, whereof HISTORY is the BASIS" (*OFB*, IV: 85). In the realm of nature, the pyramid narrowed from natural history (the accumulation of data), to physic, to metaphysic (forms; final causes), to God. A similar pyramid existed for civil knowledge. But such philosophy only made sense through harmony. Seeing the pieces as separate and dissonant meant that the underlying relationships disappeared from view. Again, Bacon took refuge in an historical analogy:

> For as, when I read in *Tacitus,* the Actions of *Nero,* or *Claudius*, with circumstances of times, inducements and occasions, I finde them not so strange: but when I reade them in *Suetonius Tranquillus* gathered into tytles and bundles, and not in order of time, they seeme more monstrous and incredible; So is it of any Philosophy reported entier, and dismemberd by Articles. (*OFB*, IV: 92)

The processes of understanding history and understanding nature were at least parallel, if not identical.

When Bacon turned his attention to the individual, he raised anew the question of how to persuade men to be moral. Surprisingly, Bacon insisted that this part of philosophy had not been adequately dealt with: instead there had been too much emphasis on subtlety of argument, and elegance of presentation had eclipsed hard thought. What was needed, he thought, was a "Georgickes of the mind" (*OFB*, IV: 135).[42] Here as elsewhere in such discussions, Bacon demanded that men put the public good

---

41 Sidney, *Defence*, 89.

42 *Ibid.*, 135. On this, see Stanley E. Fish, "Georgics of the Mind: The Experience of Bacon's Essays," in *Seventeenth-Century Prose*, ed. Stanley E. Fish (New York: Oxford Univ. Press, 1971), 251–80.

before the private–by which he did not mean that the private good should be abandoned altogether; instead, it should be seen in societal rather than in individual terms. This is not yet civil knowledge, but it might well be seen as a step in that direction.

As Bacon admitted, the difficulty was that civil knowledge was of all subjects the most immersed in the mundane, the most tied to the particular, and thus the hardest to reduce to axioms. That helped explain why it was that two of the three branches of civil knowledge–conversation (the wisdom of behavior) and negotiation (the wisdom of business)–had been relatively neglected. Insofar as there were materials available for studying negotiation, histories of lives served better than most. Aphorisms, such as those called "The Wisdom of Solomon," could also be pressed into service if properly interpreted; so too could a novel form Bacon called discourses upon letters, *discorsi* in the style of Machiavelli but applied to political correspondence. As for the third branch of civil knowledge, government, history of times provided the best materials for generalization. In these discourses–and here Bacon explicitly had Machiavelli's *Discorsi* in mind–the *arcana imperii* might be revealed. To illustrate, Bacon spent a considerable amount of time on what he called the architecture of a man's own fortune, which involved the ways in which a man might advance himself in a political situation. There is an element of the self-serving here, contradicting Bacon's earlier protestations that the public should take precedence over the private. But it is also possible to read this section of *The Advancement* as an analysis of political motives, as a way of understanding the human psychology that drove politics. Bacon's own political analyses, here as elsewhere, followed Machiavelli in having a psychological basis.[43] Nevertheless, however one interprets the details of the argument, the underlying point in Bacon's discussion of civil knowledge was that the best source material derived from records of action, from history, rather than from *a priori* discussions of the nature of man.

Early in Book One of *The Advancement*, when Bacon was defending learning by attacking those who disparaged it, he challenged the claim that too much study makes irresolute those who make political decisions. The opposite was true, he insisted, for learning could teach men how to make decisions; indeed, it could even teach them "how to carrie thinges in suspence without preiudice, till they resolue." It could teach them how to distinguish between the conjectural and the definitive, how to evaluate principles and rules and locate the exceptions to them. And again, Bacon chose historical examples to make his case, carefully ranging from Clement VII in Guicciardini, to Cicero's Letters to Atticus, to the Life of Cato the Younger (*OFB*, IV: 12).

Once large quantities of information had been collected, the question became how to organize the material in such a way as to allow generalizations to be drawn from it. That question was not, in itself, a new one. Some writers had used comets and other portents as guides to the relative importance of events, and continued to do for most of the seventeenth century. Jean Bodin, in his *Methodus*, had tried out a number of schemes: perhaps, as the ancients had suggested, geography and climate could be

[43] George H. Nadel, "History as Psychology in Francis Bacon's Theory of History," *History and Theory* 5 (1966): 275–87.

used to determine how the peoples of different nations would behave; perhaps the alternation of kingdoms followed a mathematical cycle, each monarchy replacing its predecessor every four centuries; perhaps the regularities operated at the level of the individual where the numbers sixty-three (or, better, seventy) marked off a climacterical year.[44] Some of this speculation was repeated in William Harrison's *Description of England.*[45] William Camden spent part of his apprentice years drawing up astrological diagrams, perhaps in the hope that through these he could discern patterns in history.[46] Bacon moved in an entirely different direction. He insisted that the method of induction, set out in detail in the *Novum Organum*, first published 1620, could also be applied to subjects other than natural philosophy, including logic, ethics, and politics, "for I form a history and table of discovery for anger, fear, shame, and the like; for matters political; and again for the mental operations of memory, composition and division, judgment, and the rest."[47] In *The Advancement* he had implied as much by shifting constantly between examples of natural philosophy and civil knowledge. The teams of scholars Bacon hoped for would thus be spending their time not so much assembling enormous perfect histories–though these were valuable enough–but instead would treat historical data much like the natural scientists treated instances of hot and cold.

The rest of the story may be divided into two parts: the afterlife of the philosophy of history found in *The Advancement* and Bacon's own accomplishment when he finally found it expedient to turn his hand to historical narrative. It is true that he had been contemplating writing a history of the Tudors for some time: he had offered such proposals to influential friends and had identified the project as missing in *The Advancement.*[48] Then, in the altered circumstances of 1621, after his fall from power, he decided to write a history of Henry VII's reign. Presumably, this was an attempt to regain royal patronage by setting forth a model of kingship. Such a regnal history would have fitted into that part of the category of perfect history called "lives," of all the most excellent in profit and use, for "*Liues* if they be well written, propounding to themselues a person to represent, in whom actions both greater and smaller, publique & priuate haue a commixture; must of necessitie containe a more true, natiue, and liuely representation" (*OFB*, IV: 66). Bacon was convinced that more could be learned from the detailed description of a life than from the more prestigious, though superficial, history of times. He had also concluded that acts of peace were as, or

44 Bodin, *Method*, 223–36.

45 Leonard F. Dean, "Bodin's *Methodus* in England Before 1625," *Studies in Philology* 39 (1942): 160–6, at 161. Harrison's *Description* was part of Holinshed's larger compilation.

46 The diagrams appear in Camden's manuscripts: British Museum, Cotton MS. Julius F.xi., ff. 70^v–71^r, 151^v–152^v etc., but not in his published work.

47 Francis Bacon, *The New Organon*, ed. Fulton H. Anderson (Indianapolis: The Bobbs-Merrill Company, 1960), 116. (CXXVII) On Bacon and the science of history, see Stephen Pumfrey, "The History of Science and the Renaissance Science of History," in *Science, Culture and Popular Belief in Renaissance Europe*, ed. Stephen Pumfrey, Paolo L. Rossi, and Maurice Slawinski (Manchester: Manchester Univ. Press, 1991), 48–70.

48 Some of these are reprinted, along with an account of their origins, in Bacon, *The History of the Reign of King Henry VII*, ed. Vickers, 209–14.

more, instructive than acts of war, and that these were best described by an experienced statesman, someone like Bacon himself. Moreover, on the basis of his legal experience, Bacon concluded that Henry had been the greatest lawgiver since Edward I, and so he added to the usual contents of a history a great deal of information on Henry's parliaments and laws. Nor did he stop at describing the laws: he also analyzed them in terms of their potential effects. Though Bacon found the time to call on the assistance of a number of friends, aid that included searching for manuscript records in the manner that had now been deemed necessary for such historical writing, the composition of *The History of the Reign of Henry VII* was nevertheless rather hurried, and–perhaps because of his enforced exile from London–was based in large part on the narrative in John Speed's *History,* a volume to which Bacon had however earlier contributed.[49] In fact, the archival material contributed relatively little to the final product and Bacon had, in essence, succeeded in writing an history in the humanist mode, elegant, thoughtful, psychologically acute, but mainly derived from matter assembled by others. It is nevertheless true that the attempts at delving into Henry's psychology were in keeping with Bacon's insistence on the need for some sort of historical generalizing. The result was certainly very persuasive: the portrayal of Henry's greed is (for better or worse) with us still. On the other hand, and in the light of events during the remainder of the seventeenth century, it may be doubted whether the demonstration that Henry was "of an high mind, and loved his own will and his own way; as one that revered himself, and would reign indeed" was altogether a fit model for royal rule.[50]

From its publication, the book was an enormous success. John Selden praised Bacon's *History,* together with Camden's *Annales*, as books "lately set forth by learned men of most excelling abilities," arguing that no other work concerning the history of England "tastes enough either of the *Truth* or *Plenty* that may be gained from the Records of this Kingdome."[51] Degory Wheare, in his Oxford lectures, claimed the life and reign were presented "so happily and so fully, that if he hath not excelled the best Historians, he has yet at least equall'd them."[52] Bishop Nicolson's *English Historical Library*, written at the end of the century, was much more fulsome. Not only had Bacon overcome all the difficulties he had himself pointed out, but

> He has perfectly put himself into King *Henry's* own Garb and Livery, giving as spritely a View of the Secrets of his Council, as if himself had been President in it. No trivial Passages, such as are below the Notice of a Statesman, are mix'd with his Sage Remarks: Nor is any thing of Weight or Moment slubber'd over with that careless Hast and Indifferency, which is too common in other Writers. No Allowances are given to the

---

[49] Daniel R. Woolf, "John Seldon [*sic*], John Borough and Francis Bacon's *History of Henry VII*, 1621," *Huntington Library Quarterly* 47 (1984): 47–53.

[50] Bacon, *The History of the Reign of King Henry VII*, ed. Vickers, 199.

[51] Augustine Vincent, *A Discoverie of Errours,* "To my singular good Friend, Mr. *Augustine Vincent*," av.

[52] Degory Wheare, *The Method and Order of Reading Both Civil and Ecclesiastical Histories,* trans. and enlarged Edmund Bohun, 2nd ed. (London: for Charles Brome, 1694), 159. Wheare's lectures had been delivered more than half a century before.

author's own Conjecture or Invention; where a little Pains and Consideration will serve to set the Matter in its proper and true Light.[53]

As far as the seventeenth century was concerned, Bacon had written an enduring classic, and one that perfectly fulfilled the requirements for history laid down in *The Advancement*.

In a more general sense, that triumph applied only to the part of perfect history called "Lives," and it is possible that the very success of Bacon's life of Henry VII, and others like it, helped to inhibit the development of a new form of historical writing, perhaps one embodying the techniques worked out by the antiquaries, for example, by Selden in his own writings. Whatever the explanation, little changed. "Lives," as a way of explicating the past, continued to dominate the writing of the English national history. When at the end of the seventeenth century William Temple and his successor, John Hughes, set to work to produce a new history of England, all agreed that it was beyond the powers of any one person to write the whole narrative. Instead, the book was a compilation of regnal histories–Bacon's was one–almost all written in the first half of the seventeenth century, fifty or more years before. Imposing though the huge folios were, *The Complete History of England* seemed not to be enough. Barely a decade after its publication, Joseph Addison repeated the complaint, heard by then for more than a century, that "our Country, which has produced Writers of the first Figure in every other kind of Work, has been very barren in good Historians."[54]

Bacon's philosophy of history was less influential than his practice. Some of the ideas resurfaced in "Of Reading History," published in 1620 as part of *Horae subsecivae*, a collection of essays and discourses written by William Cavendish, 2nd Earl of Devonshire, and his friend and former tutor, Thomas Hobbes. As both men were friends of Bacon's, and Hobbes had played some role as Bacon's secretary, it is not surprising that there was a resemblance to *The Advancement*. "History" was divided into Baconian categories, though rather more was made of the use of history for moral instruction. Once more, the scope of poetry was strictly limited, while invented speeches were still allowed. And the "Discourse vpon the beginning of Tacitus" that followed was an example of the sort of Machiavellian reflection on history that Bacon had praised, while consigning it to the realm of reason.[55] A decade later, when Hobbes came to write the introduction to his translation of Thucydides, he had shifted his position away from Bacon's. He still held the view, reminiscent of Amyot

[53] William Nicolson, *The English Historical Library* (London: Abell Swall and T. Child, 1696), 224. See also Joseph M. Levine, *The Battle of the Books* (Ithaca: Cornell Univ. Press, 1991), 288–9.

[54] Quoted by Philip Hicks, *Neoclassical History and English Culture* (Basingstoke: Macmillan, 1996), 23, from Joseph Addison, *The Freeholder* (1716). The entire story of the Temple-Hughes-Kennett *History* has been told by Joseph Levine, *Humanism and History*, 164–71, and at greater length in his *Ancients and Moderns*, 294–319, which also mentions Addison's historical ideas, 285–6.

[55] For further discussion of *Horae subsecivae*, see my "The background of Hobbes's *Behemoth*," in *The historical imagination in early modern Britain*, eds. Donald R. Kelley and David Harris Sacks (Cambridge: Woodrow Wilson Center Press and Cambridge Univ. Press, 1997), 243–66, esp. 257–63.

and shared by almost everyone, that it was "the *principall* and *proper worke* of History, being to instruct, and enable men, by the knowledge of Actions *past*, to beare themselues prudently in the *present*, and prouidently towards the *Future*;" and he was convinced that Thucydides fulfilled that promise better than any other historian.[56] Yet the reason was not Thucydides' lucid exploration of causes and policies nor yet his subtle conjectures but precisely the opposite: Thucydides did not insert such digressions into his history–which served mainly to commend the writer rather than the story–but instead allowed the readers to deduce them from the narrative. In this he followed Bacon. When, much later, Hobbes wished to explain the English Revolution and returned to history in *Behemoth,* it was by way of a discourse-in-dialogue, not a narrative.[57] The idea that historians, modern or ancient, might provide the materials for a "scientific history" seemed to have disappeared.

Let me summarize. When Bacon set out to compose *The Advancement of Learning*, the common way of writing history was in the humanist mode. To be sure, chronicles were still being produced, but the writer who took his classical training seriously considered history to be part of rhetoric. The point was not to search out new material but to present old material convincingly. Basing themselves on the best ancient writers, these humanist historians would include speeches (usually invented) and, often enough, would add moral commentary. But, by around 1600, this humanist paradigm was coming under considerable pressure. Poets like Philip Sidney, eager to defend their chosen calling, denied history's ability to provide moral examples and, in Sidney's case, also denied the historian's claim to be telling nothing but the truth by pointing to the fictional aspects of the construction of causes to explain events. Others elaborated the point by attacking the invention of speeches. However, the humanist historians were not all in agreement about the utility of what they wrote. Some retained a belief in its ability to teach virtue; others argued that the lessons history taught best were political and prudential. For the second group, rhetoric was of relatively little importance; on the other hand, detailed accuracy was essential if the correct lesson were to be drawn. In addition, still another group of scholars, also influenced by their classical training, were extending the study of ancient history–and then modern–through the analysis of antiquities. They insisted on hard facts and, later, on primary (often archival) sources. For them, it was no longer enough to warm over old stories.

Faced with these often contradictory ways of thinking about human history, Bacon had to decide how to fit the subject into his grand scheme. Like Sidney, he divided poetry from history, though he was less concerned with the poetical aspects of history writing. But Bacon went beyond Sidney in subdividing history, in noting that even perfect history could be further subdivided, and–most important–in arguing that the subdivisions differed not only in their usefulness but in their standards of proof. In addition, Bacon removed commentary on history from the realm of memory

[56] Thucydides, *Eight Bookes of the Peloponnesian Warre*, trans. Thomas Hobbes (London: Henry Seile, 1629), A3^{r}.

[57] See my "The background of Hobbes's *Behemoth*," cited above; and see as well the essays by David Wootton and Patricia Springborg in the same volume.

altogether, and placed such discourses under the rubric of reason. And it was such ruminated history that was most useful for the politician, be he prince or counselor.

Given the context, Bacon's organizational scheme was not extraordinary, and that may help explain why its influence was limited. His inability to see the importance of antiquarianism, or of the rapidly growing insistence on the use of primary sources, even among humanist historians, must also have made *The Advancement* soon appear to be a little dated. What *was* novel in the theory was Bacon's argument, adumbrated here but not fully developed until a little later, that it might be possible to create a science of history, not through astrology or numerology, but in a way that paralleled his handling of the phenomena of nature. That notion was not picked up. Instead, humanist history continued as it had before, propelled in part by the success of Bacon's own life of Henry VII. So too did antiquarianism, though on a different track, adding learned folio to learned folio. In that divided world, there was little need for Bacon's philosophy of history.

Chapter 11

"Seated Between the Old World and the New": Geopolitics, Natural Philosophy, and Proficient Method

Timothy J. Reiss

> neque enim ritus moresque nunc tractabimus innumeros ac totidem paene quot sunt coetus humanum … in quibus prodigiosa aliqua et incredibilia multis visum iri haud dubito. quis enim Aethiopias antequam cerneret credidit? aut quid non miraculo est cum primum in notitiam venit? quam multa fieri non posse priusquam sunt facta iudicantur? naturae vero rerum vis atque maiestas in omnibus momentis fide caret si quis modo partes eius ac non totam complectatur animo.[1]

This essay tracks some of the evidence for and consequences of Bacon's alignment of new learning on the politics of an "imperial" expansionist project. I do not mean this as an ideological or moral judgment, but as a matter of fact. I argue in particular that in the politics of colonial expansion and, equally importantly, in the writings *on* that expansion–the *Histories*, as their authors so often called them, Bacon found issues of method that completed those he knew from natural philosophy, rhetoric, and jurisprudence. All these writers emphasized the need and particular nature of experience in obtaining first cognitions of a wondrous unknown. All stressed the vast importance of ways of communicating these cognitions and these (hitherto) unknown phenomena, of validating them in their modes of expression, and indeed of simply recording them in appropriate ways. Their *Histories*, like Bacon's scientific histories, sought to do this. Further, to the extent that many of these unknowns implied quite different ways of being in the world, whether one saw indigenous peoples, for example, as simply other, or as children potentially able to grow into adults, they also implied that natural knowledge was *not*, as Aristotle asserted, knowledge of unchanging things, but precisely that of *changing* things–Aristotle's prudential knowledge. Bacon's idea that

[1] "Nor shall we now treat the countless manners and customs, almost as many as there are human societies, … among which I do not doubt that some will seem unnatural and unbelievable to many. For who believed in black Africans before seeing them? What is not miraculous when first it comes to knowledge? How many things are judged impossible before they occur? Truly the power and majesty of the nature of things is constantly unbelievable if one grasps only parts of it in the mind and not the whole" (Pliny the Elder, *Natural History* VII.i.6–7). I thank Julie Robin Solomon for her generous and patient attention to earlier versions of this chapter, and Daniel Javitch for a late correction.

knowledge was the same kind of process as the "enlargement of territories" thus changes a traditional notion of natural philosophy. It also changes the notion of the state as Bacon himself presented it in the *Advancement*, since he there associated "active good" with individual endeavor and "passive good" with the common good for which the state was responsible.[2] I treat each of these particulars in turn, but before I can do so, I must show that Bacon did indeed see knowledge and its acquisition in the light of an imperial project–more specifically, that he saw them as inseparably tied, the one rooted in the other.

Expanding intellectual and material globes: methods and purposes

Ending *The Proficience and Advancement of Learning* in 1605, Bacon took pride in having "made as it were a small Globe of the Intellectuall world, as truly and faithfully as I coulde discouer, with a note and description of those parts which seeme to mee, not constantly occupate, or not well conuerted by the labour of Man" (*OFB*, IV: 192).[3] He set his science project flatly on those of settling, mastering, and making proficient "new," "savage," "unproductive" lands, whether the Ireland of Edmund Spenser, Essex, and himself, or the Americas of Walter Ralegh and his European precursors. Indeed, Bacon's phrase may purposely recall Ralegh's notorious one about Guiana's "virginity" and the land's consequent availability for those who would make it fruitful. Bacon drew deeply on these latter projects for his scientific method, the "soundnesse of direction" that was first of the three aids needed to "rectifie or raise the Sciences themselues."[4] Not by chance did he name this "direction" (the method that "preuenteth error," "the Inuention or election of the Meane" to find new knowledge) just after likening the quest to expand the sciences to sailing from the Mediterranean into the Atlantic in the *direction* of a new world (55–6). His means to science owed a lot to those who had done. Not by chance did he say of the "INVENTION … of ARTS and SCIENCES," the first kind of the first of the "ARTS INTELLECTVALL," that "as money will fetch all other commodities; so this knowledge is that which should purchase all the rest," just as the "small motion" of the "Mariners Needle" enabled discovery of the "vast Regions" of "the *West Indies*" (107).[5] It matters, too, that he sees language, vital factor of discovery and experience, as "*Words*, and *Writings by letters*" that are "the Mint of knowledge (for wordes, are

---

[2] Julie Robin Solomon, *Objectivity in the Making: Francis Bacon and the Politics of Inquiry* (Baltimore: Johns Hopkins Univ. Press, 1998), 125–6.

[3] All subsequent non-identified page references in my text are to *The Advancement of Learning* in this edition.

[4] The others are "amplitude of reward" and "coniunction of labors."

[5] The four intellectual arts are "ENQVIRIE or INVENTION," "EXAMINATION or IVDGEMENT," "CVSTODIE or MEMORIE" and "ELOCVTION or TRADITION." For "mans labour is to *inuent* that which is *sought* or *propounded*: or to *iudge* that which is *inuented*: or to *retaine* that which is *iudged*: or to *deliuer* ouer that which is *retained*." The two kinds of the first of these are "of arts and sciences" and "of speech and argvments" (107). Discovery in the "arts and sciences" most concerns me in this essay.

the tokens currant and accepted for conceits, as Moneys are for values ...)" (120). Fellows, even instruments, of scientific discovery and method, expansive trade, and settlement signally inflected both.

I do not say that Bacon owed nothing to the acid sixteenth-century debates on *methodus*, whose issue made mathematics a seat of scientific knowledge. Rather, he saw the "*Mathematicke*" of "*Quantitie determined, or proportionable* ... to bee one of the *essentiall formes* of things; as that, that it is causatiue in Nature of a number of Effects" (87). Its abstraction satisfies human reason's "delight in the spacious libertie of generalities," but at the same time "many parts of Nature can neither be inuented with sufficient subtiltie, nor demonstrated with sufficient perspicuitie, nor accommodated vnto vse with sufficient dexteritie, without the aide and interueyning of the Mathematicks" (88). If, he added later, "you will haue Sciences growe," you must "looke well to the takinge vp of the Rootes. Of which kinde of deliuerie the *Methode* of the *Mathematiques*, in that Subiect, hath some shadowe"–even if it is yet "deficient" (123).[6] These remarks, like the direction taken in the earlier debates, may themselves have owed much to expansive navigation, for which a good mathematical head had for more than a century been recognized as essential. Indeed "mathematics" was often a synecdoche or metonymy for "navigation" (see note 29 below).

Even less do I propose that he owes nothing to debates in rhetoric and law. Whatever his acquaintance with the "science" debates, scholars have amply shown that he is indebted to these linguistic ones. Trained in them as what Lisa Jardine calls "a well-educated English gentleman," Bacon well knew "the limitations of that education as the basis for any growth of understanding of the natural world."[7] Indeed, a distinction between disputes *over* language and language's use properly to express nature and "works" is a *Leitmotiv* of *The Advancement of Learning*: one that again, we shall see, echoed dilemmas constantly expressed by the "historians" of the Americas. While he had left Cambridge before the vehement 1579–81 quarrel between Everard Digby (*circa* 1550–92) and William Temple (1555–1627), he referred to it as "a Controuersie in our time" (122) and knew Digby's defense of a confused dialectical method against Temple's arguments that "all knowledge of

[6] Such remarks belie Bacon's putative disregard of mathematics (however unmathematical he was himself). On this, see Graham Rees, "Mathematics and Francis Bacon's Natural Philosophy," *Revue Internationale de Philosophie* 40 (1986): 399–426.

[7] Lisa Jardine, *Francis Bacon: Discovery and the Art of Discourse* (Cambridge: Cambridge Univ. Press, 1974), 7. This is explored as a crucial issue from the 1520s, debated through the century as a shift in matters of discovery from language to the mathematical arts, in Timothy J. Reiss, *Knowledge, Discovery and Imagination in Early Modern Europe: The Rise of Aesthetic Rationalism* (Cambridge: Cambridge Univ. Press, 1997). Bacon still stressed this, inveighing against the "cobwebs" of verbal dispute, unable to fix proper axioms, well-found inductions, or any "*Subiect of Nature*": "the subtiltie of Nature and Operations will not be inchayned in those bonds: For *Arguments* consist of *Propositions*, and *Propositions of Words*, and *Wordes* are but the *Current Tokens or Markes of popular Notions of thinges*; which Notions if they bee grossely and variably collected out of Particulars; It is not the laborious examination either *of Consequences of Arguments, or of the truth of Propositions*, that can euer correct that Errour; being (as the Phisitians speake) in the first digestion" (*OFB*, IV, 107–11, here 110).

general precepts is formed by observation and induction."[8] Brian Vickers shows that rhetoric and dialectic were still a major source of Bacon's thinking.[9] But language could only go so far if it and its objects were not grounded in mutually congruent rules, something, again of which those seeking to understand and express unknown wonders were fully aware.

By themselves, neither mathematics, rhetoric, or law sufficed. Uncovering legal facts (work of observation and induction) did need analysis whose tools, besides gathering apt witnesses, were rhetoric and dialectic acting on a body of law "structured in accord with reason."[10] But Bacon went beyond this to the belief that this body of law was permeated by "rules and grounds" that careful study could extract and that matched nature's rules.[11] As he wrote to King James in 1603: "There is a great affinity and consent between the rules of nature, and the true rules of policy: the one being nothing else but an order in the government of the world, and the other an order in the government of an estate … I conclude that there is … a congruity between the principles of Nature and Policy" (*LL*, III: 90–1). This was as new in legal reason as *The Advancement* was in philosophy, where he repeated the thought, urging that "naturall Philosophie" and knowledge of "manners, and policie" had to work together, the one providing "a rich Store-house for the glorie of the Creator," the other a practice for "the reliefe of Mans estate" (32), the second "repressing the inconueniences which grow from man to man," the first "relieuing the necessities which arise from nature" (39). Such fast kinship explains why their practices could so readily overlap and learn from one another, for in these claims of the ruled congruency of law, natural philosophy and state policy we are back at our starting point: the articulation of expansive polity and expansive knowledge.

So it is natural that Bacon admired one great polity that expressed in its very organization the primacy of "philosophia prima," the "vniuersal Science" containing these principles as "the Maine and common way" to all knowledge (76), and desired, he wrote in *Valerius Terminus*, "an administration of knowledge of such order and policy as the king of Spain in regard to his great dominions useth in state, who though he hath particular councils for several countries and affairs, yet hath one council of state or last resort, that receiveth the advertisements and certificates from all the rest" (*SEH*, III: 231).[12] Admiration was muted, Spain being England's enemy, but this did

---

8 Jardine, 64. See, too, Kiernan's commentary, *OFB*, IV: 319–20.

9 *Francis Bacon and Renaissance Prose* (Cambridge: Cambridge Univ. Press, 1968).

10 Julian Martin, *Francis Bacon, the State, and the Reform of Natural Philosophy* (Cambridge: Cambridge Univ. Press, 1992), 77–9, here 79.

11 Martin 117, 142; cf. Daniel R. Coquillette, *Francis Bacon* (Stanford: Stanford Univ. Press, 1992) and Mark S. Neustadt, "The Making of the Instauration: Science, Politics, and Law in the Career of Francis Bacon," Ph.D. diss., The Johns Hopkins University, 1987.

12 This thought may not have been unusual. In 1622, Sir Richard Hawkins praised "Spanish discipline, and manner of their government in generall; which is in many things different to ours." Its main difference is in only ever giving one source "absolute authoritie and commaund … for the Spaniards never give absolute authoritie to more than one. A custome that hath beene, and is approoved in all empires, kingdomes, common-wealths, and armies, rightly disciplined: the mixture hath been seldome seene to prosper,

not stop him, twenty years' later, from making things Spanish central to his exploration of new knowledge in *New Atlantis*: not only were his sailors Spanish-speaking, but when the people of Bensalem first offered a written text to communicate with these strangers before they landed, the "scroll" was in Hebrew, Greek, Latin, the three great languages of European humanism–and Spanish: the new "augmentation" is signaled not by Italy, France, or England, but Spain. But if Spanish was shown as the *lingua franca* of the Americas, the "notary"-witnessed oath-taking, by which the sailors swore their peaceable intentions, parodied the *requerimiento* with which the Spanish justified their warlike impositions (*SEH*, III: 130–1). For Bacon had also always urged that England should prevent and be revenged on Philip's depredations by attacking Spain where lay her wealth, since she was otherwise, he said in 1592, "much drained and exhausted by the Indies and by continual wars," "notwithstanding the pioners [*sic*] do still work in the Indian mines." Drake's despoiling of "Carthagena and San Domingo in the Indies," which Bacon elsewhere called "the memorable voyages into the Indies," were elements of a policy which advised Queen Elizabeth in 1584 that she "should not spare [attacks] thoroughly and manifestly both upon his Indies and Low Countries, which would give themselves to you."[13]

This policy would reduce Spain's power and wealth, even as it increased England's. At present, he said, the queen has "for her part no Indies nor mines" (*LL*, I: 176). This policy might acquire them. But it could bear moral as well as political, military, and commercial fruit, since Spain's actions have only brought harm: "The poor Indies, whereas the Christian religion generally brought enfranchisement of slaves in all places where it came, in a contrary course by it men are brought from freemen to be slaves, and slaves of the most miserable condition."[14] In this period, Bacon seems to have written a "Device of the Indian Prince" for courtly performance, showing a blind "Indian Prince" led to Elizabeth by the attendant of a "Master" in the Americas (presumably Ralegh) to be cured so that he could fulfil "an ancient prophecy that it should be he that should expel the Castilians, a nation of strangers, which as a scourge hath wound itself about the body of that continent, though it hath not pierced near the heart thereof" (*LL*, I: 385–91, here 388, 390). Taken to England, "Seated between the Old World and the New, / A land … no other land may touch" (cf. "this Iland of *Brittany* deuided from all the world": *OFB*, IV:

---

as will manifestly appeare, if we consider the issue of all actions and journeys committed to the government of two or more, generally"(*The Observations of Sir Richard Hawkins Knight, in His Voiage into the South Sea. Anno Domini 1593* [London: I. D. For Iohn Iaggard, 1622], in *The Hawkins Voyages during the Reigns of Henry VIII, Queen Elizabeth, and James I*, ed. Clements R. Markham [London: Hakluyt Society, 1878] 83–329, here 279]). Of course, Bacon could have been source of Hawkins' idea, derived in general, if not in this particular, from Jean Bodin.

13 The first three quotations are from "Observations on a libel published this present year, 1592" (*LL*, I: 169, 177, 195); the next from "Of Tribute" (*Francis Bacon: A Critical Edition of the Major Works*, ed. Brian Vickers [Oxford: Oxford Univ. Press, 1996], 50); the last from "Letter of Advice to Queen Elizabeth" (*LL*, I: 55).

14 "Of Tribute," Vickers ed., 46.

68), cured by Elizabeth, he would eject the Spanish and lead England to the continent's heart.[15]

Bacon thus had major policy reasons to focus on the Americas. Doing so, he found issues of method that completed those he knew from natural philosophy, rhetoric, and law. They concerned, as I said at the outset, experience; ways of communication, validation, and recording; even the very idea of what constituted scientific knowledge. The colonizing and commercial tradition provided him with a matrix for thought at least as crucial as those others named. As Julie Solomon notes: "Bacon's rejection of the authority of tradition for that of experience is itself the scientific articulation of the pragmatic experience of merchants and travelers, whose financial success depended not upon traditional book-learning, but upon their relatively privileged access and sensitivity to a variety of places and contexts in the world."[16] I am not interested here in ideological argument but in seeing how European involvement in the Americas was basic to Bacon's arguments about the possibility and forms of new learning and science. The difficulties of understanding, explanation, and communication incurred by transatlantic expansion were *exactly* his. He turned naturally to these "works" (in his sense of the word), even had he lacked political and still deeper cause. To tie state expansion to that of knowledge was neither metaphor nor formal symmetry. They coincided, like all acting and knowing, whose "essentiall fourme … is nothing but a representation of truth; for the truth of being, and the truth of knowing are one, differing no more than the direct beame, and the beame reflected" (25–6). "[I]t is the perfite lawe of enquiry of trueth," he added, "*That nothing bee in the globe of matter, which should not be likewise in the globe of Crystall, or Fourme*, that is, that there be not anything in being & action, which should not bee drawne and collected into contemplation and doctrine" (165). That was why philosophy teaching was as important as political doing (57), and why "*Veritas*, and *Bonitas* differ, but as the Seale and the Print: for Truth prints Goodnesse" (50)–the acmes of knowledge and action echoed and guided each other. So it was utterly natural to set "enlargement of territories and ennobling of country" (I reword very slightly) beside philosophical "conquest of the works of nature."[17] Doing and knowing were inseparable, their expansions to be elided: "And this Proficience in Nauigation, and discoueries, may plant also an expectation of the

[15] Vickers cites James Spedding as saying there is "nothing to suggest … that [this device] was written by Bacon" ("The Authenticity of Bacon's Earliest Writings," *Studies in Philology* 94.2 [Spring 1997]: 248–96, here 265). Spedding actually says that if it was *not* by Bacon (rather than Essex, e.g.), he would "despair of distinguishing their several works by examination of the workmanship" (I, 391). Vickers' view differs: "it is unlike Bacon, being too tidy in its sequences of rhetorical symmetries, correct and neatly balanced but without point or surprise. It's diligent, careful, but not looking beyond the next balance. The 1597 *Essays* are the work of a craftsman, this a beginner's, it seems to me" (personal communication, 12/16/03). I deeply respect this judgment, and thought of removing all reference to *The Indian Prince* (although that a writer must always be at top form seems dubious, especially with regard to unpublished writing), but if it is not *by* Bacon, it is certainly by someone in his circle who shared his views.

[16] "Between Magic and Travel: Francis Bacon and the Theaters of Human Knowing." Ph.D. diss., University of Pennsylvania, 1987, 272.

[17] "A Device for the Gray's Inn Revels" [1594–5], in Vickers ed., 53–4.

furder proficience, and augmentation of all Scyences" (71). On this bond, Bacon was himself endlessly expansive.

He thus began *The Advancement*'s second book by again tying geopolitical expansion to "the further endowement of the world with sound and fruitfull knowledge," asking the king: "For why should a fewe receiued Authors stand vp like *Hercules Columnes*, beyond which, there should be no sayling, or discouering, since wee haue so bright and benigne a starre, as your Ma[jesty] to conduct and prosper us?" (55). "To circle the Earth, as the heauenly Bodies doe," he said, was something humans had managed only in "these later times," and this modern

> Proficience in Nauigation, and discoueries, may plant also an expectation of the furder proficience, and augmentation of all Scyences, because it may seeme they are ordained by God to be *Coevalls*, that is, to meete in one Age. For so the Prophet *Daniel* speaking of the latter times foretelleth: *Plurimi pertransibunt, & multiplex erit Scientia*, as if the opennesse and through-passage of the world, and the encrease of knowledge were appointed to be in the same ages, as we see it is already performed in great part. (71)

Growth of knowledge and expansive polity were elided under a monarch who was, in his own eyes and Bacon's first address (3–5), the philosopher-prince nonpareil.[18] As he licensed sea captains and merchant venturers, so he would warrant seekers after the knowledge which, *The New Atlantis* still showed twenty years later, was intrinsic with geopolitical invasion, trade, and settlement.

In 1612, reworking *The Advancement* toward the first part of the *Magna instauratio*, Bacon made a vital phrase of the passage with which I began this essay into a title: *Descriptio globi intellectualis* (*OFB*, VI: 95–179). In 1620, he repeated it in the *Distributio operis* section of the *Novum organum*, where picking up his earlier thought, he wrote that he had not "failed to sail along the shores of received sciences, and even to import some useful things into them in our passage." "For," he went on, "the divisions of the sciences that we use include not only things already found and noted but things omitted till now that should be there. Since in the intellectual world as in the terrestrial [in globo intellectuali, quemadmodum in terrestri], both cultivated and deserted places are found" (*SEH*, I: 134).[19] *The New Organon*'s frontispiece pictured the alignment, famously showing two warships leaving an old world of knowledge, one nearing the Atlantic horizon, another passing the Pillars of Hercules into that ocean (*SEH*, I: 119).

With Daniel's prophecy as motto, the engraving took up *The Advancement of Learning* and a focal statement of the Preface to the *Magna instauratio* preceding the *Novum organum*:

---

[18] See, too, Kiernan in *OFB*, IV: 249–50, 272. Bacon's usual idea of war as exercise (e.g., "Exercise of War" in the Gray's Inn Revels: Vickers ed., 52–3) shares the elision. In *Advancement* he advised: "if you will haue Scyences flourish, you must obserue *Dauids* military lawe," teachers of a profession being to practioners as guards of the "Carriage" are to those "*in the Action*" (58).

[19] I give the *SEH* reference for convenience, but have also used the text and translation of *The New Organon*, ed. Lisa Jardine and Michael Silverthorne, trans. Michael Silverthorne (Cambridge: Cambridge Univ. Press, 2000), here 14–15 (translation slightly changed).

> as in earlier times when people set their course in sailing only by observations of the stars, they could certainly follow the shores of the old continent and cross some rather small inland seas, but before the ocean could be crossed and lands of the new world revealed, it was necessary to have knowledge of the nautical compass as a more faithful and sure guide; by exactly the same reasoning, things so far found in the arts and sciences are such as could be lit on by use, thought, observation and argument [ut usu, meditatione, observando, argumentando], in that they are closest to the senses and fall under common notions; just so, before one can sail to more remote and secret places of nature, it is absolutely necessary to introduce a better and more perfect use and application of the mind and understanding. (*SEH*, I: 129–30; Jardine and Silverthorne, 10–11)

This cast of mind, as well as the expression and experience of knowing, being and acting, is fundamental, always articulating the very idea of knowledge on an expansionist globalizing experience taken as essentially woven into it. It is natural for Bacon to offer "*Columbus* in the westerne Nauigation" (along with Alexander) as his example of not yielding to the idea that no new action or knowledge is possible. He adds that once this action and knowledge have been achieved and no longer "seeme strange to our assent," "our mind accepteth of them … as if we had knowne them before" (29). The explanation, almost an expansion of my Pliny epigraph, argues that such novelties *can* be familiarly expressed and communicated. Throughout *The Advancement*, Bacon looks to the same source for the method and matter of the inquiry whose results will need this expression and communication.

Toward method and histories of new knowledge

First, the "*Methode*" setting the limits of each science teaches us that "euerie Knowledge may be fitly sayd, besides the *Profunditie* (which is the truth and substance of it, that makes it *solide*) to haue a *Longitude*, and a *Latitude*: accounting the latitude towardes other Sciences: and the Longitude towards Action: that is, from the greatest Generalitie, to the most particular Precept." The first rules relations between disciplines, the second the "degree of particularitie [to which] a knowledge should descend" (126). Not to follow such methodical rule promotes idle speculation, the "hidden Rocke wherevppon … many … barques of knowledge haue been cast away, which is, that men haue despised to be conuersant in ordinary and common matters, the iudicious direction whereof neuertheless is the wisest doctrine" (134). Just so, in natural philosophy, to offer "final causes" of natural phenomena before "seuere and diligent enquirie of all *reall* and *phisicall causes*," is to allow "*Remoraes* and hinderances to stay and slugge the Shippe from furder sayling," when proper "search of the *Phisicall Causes*" would speed it along (86).

Second, just as "the inuention of the Shippe was thought so noble, which carryeth riches, and commodities from place to place, and consociateth the most remote regions in participation of their fruits," so the disciplines of knowledge properly ordered and lettered "as Shippes, passe through the vast Seas of time, and make ages so distant, to participate of the wisedome, illuminations, and inuentions the one of the other" (53). The globalizing experience provided not just method, but much new matter: the "opennesse of the world by Nauigation … hath disclosed multitudes of experiments, and a Masse of Naturall History" (181). The very novelty of these "experiments" and phenomena inevitably raised questions about methods of understanding (including relations between experience and instrument and between sense and reason), judgment, tying them to the known and communicating them. All these were central to Bacon's project. The passages quoted show how he always equated the project of reordering knowledge and discovery in the realm of learning with those of European geographical, economic, and colonial expansion. To align learning on an "imperial" project was altogether familiar to Spanish, Portuguese, French, and English *historians* of invasion, settlement, and trade in Africa and America. Their *Histories* always emphasized not just the novelty and wonder of what they described, but the need carefully to define its experience, its utterance, and its relation to the familiar. Bacon gave the alignment retrospective weight by his political role and philosophical authority, an authority he also based on "histories," calling "NATVRAL HISTORY," for example, the "BASIS" of the "NATVRAL PHILOSOPHY" that was his main philosophical interest (85). Given that his "natural history" was also deeply concerned with finding and expressing the unknown, it is no stretch to see an essential tie between this and the many colonial "histories" that swamped the sixteenth-century reader.

The vital idea of science as progressive discovery of the new has a stand in the now deep-rooted history of European expansion to the Americas. Solomon adds that by Bacon's time, "most if not all treatises of instruction to travellers urged the voyager to bring back an understanding of foreign technology, geography, arts, and sciences to benefit his native land."[20] Texts of advice to missionaries urged that without this knowledge their mission would fail. How this startlingly new history, its peoples and phenomena were recorded, analyzed, made *true*, and told convincingly to those who knew none of them matched Bacon's thinking about a method to find and convey things just as new: "Wonders of Nature," for instance, whose knowledge would "correct the parcialitie of Axiomes, and Opinions: which are commonly framed onely vppon common and familiar examples" and lead to true knowledge of nature and works that could correct "her wandrings" (63). The *Histories* of the Americas may have been as basic to his scientific method for *how* and *what* they argue as they were for their cast of mind–to say nothing of their form and title as "histories," on which he was expansive (62–74). From the start Bacon saw major difficulty in fixing learning on new foundations. How to ensure a language denoting and relating truths of new "experiences"? Neither the rules of logic nor the words expressing them might apply to or satisfactorily express these truths, he says in several early works. How to be sure of *any* adequate tie between concepts, language and experience? How to know that "*Experientia literata*," ruled experiment, "degree and rudiment" of "*Interpretatio Naturae*" (111), at all matched worldly phenomena? How even to record these new phenomena? How to "translate" this new "Masse of Naturall History"?

---

[20] "Between Magic and Travel," 279.

To answer such questions required a firm method of discovery and equally firm means of communicating it and its matter. He asked them in works like *Temporis partus masculus*, *Cogitata et visa, Redargutio philosophiarum*, as well as in *The Advancement*. They less echoed inquiry into scientific *Methodus* or rhetoric and law than questions posed by those telling *Historias* of the "new world."[21] As did their solution. For, as Thomas Kuhn said of the uses of experiment, unlike his precursors, Bacon did not seek to demonstrate conclusions already known by other means or find "concrete answers to questions posed by existing theory." Like later experimentalists, he wanted to know how "nature … behave[d] under previously unobserved, often previously nonexistent, circumstances."[22] This project resulted in "vast natural or experimental histories," whose most dramatic models were the many histories recounting natural (and social) discoveries in a world previously unknown to Europeans. Further, these histories of the Americas tended to suggest that for all the differences, what they told involved things in many ways familiar, but seen through strange "new" eyes and in new contexts; just as "the transformation of the classical sciences" may be as much "ascribed to new ways of looking at old phenomena" as "to a series of unanticipated experimental discoveries."[23] A first issue concerned what such "new ways" might even be. To get at them presupposed that one could somehow strip phenomena of old epistemological accretions, know "bare" facts. And what could that mean? Could any fact free of theoretical assumptions even be perceptible? What, too, if one seemed not, in fact, to be looking "at old phenomena" but at something new and "wondrous"? These are actually much the same problem, and, like many others of his time, Bacon was concerned to know what an experience of facts laid bare might even *mean*, itself a difficult and not obvious question, much discussed in the sixteenth century, especially by those who thought they were experiencing things quite new and unheard-of.

Experiencing the unknown

In his first letter to Lorenzo di Pierfrancesco de' Medici (dated 18 July 1500, but published only in 1745, Amerigo Vespucci wrote that his travels had given the lie to ancient theories of the torrid zone, showing that "più vale la pratica che la teorica," practice is worth more than theory.[24] Soon after, in his *Esmeraldo de situ orbis*

[21] On the long *Methodus* debates, see, e.g., Timothy J. Reiss, "Neo-Aristotle and Method: Between Zabarella and Descartes," in *Descartes' Natural Philosophy*, eds. Stephen Gaukroger, John Schuster, and John Sutton (London and New York: Routledge, 2000), 195–227.

[22] Thomas S. Kuhn, "Mathematical versus Experimental Traditions in the Development of Physical Science," in *The Essential Tension: Selected Studies in Scientific Tradition and Change* (Chicago: Univ. of Chicago Press, 1977), 31–65, here 43. Cf. Vickers' note on Bacon's efforts to found academic centers quite unlike the norm, for whom the "notion that they should also devote themselves to research in order to generate new knowledge was unheard of" (Vickers ed., 581).

[23] Kuhn, "Mathematical versus Experimental Traditions," 43, 46.

[24] Américo Vespucio, *El nuevo mundo: Cartas relativas a sus viajes y descubrimientos*, textos en italiano, español e inglés, intro. Roberto Levillier (Buenos Aires: Editorial Nova, 1951), 94n, 106.

(1505–08), the fabled Portuguese naval commander and navigator, Duarte Pacheco Pereira, often calling experience "mother of things," and teacher "of the truth of everything," agreed that of "land very close to the equinoctial circle, which the ancients said was uninhabitable, we have learned the contrary by experience."[25] Experience has undeceived us of "the errors and fables that some ancient cosmographers wrote about the description of earth and sea." "Because experience is the mother of things," he insisted, "by it we discovered the very root of truth [por ella soubemos rradicalmente a verdade]."[26] Pereira's work, written for King Manuel, was not recorded until the eighteenth century and not published until 1892.[27] But to think experience mother of all things was not rare. Joaquim Barradas de Carvalho, in whom I found this last case, finds the phrase in a 1438 *cédula* of Alfonso V of Aragon and a like one in Leonardo da Vinci's 1508 notebooks. He finds it a little later in an unnamed "Philosopher" and in Jacques Cartier's 1545 narrative, where it is referred to "le prince d'iceulx philosophes," presumably Aristotle. The maxim, Carvalho judges, was probably "current at the time."[28] Care is needed, though, because for Leonardo and others, the experience or "practice" that was mother of theory was grounded in a "Florentine notion of *disegno*," echo of the Divine Maker's conceptual and practical skill by which he crafted the world. Such practice always had a preordered plan.[29] This seems to jibe with the thesis that Bacon was concerned "with the unveiling of latent structures and processes which could enable man to (re)produce

---

25 Duarte Pacheco Pereira, *Esmeraldo de situ orbis*, ed. Augusto Epifânio da Silva Dias (Lisbon, 1905; facs. rpt. Lisbon: Sociedade de Geografia de Lisboa, 1975), 23, ll.11–12: I.2; 188, l.29: IV.6; 125, ll.30–2: II.10. The second of these, that "experience teaches us the truth of everything we said before" (IV.6), is the last line of the book's final chapter.

26 *Esmeraldo*, 127, ll.5–7: II.11; 152, ll.24–5: IV.1.

27 Besides being a highly successful naval commander in Europe and the East (and later governor of the Portuguese factory/fort at São Jorge da Mina), he was a delegate to the debates that established the line of Tordesillas and the Treaty of its name, and a "scientist and technician who, two centuries in advance, established the degree of meridian to within an error of just 4°," a figure not otherwise achieved until 1762 (Joaquim Barrados de Carvalho, *A la recherche de la spécificité de la renaissance portugaise: L''Esmeraldo de situ orbis' de Duarte Pacheco Pereira et la littérature portugaise de voyages à l'époque des grandes découvertes. Contribution à l'étude des origines de la pensée moderne*, 2 vols. [Paris: Fondation Caliste Gulbenkian, Centre Culturel Portugais, 1983], 2.776).

28 Carvalho, 1.494–502; 2.677–8.

29 Catherine Wilkinson-Zerner, *Juan de Herrera: Architect to Philip II of Spain* (New Haven: Yale Univ. Press, 1993), 17–18. Carvalho read hugely in Portuguese works from the tenth through sixteenth centuries. He found the word "experience" (in any sense) only from 1379, and only after 1453 did he find it tied to maritime discovery and such. The mathematician-sailor João de Castro used it in his 1548 *Tratado da Sphaera* especially interestingly: "How do we know that to each degree corresponds 17 and a half leagues? We have learned it by mathematicians' great experience, of which the principal and most probative experience undertaken in our time is that done by the Florentine Amerigo Vespucci, a nobleman and great mathematician, when he was sent by king Manuel to discover the new world by the eastern part" (Carvalho, 2.675–9; 758–9n154). To elide "mathematics" and "navigation" was still common usage in the seventeenth century: see, Richard Jobson, *The Discovery of River Gambra (1624)*, eds. David P. Gamble and P. E. H. Hair (London: Hakluyt Society, 1999), 79n2.

Nature's effects."[30] Certainly Bacon cared about underlying rules, but he never saw them as being first, for us, humans. The "Notions of the minde" were necessary for "light of Nature" but "the Reports of the Sences" were so first (*OFB*, IV: 76). The mind could only work from accumulations and lists of "instances."

A more pragmatic idea of experience did go back to Aristotle and was widely familiar from the opening sentence of the *Metaphysics* (980a22): "There is no desire more natural than the desire for knowledge," Michel de Montaigne quoted at the start of "De l'expérience," going on to deny Aristotle's argument that this knowledge came first from experience: "We try all means that can bring us to it. When reason fails us, we use experience for it … which is a weaker and less worthy way; but truth is so great a matter that we must scorn no intermediary that may lead us to it."[31] Montaigne's criticism was in line with the search for a logico-mathematical *method* that by his time was increasingly thought to match the divine *disegno* ordering experienced actuality, and whose trail was leading to Galileo's mathematical language of nature, Kepler's divine numbers and forms, and Descartes' analytical geometry. In some ways, these opposed an older idea of experience whose models, Montaigne's essay shows, were law and medicine; in other ways they echo aspects of an equally old idea of syllogistic. While writings on navigation, travel, and geographical discovery increasingly assumed mathematical rule (see notes 27 and 29), they still emphasized experience as use, observation, participation, and physical familiarity.

This was certainly what Pietro Pomponazzi had in mind in his 18 March 1523 lecture on the second book of Aristotle's *Meteora* in the University of Bologna, having to explain why the lands of the torrid zone were uninhabitable. Since his 1503 *Mundus novus*, Vespucci had published the contrary, and Pereira's case shows how fast that opposite doctrine was shared. By March 1523 the contrary argument was even harder. On 8 September 1522, Juan Sebastián del Cano had sailed the surviving ship of Ferdinand Magellan's global expedition into Seville harbor. Pomponazzi was in a bind. More open than most of his colleagues for many years, he followed his obligatory gloss on Aristotle by telling his students, according to their extant notes, that a Venetian friend had written from Spain that an expedition had sailed for months far into the torrid zone and found more than three hundred islands that were not only habitable, but inhabited. "Where experience," he said, "is in conflict with reason, we must rely on experience and abandon reasons" [dove l'esperienza è in conflitto col ragionamento, dobbiamo stare all'esperienza, e lasciare andare I ragionamenti.] We must follow those who have been and returned and told us what they have seen, "quel che han visto."[32] This reveals a marked change from older ideas of what constituted sufficiently witnessed truths, where actually seeing something was one element in a far more complex process in which number, status, authority of witness and other

30 Antonio Pérez-Ramos, *Francis Bacon's Idea of Science and the Maker's Knowledge Tradition* (Oxford: Clarendon, 1988), 265.

31 Michel Eyquem de Montaigne, *Oeuvres complètes*, eds. Albert Thibaudet and Maurice Rat (Paris: Gallimard, 1962), *Essais*, 3.13.1041.

32 Bruno Nardi, *Studi su Pietro Pomponazzi* (Florence: F. Le Monnier, 1965), 41–2, quoted in the Introduction to *Il mondo nuovo di Amerigo Vespucci*, ed. Mario Pozzi (Milan: Serra e Riva, 1984), 11.

things always outweighed experience alone. Andrea Frisch shows in perspicuous detail that as the sixteenth century advanced gradual changes occurred in the relative importance of the supports taken to validate knowledge.[33] Not until very late did experience become principal. That it should bear most probative weight was not at all self-evident, any more than was what the very idea of "experience" actually meant, entailing at least, Hernán Cortés noted in 1520, that "we see with our own eyes" and "understand with the intellect [con el entendimiento]."[34]

Gonzalo Fernández de Oviedo began his 1526 *Sumario de la natural historia de las Indias* by asserting that written "historias y libros," histories and books, best kept "las obras de natura," nature's works, in people's memory, but that "those are thought most true and authentic that a man of prudent understanding who has traveled the world and seen with his own eyes took upon him to write down and said what he could see and hear of such matters."[35] "La novedad," novelty, of everything in the Indies was radical, but the truth of his writing lay in his own eyewitness and its easy verification by the many other "witnesses worthy of belief who have been in those parts, who live in these kingdoms, and others who are now presently in your majesty's court and come here, who live in those parts."[36] In his great *Historia natural y general de las Indias* of 1535–7, Oviedo accused Pietro Martire, first general historian of the American voyages, of falsification, because he only relied on books, whereas, he repeated: "I write on the authority of no historian or poet, but as eyewitness of most of what I shall treat here. And what I have not seen myself I shall relate from the account of trustworthy people, not giving credit in anything to a single witness, but to many, in those things that I have not experienced in person.[37] Bartolomé de Las Casas was also repetitively adamant of being eyewitness to the facts of the Americas that he told and to the horrors committed by invaders and settlers or, when not, of directly quoting others who were. Bacon echoed this in *The New Organon*'s "use, thought, observation, and argument" (*SEH*, 1, 130). If it had nothing to do with instruments, mathematical calculation, or theorized manipulation, it did with practical tests. A little earlier, Martire was, says Antonello Gerbi, already

33 Andrea Frisch, *The Invention of the Eyewitness: Witnessing and Testimony in Early Modern France* (Chapel Hill: Univ. of North Carolina Press, 2004).

34 Hernán Cortés, *Cartas de relación*, ed. Manuel Alcalá (1960; rpt. Mexico City: Porrúa, 2002), 76.

35 Gonzalo Fernández de Oviedo, *Sumario de la natural historia de las Indias*, ed. Nicolás del Castillo Mathieu (Santafé de Bogotá: Instituto Caro y Cuevo and Universidad de Bogotá "Jorge Tadeo Lozano," 1995), 3 (Dedication to the Emperor): "aquellas por más verdaderas y auténticos se estiman, que por vista de ojos el comedido entendimiento del hombre que por el mundo ha andado se ocupó en excribirlas, y dijo lo que pudo ver y entendió de semejantes materias."

36 *Sumario*, 5: "muchos testigos fidedignos que en aquellas partes han estado, que viven en estos reinos, y otros que al presente en esta corte de vuestra majestad hoy están y aquí andan, que en aquellas partes viven."

37 Gonzalo Fernández de Oviedo, *Historia general y natural de las Indias*, ed. Juan Pérez de Tudela Bueso, 5 vols. (1959; rpt. Madrid: Atlas, 1992), 1.13b (II.i): "no escribo de auctoridad de algún historiador o poeta, sino como testigo de vista, en la mayor parte, de cuanto aquí tratare; y lo que yo no hobiere visto, dirélo por relación de personas fidedignas, no dando en cosa alguna crédito a un solo testigo, sino a muchos, en aquellas cosas que por mi persona no hobiere experimentado."

"an empiricist, as Oviedo will be. To convince himself of the hardness of an American stone attached to a wooden handle, brought back for him by Columbus himself, he brings it down with all his might on some iron bars, and dents them, without the stone suffering the slightest scratch."[38]

Just so did Bacon reject Copernicanism, because "no system so complex, abstract and mathematical could contribute to either the understanding or the control of nature"[39]: two objectives, one may add, exactly required by the colonial and trade objectives of most historians of the Indies. For Bacon the understanding of rules could only come after observation and collected experience, to do things the other way about brought knowledge to a halt. The paid "Spyalls and Intelligencers of Nature" (59), who seek out its objects and events, must, we already saw, be close to and experience what they record: "For knowledge drawne freshly and in our view out of particulers, knoweth the waie best to particulers again" (162). For "it is plaine, that the more you recede from your grounds, the weaker doe you conclude, & as in nature, the more you remoue your selfe from particulars, the greater peril of Error you doe incur" (187). Only direct inquiry lets a knowledge advance, because only from original facts can you climb up to causes and then back down again to new facts: for only thus "all true and frutefull NATVRALL PHILOSOPHIE hath *A double Scale* or *Ladder*, *Ascendent* and *Descendent*, ascending from experiments to the *Inuention of causes*; and descending from causes, to the *Inuention of newe experiments*" (80).

How the experience of naked facts and events could be made meaningful in this way required a specific conception of its interaction with right language, whose use, in turn, required that direct experience. Eyewitness historians of the Americas were adamant on the issue, and it was exactly in talking about the voyages of European exploration to the West and the East Indies that Richard Eden (1521–76), addressing the reader of his 1553 translation of Sebastian Münster's *Cosmographia*, remarked on the wondrous new knowledge found by sailing "the hole [*sic*] globe of the world":

> A thing surely most wonderful, and in maner incredible, but that the same is proued most certayne by experience, the teacher and mestres of all sciences, for lacke of whose ayde (experience I meane) lyke as many greate wittes haue fallen into great errours, so by her ayde, many base and common wittes haue attayned to the knowledge and practice of such wonderful effectes, as could hardely be comprehended by the discourse of reason. Which thing or other lyke, I suppose was the cause why the noble Philosopher Aristoteles sayde: *Quod nihil est in intellectu quod non fuit prius in sensu*, that is, that nothing is in vnder-

---

38 Antonelli Gerbi, *Nature in the New World: From Christopher Columbus to Gonzalo Fernández de Oviedo*, trans. Jeremy Moyle (Pittsburgh: Univ. of Pittsburgh Press, 1985), 65; see Pietro Martire d'Anghiera, *De orbe nouo Petri Martyris ab Angleria mediolanensis protonotarii Cesaris senatoris decades*, facs. ed., in *Opera*, intro. Erich Woldan (Graz: Akademische Druck- u. Verlagsanstalt, 1966), 176 (*Dec.* V.iv), full translation in *De orbe novo: The Eight Decades of Peter Martyr d'Anghera*, trans. Francis Augustus MacNutt, 2 vols. (New York and London: Putnam's, The Knickerbocker Press, 1912), 2.115. Gerbi (65n69) notes another test of stones "that almost ruin hammer and anvil and blacksmith before they are broken (*De orbe nouo*, 230 [*Dec.* VII.vii], trans. MacNutt 2.297), and finds similar empiricism in other reports (66–8).

39 Kuhn, 48.

> standinge, but the same was fyrst in sense, that is to saye subiect to ye senses. Yet would I not that any rashe witte shoulde hereby take holde, as thoughe eyther Aristotle or I, meant that sence were more excellent than reason, but rather that reason vsing sense, taketh his principles and fyrst sedes of thinges sensyble, and afterwarde by his owne discourse and searching of causes, encreaseth the same from a seede to a tree, as from an acorne to an oke. Nought else to say, but that experience to be most certayn which is ioyned with reason or speculacion, and that reason to be most sure which is confirmed with experience, accordinge as the Phisicians determen in theyr science, that neyther practyse is safe without speculacion, nor speculacion without practyse. Wherfore, whereas men of great knowledge and experience, are to great affaires, theyr attemptes haue for the moste parte good successe, as doeth most playnly appeare in all histories.[40]

Bacon could have written such a passage word for word, and Edward Arber is surely right to draw attention to what he calls Bacon's "Inductive Method," and to the overlap of their lives.[41] The reference to medical science and practice is as Baconian as the acorn/oak metaphor, but so, above all, in connection with experience, knowledge, speculation, and practice, is the idea of "histories."

Only such direct experience could be truthful, and yet, if it was unheard and unknown, as Las Casas put it, it was lost. To *become* truth, it required a whole new language. And how could that be? Brute experience, Galileo and Bacon would say, was meaningless. *That* might require, said Oviedo, that one's writings "go naked [desnudos]," even as writing "the new" needed "foreign and barbaric words."[42] Oviedo's "algunos vocablos extraños y bárbaros" is a crucial phrase, not just "extraños" but "bárbaros" referring particularly to incomprehensible language and those who speak it. For what language could "give [such experience] letters," make it "*literata*," said Bacon, or join experience "with reason or speculacion," as Eden had it? Pomponazzi's opposition of the two may show an early stage of thinking, for coeval was the idea that experience needed reason to adjust it. Carvalho says that mathematicians like Pedro Nunes or Castro held them opposed. His examples, though, show that not *experience*, but misuse of the senses opposed reason. Reason worked *with* experience to correct, even replace, brute or naked sense data. Castro clarified Nunes' remark that "the senses should obey the understanding," by adding that "with *reason* and *experience*, the understanding is convinced and the senses undeceived."[43] Cortés had said little else.

To understand and communicate such unheard-of experiences–a world, peoples, fauna, and flora unknown–right expression, right comprehension, right language, and use of language were essential. To such experiences wonder was a first reaction. Tellingly, Descartes and Hobbes would later make wonder the first of the passions:

---

[40] Edward Arber, ed., *The First Three English Books on America [?1511]–1555* A.D. *Being Chiefly Translations, Compilations, &c., by Richard Eden, from the Writings, Maps, &c., of Pietro Martire ..., Sebastian Münster ..., Sebastian Cabot, with Extracts, &c., from the Works of Other Spanish, Italian, and German Writers of the Time* (Birmingham: n. p., 1885), 9.

[41] Arber, xii.

[42] Oviedo, *Historia* 1.9b, 10b: Bk. I.

[43] Last three references to Carvalho, *La recherche*, 2.679–80, 760–1nn157–60.

first, because it enabled all other passions and all knowing (views surely not foreign to Bacon's insistence on the epistemological importance of "wonders of nature"). But such wonder was useless if its originating experiences had no words to express them, no language to name and explain them. Cortés lamented to Charles V in 1522 that he had to describe things "so marvelous that they cannot be made meaningful in writing [por escrito no se pueden significar] nor can be understood if they are not seen."[44] Such experiences had to be given meanings or, as Bacon put it, "made literate." How was one to do that?

Language: speaking wonders

In a 1551 dialogue written to instruct missionaries, the Franciscan Pedro de Quiroga had his surrogate Barchilón warn his newly-arrived friend Justino to have "no dealings with the things of this land until you understand them, because they are strange affairs and a strange language which only experience will reveal to you."[45] "All things that have happened in the Indies since their marvelous discovery," Las Casas began his *Brevísima relación* in 1542, "have been in every way so admirable and so unbelievable to anyone who did not see them that they seem to have obscured and silenced and even plunged into oblivion all those things, however heroic, that were seen and heard in the world in past centuries."[46] The difficulty of matching language to new experience was one early historians and chroniclers of the Americas made much of: what (and whose) words could denote, narrate, express and communicate knowledge of "unheard of" things? How to ensure that they did not stay unheard? Or make familiar things impossible to hear–as in Las Casas' remark? We saw Cortés express the difficulty of giving unknown things meaning. Eighteen months earlier (1520) he told Charles that he could not properly describe the "extrañas y maravillosas cosas," the strange and marvelous things," they had seen: "much time would be needed and many very expert narrators/reporters [muchos relatores y muy expertos]. I will be unable to say a hundredth of what could be said of them, but as far as I can I will say some things of those I saw, which, although badly said, I well know will be so wondrous [de tanta admiración] as not to be believed, for we who see them here with our own eyes cannot understand them with the intellect."[47]

The demand for accurate and detailed narration of these things also matched how the "experimentalists scorned thought experiments and insisted upon both accurate and circumstantial reporting."[48] Properly to know and then report the "Masse" of natural phenomena, Bacon agreed, required that one see them up close and naked: "it

44 Cortés, 205.

45 Quoted in Anthony Pagden, Introduction to Bartolomé de Las Casas, *A Short Account of the Destruction of the Indies*, ed. and trans. Nigel Griffin (Harmondsworth: Penguin, 1992), xxxiv.

46 Bartolomé de Las Casas, *Brevísima relación de la destruición de las Indias*, ed. André Saint-Lu (1984; rpt. Madrid: Catedra, 1999), 69.

47 Cortés, 76.

48 Kuhn, 44.

The University of Arizona Library

The UA Library catalog will be down due to a software upgrade from 8:00 - 8:30 a.m. on Tuesday, January 27, 2009.

UA Library Catalog

AUTHOR barrera-osorio View All Collections Search

Record 1 of 2

Record: Prev Next

LOCATION Science-Engineering Library
Call # Q127.L29 B37 2006
Author Barrera-Osorio, Antonio, 1964-
Title **Experiencing nature : the Spanish American empire and the early scientific revolution / Antonio Barrera-Osorio**

Location	CALL NO.	STATUS
Science-Engineering Library	Q127.L29 B37 2006	DUE 04-13-09

Edition 1st ed
Publisher Austin, TX : University of Texas Press, 2006
Subjects Science -- Latin America -- History -- To 1830.
Science -- Spain -- History -- To 1830.
Science -- United States -- History -- To 1830.
Latin America -- History -- To 1830.
Spain -- History.
Spain -- Colonies -- America.

Contents 1. Searching the land for commodities -- 2. A chamber of knowledge : the Casa de la Contratacion and its empirical methods -- 3. Communities of experts : artisans and innovation in the new world -- 4. Circuits of information : reports from the new world -- 5. Books of nature : scholars, natural history, and the new world -- Conclusions : the politics of knowledge -- App. 1. Pilots and cosmographers at the Casa de la Contratacion -- App. 2.

The University of Arizona Library

UA Library Catalog

Your request for **Experiencing nature : the Spanish American empire and the early scientific revolution / Antonio Barr** was successful.

Science-Engineering Library	Q127.L29 B37 2006	DUE 04-13-09

'In Library' items will be on the hold shelf of the appropriate library within 24 hours.
RECALLED items will be available within 10 days.
Requests for materials with a status of 'In Library' located at the Arizona Health Science and Law Libraries will be canceled.

in the institu[illegible] "empirical" [illegible]
political cult. of
first fostered by the Spanish transatlantic
enterprise.

is the *remote standing or placing thereof*, that breedeth these Mazes and incomprehensions; for as the sence is far off, is full of mistakings, but is exact at hand" (98). Their reporting had to be of the same kind, "a continuance of the naked euents & actions, without the motiues or designes …" (65). One recalls, as Bacon surely did, Montaigne's and many others' insistence, apropos precisely of the Americas, that those reporting on its peoples, objects, behaviors, conditions, and phenomena do so in narratives of the plainest language, without rhetorical ornament "to give their interpretation legitimacy and persuade us to believe it."[49] By this time, such remarks were a cliché of histories of the Americas.

Language and interpretation was always a central care of *all* early chroniclers, not just for how they could adequately convey their knowledge but for how they could even get it: Columbus, Vespucci, Martire, Oviedo, to name the earliest writers on the Americas, wrote repeatedly of the many native tongues and of interpretation.[50] Martire noted the trouble that arose for lack of the last, telling how the brother of one of the local Hispaniola leaders tried to explain why the Spanish settlers left by Columbus on his first voyage had been killed (because of their brutality it later appeared), "but as there was no interpreter, nobody at the time understood his words."[51] Until the truth was known (as other sources tell, including Dr. Chanca's letter on Columbus' next voyage), no one knew what to do, but distrust and fear ruled. Not just knowledge, but policy, settlement, trade, and all else hinged on right language. Europeans had long known this, and Elio Antonio de Nebrija's now-famous remark, in his dedication to Queen Isabel of his 1492 Castilian *Gramática*, that "language was always the companion of empire" had merely emphasized the point.[52]

By 1454, the Portuguese had built interpretation into the very structure of the West African voyages that preceded and led to the transatlantic ones. Whenever they met a new people, they took captives to teach them Portuguese, always including such slave interpreters in their ships' crews: "each of our ships," wrote Alvise Cadamosto in 1454–56, "had negro interpreters on board brought from Portugal, who had been sold by the lords of Senega to the first Portuguese to discover this land of the Blacks." These Christianized interpreters, "Spanish"-speaking (as Cadamosto called them), had food and wages on board and were received "from their owners on the understanding that for the hire and pay of each we would give one slave to be chosen from all our captives." Further, "each interpreter who secured four slaves for his master

---

49 *Essais*, 1.31.202 ("Des cannibales").

50 "What mortal mind," cried Oviedo, "will be able to understand so great a diversity of languages, of dispositions, of customs among the peoples of these Indies?" (*Historia* 1.8: Bk. I).

51 Martire, *De orbe nouo*, *Dec*. I.ii, in *Opera*, 45 (trans. MacNutt 1.78).

52 Elio Antonio de Nebrija, *Gramática de la lengua castellana*, ed. Antonio Quilis (Madrid: Editora Nacional, 1980), 97. Nebrija did not originate the equation. Lorenzo Valla (who had close ties to the court of Aragon) perhaps did so in his 1471 *Elegantiae*, a work of very wide circulation, but he too was uttering what was already a commonplace: see Eugenio Ascensio, "La lengua compañera de imperio: Historia de une idea de Nebrija en España y Portugal," *Revista de Filología Española* 43.3–4 (1960): 399–413.

was to be given his freedom."[53] Cadamosto spoke much of these interpreters, noting, too, that it was the king who

> had enjoined them that, from the farthest land they reached, if it chanced that the people were unable to understand their interpreters, they were to contrive to bring away a negro, by force or persuasion, so that he might be interrogated through the many negro interpreters to be found in Portugal, or in the course of time might learn to speak, so that he might give an account of his country.[54]

Speaking often of the matter, the royal chronicler Gomes Eanes de Zurara wrote that it was Henry the Navigator who had ordered his sailors, on making landfall anywhere, always to "get an interpreter from these people, capturing some from whom you can find out knowledge of the land."[55] Similarly, Columbus, Vespucci, and Martire their commentator, continually recalled their efforts to capture people to make them translators.

Early in his 14 March 1493 letter to Raphael Sanchez, the Catholic Monarchs' Treasurer, about his first voyage, Columbus wrote of his inability to *speak* [*loqui*] with the people of Juana (Cuba) because they fled from him and his men; then of how he had *understood* [*intellexeram*] from some Indians he had seized that Juana was indeed an island. Finally, he adds that in fact: "as soon as I arrived in that sea, I seized some Indians by force from the first island, that they might learn [*ediscerent*] [our language] from us and teach [*docerent*] us what they knew about those parts–which succeeded as we wished."[56] This last event mattered enough for him to have used the same phrase (now in Spanish) in a 15 February letter to the court scribe Luis de Santángel:

> And as soon as I arrived in the Indies, in the first island which I found, I took by force some of [the Americans], in order that they might learn [our language] and give me information of that which there is in those parts, and so it was that they soon understood us, and we them, either by speech or signs, and they have been very serviceable.[57]

---

[53] Alvise da Ca' da Mosto, *Viagens de Luís de Cadamostro e de Pedro de Sintra*, ed. Damião Peres (Lisbon: Academia Portuguesa da História, 1948), 52–3 (ch. xxxv); *The Voyages of Cadamosto and Other Documents on Western Africa in the Second Half of the Fifteenth Century*, trans. and ed. G. R. Crone (London: For the Hakluyt Society, 1937), 55.

[54] *Viagens*, 79; *Voyages* 84 (ch. l). The occasion provoking this explanation is interesting because Cadamosto tells how the African man seized was taken back to Lisbon, where eventually they found an African woman who spoke his language, though her own was different. Did women interpreters also go on the ships? The negotiations in which these interpreters were involved could be perilous. The previous occasion cited involved taking lots to choose the interpreter who would negotiate, Cadamosto implying potential advantage for the interpreter, who was working toward gaining freedom. In this case, the interpreter was killed.

[55] Gomes Eanes da Zurara, *Crónica de Guiné*, ed. José de Bragança, [new edn.] (Barcelos: Livraria Civilização, 1973), 57.

[56] Christopher Columbus, *Four Voyages to the New World: Letters and Selected Documents*, bi-lingual edn, trans. and ed. R. H. Major, intro. John E. Fagg (New York: Corinth Books, 1961), 3, 4, 9.

[57] Cristóbal Colón, *Textos y documentos completos*, ed. Consuelo Varela, 2nd edn (Madrid: Alianza, 1992), 22–3; translation from Roland Greene, *Unrequited Conquests: Love and Empire in the Colonial Americas* (Chicago: Univ. of Chicago Press, 1999), 48.

Vespucci spoke many times of the myriad languages of the inhabitants of the southern continent, finally telling in his 1504–05 letter to Piero Soderini how, on their third voyage, they were greeted by friendly coastal people: "Accordammo in questo luogho leuare un paio di huomioni, perche cimostrassino la lingua; et uennono tre di loro uolunta per uenire a Portogallo" [We agreed in this place to take away a couple of men so that they might teach us the language; and three came of their own will to return to Portugal.][58] All this sought to establish interpretative possibilities and create a linguistic measure enabling narratives of realities from the mundane to the wondrous.

Not accidentally did Bacon echo this by having his Indian Prince accompanied by an "expert in languages." The device's theme was that of going from blindness to sight, enabled by an "Excellent Queen," in a new American world (*LL*, 1: 388). One of those whose sight it was to correct was the unnamed but implicit Ralegh, closer to Bacon in all sorts of ways than those just named. Ralegh had insisted that on his Guyana voyage he had: "with me an Indian that spake many languages, and that of *Guiana* naturally."[59] Bacon wrote in 1608 of his need to talk with Ralegh, Thomas Hariot and Northumberland, "themselves being already inclined to experim[en]ts" (*LL*, 4: 63). While Ralegh was certainly not the first to make native translators and interpreters a major part of his business, he stands out in seeing (and using) them as participants in a cultural exchange (even if he was clearly not first here either; Jean de Léry is an obvious case, but so are others), in the sheer numbers involved and in the attentive care he brought to their linguistic training in London–mostly at the hands of his polymathic linguist friend, Hariot. Alden T. Vaughan calls them "culture brokers, differing from later types most notably in the primacy of language training, the location of that training in England rather than America, and the transient nature of their broker's role." Of the six or seven from Virginia in the 1580s and the at least twelve from Guyana in the 1590s brought by Ralegh or his agents to London (where they chiefly lived in Ralegh's mansion) almost all eventually returned home, some after many years. Once home, most do seem to have served Ralegh and others as essential go-betweens, not seldom saving lives and facilitating various diplomatic and cultural exchanges.[60] Of brokers like these interpreters, or Ralegh and Hariot, Solomon notes that "to follow through with the instructions and satisfy the expectations of those … who advised [them, travelers] had to have, over and above acuity of observation, an understanding of circumstances, of the right relation of thing observed to the time and place of observation": they had to be Bacon's scientific "historian," reporters and narrators of reasoned experience.[61]

---

58 *Lettera di Amerigo Vespucci delle isole nouamente trouate in quattro suoi viaggi* (Florence 1505–06), in Vespucio, 177–264 (facsimile, with facing-page Spanish), here 254.

59 Sir Walter Ralegh, *The Discoverie of the Large, Rich and Bewtiful Empyre of Guiana*, ed. Neil L. Whitehead (Norman: Univ. of Oklahoma Press, 1997), 194.

60 Alden T. Vaughan, "Sir Walter Ralegh's Indian Interpreters, 1584–1618," *William and Mary Quarterly*, Third Series 59.2 (April 2002): 341–76, quotations from 343, 279. The essay is full of information, although its speculations on motives are dubious.

61 "Between Magic and Travel," 279.

Language: validating, recording, narrating

Bacon, like everyone else at the court, would have known of Ralegh's Indians and doubtless met them–not that this matters much, since their presence only confirmed long-familiar assumptions, but they were physical reminders of the Americas' role in these assumptions. He always stressed the importance of finding a right language to express nature's works and pass on knowledge. God had made humans' first task the proper naming of other creatures, which they did by "the pure knowledge of nature and uniuersality" (6). Humans now had to repeat this "view of Creatures, and the imposition of names" (34), again finding a language for it: as for the Atlantic historians, simple, straightforward language was essential for telling the bare facts of nature and their truths, one that would grow along with the new-developing knowledge whose discovery it narrated: "as young men, when they knit and shape perfectly, doe seldome grow to a further stature: so knowledge, while it is in Aphorismes and obseruations, it is in groweth; but when it once is comprehended in exact Methodes; it may perchance be further pollished and illustrate, and accommodated for vse and practise; but it encreaseth no more in bulke and substance" (30). Well-framed aphorisms "cannot bee made but of the pyth and heart of Sciences: for discourse of illustration is cut off … So there remayneth nothinge to fill the *Aphorismes*, but some good quantitie of Obseruation," of necessity "sound and grounded." Further, "*Aphorismes*, representing a knowledge broken, doe inuite men to enquire further" (124). Aphorisms told exactly against the disputes of language that he so frequently criticized and for the proper *use* of language to express nature and works. The observations and experiences on which aphorisms depended and which they could encapsulate eventually enabled narrative histories.

It is no surprise that Las Casas, like Oviedo, began his *Historia de las Indias* by criticizing historians who were "verbose, eloquent, rich in words," mixing "opinion, … fables and false fictions" to hide the fact that they were writing of "what they had neither seen nor experienced." His own work was bound to be written roughly, "for its penury of words, humbleness of style, lack of eloquence [would] be good witnesses of [its truth]."[62] Montaigne made the same case about his man who had lived "ten or twelve years" in "la France Antartique" and who, because he was "simple and rough," and not endowed with learning and language that would order things to his preconceptions, was of "fit condition to give true witness."[63] Francisco López de Gómara, from whom Montaigne drew much of his information about the invasion of the Americas, had used the same trope in his 1553 *Historia general de las Indias*, a work translated and read across Europe. Although he, in fact, had *not* been to the Americas or seen anything firsthand, he prided himself on having written a history that was "plain and by today's norms; its order harmonious and smooth; its

[62] Bartolomé de Las Casas, *Historia de las Indias*, ed. Miguel Angel Medina, Jesús Angel Barreda and Isacio Pérez Fernández, 3 vols., in *Obras completas*, [published] by la Fundación 'Instituto Bartolomé de Las Casas', de los Dominicos de Andalucía, vols. 3–5 (Madrid: Alianza Editorial, 1994), 3.328, 337 (Prólogo).

[63] *Essais*, 1.31.200, 202 ("Des cannibales").

chapters short to save words; its sentences clear, though brief. I have striven," he said, "to say things as they happened." He even added a note "to translators" that they faithfully copy his simple style, and neither "add nor subtract a letter."[64] This same concern with exactitude of language and what it denoted remains evident in Bacon's *New Atlantis*, a work still tying geographic, economic, military, and colonial "discovery" to that of knowledge. The fact that "the people had languages" was not just a mark of their being "full of humanity" (*SEH*, III: 130), but a promise of that record of validated ("true") knowledge of which the narrator was later to be beneficiary, learning not just the history of Bensalem and the western hemisphere, but of how to set about "the finding out of the true nature of all things" (*SEH*, III: 146). Here, too, continuous sea voyages–of the narrator and his companions, and of the members of Salomon's house–are the standard of such discovery and verification, echoing Cortés' narrative record in form *and* content.

Vespucci began his *Lettera* to Soderini by excusing his offering "things unsuitable to your state nor pleasing and written in barbarous style and out of all human order; save the confidence I have in your virtue and in the truth of my writing, which are things not written before either by ancient or by modern writers."[65] He underscored here the need for a language that could be heard and meaningful to its listeners (conuenienti), whose roughness (barbaro stile, the idea repeated by Oviedo) matched the wildness and foreign human realities it was expressing (fuora dogni ordine de humanita). Only such language could communicate with its European recipients' reason (le uostre uirtu) and speak the truth (uerita) of the experience it related, a wondrous experience unknown to ancients and moderns. Thomas More took up these concerns in *Utopia*, but a slightly later tale tied the difficulty of deciphering language, here written, with colonizing voyages and Utopia. I mean François Rabelais's 1532 *Pantagruel*, whose Chapter 24 first has Panurge trying many ways to decipher a blank letter finally understood to accuse Pantagruel of falsity and abandonment, and which leads directly into an account of their voyage down the western coast of Africa, around the Cape of Good Hope, past Zanzibar and into unknown lands, finally disembarking in Utopia.[66] Rabelais directly tied questions of true and false writing, meaning, and understanding to colonial voyaging. So he did again in the *Quart livre* (1548, 1552), a work that I read as, among other things, anti-colonizing, whose transatlantic voyage traces growing moral decay of the many islands visited (being finally so bad that Pantagruel will not even risk going ashore) and also shows language breaking down, its thawing words revealing not philosophical or other truth, but just cries and groans of war.[67] The novel, then, shows how collapse of the experiential truths enabled by the voyages and colonial ventures entails that of language. The issue here was one of validation, in this last case, of its failure.

64 Francisco López de Gómara, *Historia general de las Indias*, 2 vols. (Madrid: Espasa-Calpe, 1941), 1.1, 3.

65 "cose non conuenienti a uostro stato ne dilecteuoli & con barbaro stile scripte & fuora dogni ordine di humanita; ma la confidentia mia che tengho nel le uostre uirtu & nella uerita del mio scriuere che son cose non sitruouano scripte ne per li antichi ne per moderni scriptori" (Vespucci, *Lettera* 200).

66 François Rabelais, *Oeuvres complètes*, ed. Pierre Jourda, 2 vols. (Paris: Garnier, 1962), 1.337–40.

67 Rabelais, 2.203–8 (Chs. 55 and 56).

The point is not that Bacon drew directly on these writings, any more than he may have on Léry's many remarks on language, writing, and music in his 1578 *History* of his Brazilian voyage,[68] or Ralegh's in the 1596 *Discoverie*. As William Aldis Wright has shown, he may, however, have drawn on José de Acosta's extensive scrutiny in his 1590 *Historia natural y moral de las Indias*, which was a Europe-wide bestseller, immediately the most read book on the Americas, with many Spanish and Latin editions and translations into all European languages–one into English in 1604. Wright noticed that Bacon's mistaken remarks on Chinese ideograms in *The Advancement* (*OFB*, IV: 120) echo Chapter 5 of Acosta's several on language, writing, and other forms of expression in the Americas in his Sixth Book.[69] Acosta aimed not just to note the variety of Indian languages, but, like everyone else, to remind his readers that language was crucial to the whole enterprise. It was this, among other things, that made this tradition of discovery and its histories (tradition as it now was) so crucial, too, to Bacon. As its writers did (in an even longer tradition to which Galileo still clung), he asserted that humans had to read not just in "the Booke of Gods word," but "in the Booke of Gods workes" (9), for while the one opens "our vnderstanding … by the general notions of reason and rules of speech," the other has us meditate on what is "signed and ingrauen vppon his workes" (37–8). The assertion makes words and works, the language of reason and that of the world, interdependent. They were so in equally scientifically urgent ways.

American "Histories," then, made experience and language *essentially* interdependent as far as the former's meaning, and so usefulness, was concerned. So they are for Bacon. God's book of works is written in an alphabet that humans must decipher and understand. To start by seeking complex relations leads to error, confusion, and obfuscation:

---

68 Besides its several French editions, this work appeared in Latin in 1592 as the second part of part three of Theodor de Bry's *India occidentalis*, another European best-seller.

69 *Advancement*, ed. William Aldis Wright, 5th ed. (Oxford: Clarendon, 1926), 309–10; José de Acosta, *Historia natural y moral de las Indias*, ed. José Alcina Franch (Madrid: Historia 16, 1987), 395–6. Wright urged that Bacon also got from Acosta a remark about Americans not using flint for fire, like Europeans, because "of the rarenesse with them of flint"–although Acosta himself mentions no such rarity. The point is intriguing. Wright notes that the 1604 English text has Indians "rubbing two stones one against another," which might imply flint, where the Spanish has "fregando unos palos con otros," rubbing sticks with others (Wright ed. 305; Acosta 154 [Bk.3, Ch.2]; *OFB*, IV: 108). If Bacon's source was Acosta, he must have read a Spanish, Latin, or other edition (the entries of his commonplace book in these languages and others show that he read them). But the friction method was familiar. Oviedo described it somewhere in his *Historia*, as did, in other hugely popular works, André Thevet in 1558 (*Les singularités de la France Antarctique autrement nommée Amérique et de plusieurs terres et les découvertes de notre temps* … [facs. rpt., intro. Jean Baudry] [Paris: Le Temps, 1982], 101–2), Léry in 1578 (*Histoire d'un voyage faict en terre du Brésil (1578)*, 2nd ed., 1580, ed. Frank Lestringant [Paris: Livre de Poche, 1994], 458), and perhaps most importantly here, John Sparke in his account of John Hawkins' "Second Voyage" (1564), printed by Richard Hakluyt in his 1589 *Principal Navigations* (*The Hawkins' Voyages during the Reigns of Henry VIII, Queen Elizabeth, and James I*, ed. Clements R. Markham [London: Hakluyt Society, 1878], 52). Acosta's work was offered for travelers and missionaries (for whom he later wrote a manual).

> Noe more then it were either possible or to purpose to seeke in grosse *the formes of those Soundes which make wordes*, which by composition and transposition of Letters are infinite. But on the otherside, to enquire *the forme of those Soundes or Voices which make simple Letters* is easily comprehensible, and being knowen, induceth and manifesteth *the formes of all words*, which consist, & are compounded of them; in the same maner to enquire *the forme* of a Lyon, of an Oake, of Gold: Nay of Water, of Aire, is a vaine pursuite: But to enquire *the formes* of Sence, of voluntary Motion, of Vegetation, of Colours, of Grauitie and Leuitie, of Densitie, of Tenuitie, of Heate, of Cold, & al other Natures and qualities, which like an *Alphabet* are not many, & of which the essences (vpheld by Matter) of all creatures doe consist. (84)

That the elements of the physical world are an "alphabet" matters enormously for our ability to understand and communicate that world. It is first of all what links language and experience, and makes the world *literally* a book. It is also, of course, what made language so crucial in the communicating, validating, recording, and narrating of new experience.

For human reason also expresses itself in language or signifying gestures, these being "*competent to expresse Cogitations*" which are best expressed in "*Writings by letters* [which] doe far excell all the other wayes" of communicating rational knowledge (120). This is why knowledge of the natural world (and everything else) must be set into written "histories" (of which we saw a little in the second section of this chapter). The alphabet of the world composes "the common principles and Axiomes which are promiscuous and indifferent to seuerall Sciences" (82), which are, too, "the same footsteppes of Nature, treading or printing vppon seuerall subiects or Matters" (78). These principles can be brought to match in some way "the power and Nature of Wordes, as they are the foot-steppes and prints of Reason" (121), the "pictures of wordes" imaging "the life of reason" (46). That was why the first "great iudgement of God vppon the ambition of man, was the confusion of tongues; whereby the open Trade and intercourse of Learning and knowledge, was chiefely imbarred" (34). But it is also why, now, "the contemplation of Nature, and the obseruations of experience" (30) require, we saw, that they be set first in the open-ended simple language of aphorisms, and then into the lettered order of histories, avoiding recording anything not directly experienced or seen, "wherein the wisedome and integritie of *Aristotle* is worthy to be obserued, that hauing made so diligent and exquisite a Historie of liuing creatures, hath mingled it so sparingly with any vaine or fayned matter" (26).

For while one should not *start* with complex relations, one must certainly *get* to them: from aphorisms to the narratives of extended phenomena, Cortes' expert *relatores* narrating experiences and observations adjusted to reason. These narrative "histories" knit up the broken knowledge of aphorisms, make literate the alphabet of the footsteps of nature, reveal interrelations in space and time, beginnings and ends, causes and effects, in an operational sequence enabling the production of "works." Well-ordered writing can indeed provide an exact rendition and analysis of even the newest parts of the world's alphabet, and that is why we have to be as attentive to language as to reason and experience, since language, we saw, is "the Mint of knowledge (for wordes are the tokens currant and accepted for conceits, as Moneys are for values …)" (120).[70] Too, because of the concord between reason and nature, we can much rely on experiential accuracy: "hee that shall attentiuely obserue how the minde doth gather this excellent dew of Knowledge … distilling and contryuing it out of particulars naturall and artificiall, as the flowers of the field and Garden: shall finde that the mind of her selfe by Nature doth mannage, and Acte an Induction, much better than [logicians] describe it" (109).

The state's knowledge of the good

Having thus "settled" how language and reason could properly render utterly new matter, Bacon might hope to be able to say of knowledge what he said in 1592(?) of his queen's most signal achievement: "And lastly, to make an end where no end is, the shipping of this realm [is] so far advanced and made so mighty and potent, as this island is become (as the natural site thereof deserved) the lady of the sea; a point of so high consequence, as it may be truly said that the commandment of the sea is an abridgement or quintessence of an universal monarchy."[71] Here, the good of the state is most certainly not a passive one. On the contrary, it is by definition expansive–as is the good pursued by those in pursuit of new knowledge. By the time of *The Advancement* the two have become indivisible. Settlement of the relation between experience, language, and reason enabled true narration of phenomena (rather than its precedent "*Enumeration of particulars*" from which nothing can yet be concluded, since who can say "that there are not other on the contrarie side, which appeareth not": *OFB*, IV: 109–10), their proper ordering, the establishment of correct axioms about the natural universe, and from them descent to new particulars, which can now be tested, manipulated, and made "operatiue to the endowment, and benefit of Mans life" (*OFB*, IV: 65). This held, we saw him write to the new king in 1603, as much for the state as for nature: "There is a great affinity and consent between the rules of nature, and the true rules of policy: the one being nothing else but an order in the government of the world, and the other an order in the government of an estate … I conclude that there is … a congruity between the principles of Nature and Policy" (*LL*, III: 90–1). This was why he so admired the Spanish polity, built, he said, on recognition of that "affinity." This achievement of theoretical and practical knowledge drew on and always matched the paths and histories that circulated throughout the Atlantic. If one "will be content to beginne with doubts," Bacon wrote, "he shall end in certainties." By that scale, "rough and troublesome" as the doubts and interrogations of historians and thinkers had made sailing "in the entrance," their resolution piloted these mariners toward "faire and euen" waters (31).

---

70 Vickers has written much on Bacon's attention to language: see, e.g., *Francis Bacon* and, more recently, specifically on the aphorism, Vickers ed. 544. I have written on the alphabet of the world and *experientia literata* at more length and from another angle in *The Discourse of Modernism* (Ithaca, NY: Cornell Univ. Press, 1982), 201–16.

71 "Of Tribute," Vickers, ed. 42.

Select Bibliography

Editions of Bacon's Works

The Works of Francis Bacon. eds. James Spedding, Robert Leslie Ellis, and Douglas Denton Heath. 14 vols. (London, Longman, 1857–74).

The Oxford Francis Bacon. eds. Graham Rees, Lisa Jardine, Michael Kiernan, Julian Martin, Marta Fattori, Michael Edwards et al., 15 vols. forthcoming (Oxford: Clarendon Press, 1996–).

Francis Bacon: A Critical Edition of the Major Works. ed. Brian Vickers (Oxford: Oxford University Press, 1996).

The Advancement of Learning. ed. William Aldis Wright. 5th edn. (Oxford: Clarendon, 1926).

"Aphorisms on the greater Law of nations or the fountains of Justice and Law." ed. Mark S. Neustadt in *The making of the instauration: science, politics, and law in the career of Francis Bacon* (Ph.D. diss., John Hopkins University, 1987).

De Augmentis in *Philosophical Works of Francis Bacon*. ed. John M. Robertson (New York: E. P. Dutton, 1905).

The History of the Reign of King Henry VII. ed. Brian Vickers (Cambridge: Cambridge Univ. Press, 1998).

New Atlantis and Great Instauration. ed. Jerry Weinberger (Arlington: Harlan Davidson, 1989).

The New Organon. ed. Fulton H. Anderson (Indianapolis: Bobbs-Merrill, 1960).

The New Organon. eds. Lisa Jardine and Michael Silverthorne (Cambridge: Cambridge Univ. Press, 2000).

The Philosophy of Francis Bacon. ed. Benjamin Farrington (Liverpool: Liverpool Univ. Press, 1964).

Francis Bacon: A Selection of His Works. ed. Sidney Warhaft (New York: Macmillan, 1982).

Sylva Sylvarum (Montana: Kessinger Publishing Co., n.d.).

Biographical Studies

Aubrey, John. *Brief Lives*. ed. Oliver Lawson Dick (London: Secker & Warburg, 1960).

Bowen, Catherine Drinker. *Francis Bacon: The Temper of a Man* (Boston: Little Brown, 1963).

Du Maurier, Daphne. *Golden Lads* (London: Gollancz, 1975).

________. *The Winding Stair: Francis Bacon, His Rise and Fall* (Garden City, NY: Doubleday 1977).

Jardine, Lisa and Alan Stewart. *Hostage to Fortune: The Troubled Life of Francis Bacon* (London, Victor Gollancz, 1998).

Occasional Publications

Francis Bacon: The Commemoration of his Tercentenary at Gray's Inn (London: Chiswick, 1912).

The Unveiling of the Statue of Francis Bacon (London: Chiswick 1912).

Critical and Historical Studies

Archer, John Michael. "Surveillance and Enlightenment in Bacon's *New Atlantis.*" *Assays* 6 (1991): 111–27.

Aughterson, Kate. "'Strange things so probably told': gender, sexual difference, and knowledge in Bacon's *New Atlantis*" in *Francis Bacon's New Atlantis: New Interdisciplinary Essays.* ed. Bronwen Price (Manchester: Manchester Univ. Press, 2002), 156–79.

Benjamin, Edwin B. "Bacon and Tacitus." *Classical Philology* 60.1(1965): 102–10.

Bergeron, David M. "Francis Bacon: An Unpublished Manuscript." *Papers of the Bibliographical Society of America* 84.4 (1990): 397–404.

Briggs, John Channing. *Francis Bacon and the Rhetoric of Nature* (Cambridge: Harvard Univ. Press, 1989).

Bundy, Murray W. "Bacon's True Opinion of Poetry." *Studies in Philology* 27 (1930): 244–64.

Cohen, Jonathan. "Some Remarks on the Baconian Conception of Probability." *Journal of the History of Ideas* 41 (1980): 219–31.

Clucas, Stephen. "In Search of 'The True Logick': Methodological Eclecticism among the 'Baconian Reformers.'" *Samuel Hartlib and Universal Reformation: Studies in Intellectual Communication.* eds. Mark Greengrass, Michael Leslie, and Timothy Raylor (Cambridge: Cambridge Univ. Press, 1994), 51–74.

Coquillette, Daniel R. *Francis Bacon* (Stanford: Stanford Univ. Press, 1992).

Donnelly, M. L. "Francis Bacon's Early Reputation in England and the Question of John Milton's Alleged 'Baconianism.'" *Prose Studies* 14 (1991): 1–20.

Faulkner, Robert K. *Francis Bacon and the Project of Progress* (London: Rowman & Littlefield, 1993).

Ferguson, A. B. "The Non-Political Past in Bacon's Theory of History." *Journal of British Studies* 14 (1974): 4–20.

Fish, Stanley E. "Georgics of the Mind: The Experience of Bacon's Essays" in *Seventeenth-Century Prose.* ed. Stanley E. Fish (New York: Oxford Univ. Press, 1971), 251–80.

Gaukroger, Stephen. *Francis Bacon and the Transformation of Early-Modern Philosophy* (Cambridge: Cambridge Univ. Press, 2001).

Harrison, John L. "Bacon's View of Rhetoric, Poetry, and the Imagination." *Huntington Library Quarterly*, and 20 (February 1957): 107–25.
Hogan, John, C. Mortimer, and D. Schwartz. "On Bacon's 'Rules and Maximes' of the Common Law." *Law Library Journal* 76 (Winter 1983): 48–77.
Hunter, Michael and Paul B. Wood. "Towards Solomon's House: Rival Strategies for Reforming the Early Royal Society." *History of Science* 24 (1986): 49–108.
Hutton, Sarah. "Persuasions to Science: Baconian Rhetoric and the *New Atlantis*" in *Francis Bacon's* New Atlantis: *New Interdisciplinary Essays*. ed. Brownen Price (Manchester: Manchester Univ. Press, 2002), 48–59.
Jardine, Lisa. *Francis Bacon: Discovery and the Art of Discourse* (Cambridge: Cambridge Univ. Press, 1974).
Keller, Evelyn Fox. "Baconian science: an hermaphroditical birth." *Philosophical Forum* 11.3 (1980): 299–308.
Landau, Iddo. "Feminist criticisms of metaphors in Bacon's philosophy of science." *Philosophy* 73 (1998): 47–61.
Levao, Ronald. "Francis Bacon and the Mobility of Science." *Representations* 40 (1992): 1–32.
Lynch, William T. *Solomon's Child: Method in the Early Royal Society of London* (Stanford: Stanford Univ. Press, 2001).
Martin, Catherine Gimelli. "The Ahistoricism of the New Historicism: Knowledge as Power versus Power as Knowledge in Bacon's *New Atlantis*" in *Faultlines in the Field*, eds. Claude Summers and Ted-Larry Pebworth (Columbia: Univ. of Missouri Press, 2002), 22–49.
Martin, Julian. *Francis Bacon, the State, and the Reform of Natural Philosophy* (Cambridge: Cambridge Univ. Press, 1992).
Nadel, George H. "History as Psychology in Francis Bacon's Theory of History." *History and Theory* 5 (1966): 275–87.
Olivieri, Grazia Tonelli. "Galen and Francis Bacon: Faculties of the Soul and the Classification of Knowledge" in *The Shapes of Knowledge from the Renaissance to the Enlightenment*. eds. Donald R. Kelley and Richard H. Popkin (Dordrecht: Kluwer Academic Publishers, 1991). 61–92.
Peltonen, Markku, ed. *The Cambridge Companion to Bacon* (Cambridge: Cambridge Univ. Press, 1996).
Pérez-Ramos, Antonio. *Francis Bacon's Idea of Science and the Maker's Knowledge Tradition* (Oxford: Clarendon Press, 1988).
Pesic, Peter. "Wrestling With Proteus: Francis Bacon and the 'Torture' of Nature." *Isis* 90.1 (1999): 81–94.
Price, Brownen, ed. *Francis Bacon's* New Atlantis: *New Interdisciplinary Essays* (Manchester: Manchester Univ. Press, 2000).
Pumfrey, Stephen. "The History of Science and the Renaissance Science of History" in *Science, Culture and Popular Belief in Renaissance Europe*. eds. Stephen Pumfrey, Paolo L. Rossi, and Maurice Slawinski (Manchester: Manchester Univ. Press, 1991), 48–70.
Rees, Graham. "Bacon's Speculative Philosophy" in *The Cambridge Companion to Bacon*. ed. Markku Peltonen (Cambridge: Cambridge Univ. Press, 1996), 121–45.

________. "Mathematics and Francis Bacon's Natural Philosophy." *Revue Internationale de Philosophie* 40 (1986): 399–426.

________. "Francis Bacon's Semi-Paracelsian Cosmology and the Great Instauration." *Ambix* 22 (1975): 161–73.

________. "The Fate of Bacon's Cosmology in the Seventeenth Century." *Ambix* 24 (1977): 27–38.

Paolo Rossi. *Francis Bacon: From Magic to Science*. trans. Sacha Rabinovitz (Chicago: Univ. of Chicago Press, 1968).

Sargeantson, Richard. "Natural Knowledge in the *New Atlantis*" in *Francis Bacon's* New Atlantis: *New Interdisciplinary Essays*. ed. Bronwen Price (Manchester: Manchester Univ. Press, 2002), 82–105.

Sessions, William A., ed. *Francis Bacon's Legacy of Texts: 'The Art of Discovery Grows with Discovery'* (New York: AMS, 1990).

Snider, Alvin. "Bacon, Legitimation, and the 'Origin' of Restoration Science." *The Eighteenth Century* 32 (1991): 123–5.

Solomon, Julie Robin. "Between Magic and Travel: Francis Bacon and the Theaters of Human Knowing" (Ph.D. diss., University of Pennsylvania, 1987).

________. *Objectivity in the Making: Francis Bacon and the Politics of Inquiry* (Baltimore: Johns Hopkins Univ. Press, 1998).

________. "'To know, to fly, to conjure': Situating Baconian Science at the Juncture of Early Modern Modes of Reading." *Renaissance Quarterly* 44 (Autumn, 1991), 513–58.

Vickers, Brian. *English Science, Bacon to Newton* (Cambridge: Cambridge Univ. Press, 1987).

________. *Francis Bacon and Renaissance Prose* (Cambridge: Cambridge Univ. Press, 1968).

________. "The Authenticity of Bacon's Earliest Writings." *Studies in Philology* 94 (Spring 1997): 248–96.

________. "The Myth of Francis Bacon's 'Anti-Humanism'" in *Humanism and Early Modern Philosophy*, eds. Jill Kraye and M.W. F. Stone (New York: Routledge, 2000), 135–58.

Webster, Charles. *The Great Instauration: Science, Medicine and Reform 1626–1660* (London: Duckworth, 1975).

Weinberger, Jerry. "Metaphysics and Religion: Francis Bacon's Critique of the Ancients" in *Educating the Prince: Essays in Honor of Harvey Mansfield*, eds. Mark Blitz and William Kristol (Lantham: Rowman & Littlefield, 2000), 62–73.

Wheeler, Harvey. "Francis Bacon's *New Atlantis*: The 'Mould' of a Law Finding Commonwealth" in *Francis Bacon's Legacy of Texts*. ed. W. A. Session (New York: AMS, 1990), 291–310.

________. "Francis Bacon's 'Verulamium': the Common Law Template of the Modern in English Science and Culture." *Angelaki: Journal of the Theoretical Humanities* 4 (May, 1999): 7–26.

Wilson, E. O. *Consilience: The Unity of Knowledge* (New York: Alfred A. Knopf, 1998).

Woolf, Daniel R. "John Seldon [*sic*], John Borough and Francis Bacon's *History of Henry VII*, 1621." *Huntington Library Quarterly* 47 (1984): 47–53.
Yeo, Richard. "An Idol of the Market-Place: Baconianism in Nineteenth Century Britain." *History of Science* 23 (1985): 251–98.
Zagorin, Perez. *Francis Bacon* (Princeton: Princeton Univ. Press, 1998).
Zetterburg, J. Peter. "Echoes of Nature in Salomon's House." *Journal of the History of Ideas* 43 (1982): 179–93.

Index